C0-CDB-692

ACS SYMPOSIUM SERIES 772

Persistent, Bioaccumulative, and Toxic Chemicals I

Fate and Exposure

Robert L. Lipnick, EDITOR
U. S. Environmental Protection Agency

Joop L. M. Hermens, EDITOR
Utrecht University

Kevin C. Jones, EDITOR
Lancaster University

Derek C. G. Muir, EDITOR
National Water Resource Institute

American Chemical Society, Washington, DC

Library of Congress Cataloging-in-Publication Data

Persistent, Bioaccumulative, and toxic chemicals I: fate and exposure / Robert L. Lipnick, editor...[et al.].

   p. cm.—(ACS symposium series ; 772)

Papers presented at a symposium held Mar. 21–25, 1999 in Anaheim, Calif.

Includes bibliographical references and index.

ISBN 0–8412–3674–7

   1. Persistent pollutants—Environmental aspects—Congresses. 2. Persistent pollutants—Bioaccumulation—Congresses. 3. Persistent pollutants—Toxicology—Congresses.

   I. Lipnick, Robert L. (Robert Louis), 1941–   . II. Series.

TD196.C45 P465   2000
628.5´2—dc21   00–56557

The paper used in this publication meets the minimum requirements of American National Standard for Information Sciences—Permanence of Paper for Printed Library Materials, ANSI Z39.48–1984.

Copyright © 2001 American Chemical Society

Distributed by Oxford University Press

All Rights Reserved. Reprographic copying beyond that permitted by Sections 107 or 108 of the U.S. Copyright Act is allowed for internal use only, provided that a per-chapter fee of $20.00 plus $0.50 per page is paid to the Copyright Clearance Center, Inc., 222 Rosewood Drive, Danvers, MA 01923, USA. Republication or reproduction for sale of pages in this book is permitted only under license from ACS. Direct these and other permission requests to ACS Copyright Office, Publications Division, 1155 16th St., N.W., Washington, DC 20036.

The citation of trade names and/or names of manufacturers in this publication is not to be construed as an endorsement or as approval by ACS of the commercial products or services referenced herein; nor should the mere reference herein to any drawing, specification, chemical process, or other data be regarded as a license or as a conveyance of any right or permission to the holder, reader, or any other person or corporation, to manufacture, reproduce, use, or sell any patented invention or copyrighted work that may in any way be related thereto. Registered names, trademarks, etc., used in this publication, even without specific indication thereof, are not to be considered unprotected by law.

PRINTED IN THE UNITED STATES OF AMERICA

628.52
P466
v.1

Foreword

THE ACS SYMPOSIUM SERIES was first published in 1974 to provide a mechanism for publishing symposia quickly in book form. The purpose of the series is to publish timely, comprehensive books developed from ACS sponsored symposia based on current scientific research. Occasionally, books are developed from symposia sponsored by other organizations when the topic is of keen interest to the chemistry audience.

Before agreeing to publish a book, the proposed table of contents is reviewed for appropriate and comprehensive coverage and for interest to the audience. Some papers may be excluded in order to better focus the book; others may be added to provide comprehensiveness. When appropriate, overview or introductory chapters are added. Drafts of chapters are peer-reviewed prior to final acceptance or rejection, and manuscripts are prepared in camera-ready format.

As a rule, only original research papers and original review papers are included in the volumes. Verbatim reproductions of previously published papers are not accepted.

ACS BOOKS DEPARTMENT

Contents

Fate and Behavior

Food Chain Transfer and Exposure

Preface

This monograph, *Persistent, Bioaccumulative, and Toxic Chemicals I: Fate and Exposure*, represents one of two books derived from a symposium sponsored by the American Chemical Society (ACS) Division of Environmental Chemistry, Inc., March 21–25, 1999 in Anaheim, California. The symposium was originally inspired by the U.S. Environmental Protection Agency's 1998 program on persistent, bioaccumulative, and toxic chemicals (PBTs). The EPA's national PBT program based its priority chemical list on those previously agreed to by the United States and Canada in 1997 for "virtual elimination" of PBTs in the Great Lakes. This action was preceded by a 1995 agreement signed by Canada, Mexico, and the United States to develop and implement North American Regional Action Plans (NARAPs) under the North American Free Trade Agreement (NAFTA) to phase out or reduce emissions of PBTs.

Other regional agreements that are responsible for the surge of interest in PBTs include the protocol for Long Range Transport of Air Pollutants (LRTAP) under the U.N. Economic Commission for Europe (UN-ECE). Globally, such reduction and phaseout of PBTs is being led by the U.N.'s Environmental Program (UNEP) through global persistent organic pollutants (POPs) negotiations.

The Anaheim symposium provided a much-needed means for research scientists and policy-makers to exchange information regarding the state of scientific knowledge and theory on this subject, and how these findings can be used in the decision-making process. With the encouragement of ACS Books Department, the Editors also recruited additional papers for the monographs to fill gaps. These monographs provide a means of disseminating this information to a larger audience.

This volume includes contributions from Israel, Netherlands, Poland, the United Kingdom, and the United States, covering fate and exposure. The fate section includes chapters on behavior in the environment covering the theory of movement of such hydrophobic chemicals through sediments, uptake by plants, bioconcentration (theoretical framework, and bioaccumulation of toxaphene, DDT, and dieldrin), soil degradation, and remediation. Chapters are included related to emissions at the local and regional level for polychlorinated biphenyls, butyltin, toxaphene, and tetrachlorodioxin. These are followed by chapters pertaining to human exposure for PCBs (generic model, soil concentrations, and concentrations in breast milk), dioxin (food from animal sources and body burdens in California populations), and mercury (hair analysis and atmospheric deposition studies).

This monograph contains an introductory chapter that puts the discovery, use, and findings of adverse environmental problems with PBTs into historical perspective. This chapter details how chlorinated pesticides such as DDT initially provided a great service to humankind, especially to combat malaria and other diseases; the industrial utility of PCBs and other such substances; and the unintended production, disposal, and bioaccumulation of highly toxic polychlorinated dibenzo-*p*-dioxins, polychlorinated dibenzofurans, and other industrial by-products.

The companion volume to this monograph, entitled *Persistent, Bioaccumulative, and Toxic Chemicals II: Assessment and New Chemicals*, includes chapters on characterizing PBT properties for selecting candidate chemicals, including their long-range environmental transport. These are followed by coverage of estimation of bioavailability and toxicity of individual PBTs and mixtures from physicochemical properties; case studies of emissions and fate of polychlorinated biphenyls; PBT partitioning between air and vegetation; PBT hazard and risk assessment; and new categories of PBTs including chlorinated paraffins, musk fragrances, polychlorinated naphthalenes, polybrominated diphenyl ethers, and a recently isolated chlorinated contaminant in Antarctic seals.

The Editors gratefully acknowledge the continuing support and encouragement provided by Anne Wilson, Senior Product Manager, Kelly Dennis, Editorial Assistant, and Margaret Brown, Senior Production Specialist, ACS Books Department, in preparing this ACS Symposium Monograph.

ROBERT L. LIPNICK
Office of Pollution Prevention and Toxics
U.S. Environmental Protection Agency
401 M Street, S.W., Mail Stop 7403
Washington, DC 20460

JOOP L. M. HERMENS
Research Institute of Toxicology
Utrecht University
P.O. Box 80176
NL–3508 TD Utrecht, the Netherlands

KEVIN C. JONES
Environmental Science Division
Lancaster University
Lancaster LA1 4YQ, United Kingdom

DEREK C. G. MUIR
National Water Research Institute
Environment Canada
Burlington, Ontario L7R 4A6, Canada

Chapter 1

History of Persistent, Bioaccumulative, and Toxic Chemicals

Robert L. Lipnick[1] and Derek C. G. Muir[2]

[1]Office of Pollution Prevention and Toxics, U.S. Environmental Protection Agency, 401 M Street, SW, MS 7403, Washington, DC 20460
[2]National Water Research Institute, Environment Canada, Burlington, Ontario L7R 4A6, Canada

The history of persistent, bioaccumulative, and toxic chemicals (PBTs) is traced in this chapter from the 1825 synthesis of technical "benzene hexachloride (BHC)" by Faraday to the current international efforts to phase out 12 "persistent, organic pollutants" (POPs) under the auspices of the United Nations Environmental Program (UNEP). In the 1930s, new uses were sought for chlorine, leading to the discovery of a number of chlorinated insecticidal chemicals, including DDT, which earned Müller the Nobel Prize in Medicine for its great success during and following the War to control typhus, malaria, typhoid fever, and cholera. Beginning in the 1960s, however, concerns were identified for the persistence, bioaccumulation, and food chain biomagnification of DDT and other PBTs, and their effects on reproduction in grebes, falcons, and eagles. In addition, these compounds were being detected in geographical locations far from their production and use. PBT properties are found not only in chlorinated pesticides including DDT, but also in halogenated industrial chemicals such as polychlorinated biphenyls. In addition, PBTs, including polychlorinated dibenzo-*p*-dioxins and dibenzofurans, arise as unintended products of industrial processes or combustion. Although, a large number of chlorinated and brominated compounds have been isolated from natural sources, including marine organisms, it is their production and widespread distribution by anthropogenic sources that is of concern.

© 2001 American Chemical Society

1

Background

Persistent, bioaccumulative, and toxic chemicals (PBTs) are the subject of a concerted national, regional, and international effort to limit their production, and use, as well as the control and disposal of no longer used substances. PBTs or POPs (persistent organic pollutants) share low water solubility and high lipid solubility leading to their high potential for bioaccumulation. In addition, multimedia releases and volatility lead to long range environmental transport both via water and the atmosphere, resulting in widespread environmental contamination of humans and biota at sites distant from their use. An international working group was convened in May, 1995 by the United Nations Environmental Program (UNEP) Governing Council, to develop assessments for 12 POPs. The 12 POPs consist of aldrin, chlordane, dieldrin, DDT, endrin, heptachlor, hexachlorobenzene, mirex, toxaphene, PCBs, polychlorinated dibenzo-*p*-dioxins (PCDDs), and polychlorinated dibenzofurans (PCDFs). This working group determined that data were adequate for these 12 to eliminate or reduce emissions, and even, in some cases, halt production and use *(1)*. In 1998, 36 member states of the United Nations Economic Commission for Europe (UNECE), consisting of European countries, Russia, Canada and the United States, agreed on a Protocol which bans the production and use of some POPs. An outright ban was proposed for aldrin, chlordane, chlordecone, dieldrin, endrin, hexabromobiphenyl, mirex and toxaphene, while DDT, heptachlor, hexachlorobenzene (HCB), and PCBs were scheduled for elimination at a later stage. Uses of HCH (hexachlorocyclohexane), including lindane, were restricted. The protocol also obliges signatory Parties to reduce their emissions of PCDD, PCDFs, polycyclic aromatic hydrocarbons (PAHs) and HCB below their levels in 1990 *(2)*. Although most chemicals on the UNEP and UNECE lists contain chlorine or bromine, some chemicals which are classified as POPs such as polycyclic aromatic hydrocarbons *(2)* have no halogen substituents. There are also natural sources of halogenated organics. Thus about 3200 organohalogen compounds from non-industrial sources have been identified in the environment. More than 1600 of these such as brominated diphenyl ethers, bromomethane, bromoethane, bromoform, brominated dioxins, and brominated pyrroles have been identified, many of which are produced by marine algae and marine acorn worms *(3,4)*. In addition there are a large number of chemicals in use as pesticides, flame retardants, plasticizers and even fragrances which have some similarities to the 12 POPs in terms of chemical structure, persistence, environmental transport to remote regions and bioaccumulation, although typically less is known about their levels in the environment or their toxicity. The synthetic (nitro and polycyclic) musks represent a example of widely used products recently discovered to bioaccumulate in aquatic food webs. The chlorinated paraffins and brominated diphenyl ethers are other examples of polyhalogenated compounds with some similarities to POPs as discussed in Vol 2 of this series.

Discovery of chlorine and bromine

The history of persistent, bioaccumulative, toxic chemicals (PBTs), which are also referred to as POPs (persistent organic pollutants) and PLCs (persistent lipophilic contaminants), can be traced to the discovery by the Swedish apothecary Karl Wilhelm Scheele in 1774 of the element chlorine, which he prepared by the reaction of hydrochloric acid and manganese dioxide *(5)*. Bromine was identified as a new element in 1826 by Antóine-Jérôme Ballard in France, who isolated it from Mediterranean brine concentrates *(6)*. Chlorine and bromine substituents provide PBT molecules with properties of both persistence and biological activity, which are important with respect to their intended performance characteristics.

Chlorinated Insecticidal and other Agricultural Chemicals

Interest arose, particularly in the 1930s, to investigate the synthetic possibilities of chlorine, leading to a variety of valuable insecticidal and agricultural chemicals, which quickly replaced compounds of arsenic, lead, and copper *(7)*.

Lindane and other isomers of hexachlorocyclohexane

Faraday's Preparation of "Benzene Hexachloride"

The history of chlorinated pesticides can be traced to Michael Faraday, who reported on June 16, 1825 before the Royal Society of London, the formation of "benzene hexachloride" by the reaction of benzene (bi-carburet of hydrogen) with chlorine in the presence of sunlight. Faraday describes his findings as follows *(8)*:

"Chlorine introduced to the substance in a retort exerted but little action until placed in sun-light, when dense fumes were formed, without the evolution of much heat; and ultimately much muriatic acid [hydrochloric acid] was produced, and two other substances, one a solid crystalline body, the other a dense thick fluid. It was found by further examination that neither of these were soluble in water; that both were soluble in alcohol – the liquid readily, the solid with more difficulty. Both of them appeared to be triple compounds of chlorine, carbon, and hydrogen; but I reserve consideration of these, and of other similar compounds to another opportunity."

Discovery of the Insecticidal Properties of Hexachlorocyclohexane

Unknown to Faraday at the time, this reaction product actually consisted of a mixture of various isomers of hexachlorocyclohexane (HCH). During 1932-1933,

Harry Bender, Great Western Electrochemical Company of California, was investigating new uses for chlorine. Part of a reaction product of benzene and liquid chlorine spilled on the ground and "attracted and killed flies and bees". Unfortunately, prior to toxicity testing, the recrystallized samples consisted almost exclusively of the less soluble and much less active α- and β-isomers. The discovery of insecticidal properties was also missed during initial screening by Imperial Chemical Industries (ICI) of technical BHC (benzene hexachloride) in the early 1930s. In 1942, at ICI, the α- and β-isomers were found to be inactive, but the activity was linked to the γ-isomer the following year. The name lindane was subsequently assigned to this active γ-isomer after Van Linden, discoverer of the δ- and γ-isomers (9).

Regulation and Control of Use of Technical Hexachlorocyclohexane and Lindane

The use of a mixture of HCH isomers, known as technical HCH, began in 1943, and global consumption has been estimated to be about 6 million tonnes (metric tons) since then (10). Technical HCH use was banned in most western countries and Japan in the 1970s but continued in India, Russia and China. China is reported to have been the major world producer accounting for about 4.5 million tonnes between 1945 and 1983 (11). In 1983 China switched to lindane and Russia banned technical HCH in the late 1980s (12). In 1990 India banned the technical product for agricultural use but kept it for public health uses (10). Levels of α-HCH, the major component of technical lindane, have declined in Arctic air over the period 1979 to 1994 in concert with reductions of technical HCH use in China and Russia. Lindane remains an important soil insecticide in North America, Europe and Asia (10).

DDT

Discovery of Insecticidal Activity and Much Earlier Chemical Synthesis

Just prior to World War II, the first modern synthetic insecticides were developed. In 1939, Paul Müller, working in the laboratory of the Swiss company J.R. Geigy, discovered the insecticidal activity of DDT (dichlorodiphenyltrichloroethane). DDT was found shortly thereafter to be extremely effective against crop and household pests, and was even used for direct application to humans by the Allies during the War to fight typhus in Naples (12). DDT was also employed during and following the War to control malaria, typhoid fever, and cholera by control of mosquitoes, lice, and flies (13). The utility of DDT was so remarkable that during the first decade of use, it "...saved five million lives and prevented 100 million incidents of serious disease *(14)*." Müller was awarded the Nobel Prize in medicine in 1948 for his research on DDT *(15)*. Remarkably, DDT, like lindane, was first synthesized long before its insecticidal activity was found. Othmar Zeidler *(16)* in 1873, working in the laboratory of Adolph von Bayer at the University of Strasbourg *(17)*, prepared DDT by reaction of chloral and chlorobenzene in the presence of sulfuric acid, the same reaction subsequently employed in the large scale industrial preparation of the insecticide.

Discovery of Environmental Problems of DDT

Unfortunately, the low volatility, chemical stability, lipid solubility, and slow rate of transformation and degradation of DDT and other organochlorine insecticides that made them so successful, ultimately led to their problems due to environmental persistence, bioconcentration, and food chain biomagnification. In 1962, in her book *Silent Spring (18)*, Rachel Carson focused public and scientific attention on these problems, particularly with respect to DDT and related substances affecting reproduction of grebes, falcons, and eagles at the top trophic level of their food chains due to accumulation in fat and egg shell thinning *(19)*. Because of this, restrictions and bans on the use of DDT were introduced in Canada, US, Japan, and in most western European countries in the early 1970s *(20)*. Indeed, the increased awareness of the adverse effects of such persistent, bioaccumulative, toxic chemicals has produced more rigorous screening by industry and government in the development of subsequent insecticides. However, DDT production in China, India, Russia and possibly other countries continued during the 1970s and 1980s. DDT production peaked in China during the late 1970s and was phased out in 1984 *(21,22)*, but DDT is still produced in large quantities.

Results of DDT Regulation and Continuing International Control Efforts

DDT is still in use in Mexico, tropical Africa, India, and Brazil, and other tropical countries for control of the malaria vector. The September, 1999 meeting of delegates from about 110 countries to the United Nations Environmental Program (UNEP) negotiations on controlling 12 persistent organic pollutants (POPs), agreed to phase out aldrin, endrin, and toxaphene, and severely restrict the use of chlordane, dieldrin, heptachlor, mirex, and hexachlorobenzene. While agreement was reached on stopping all agricultural uses of DDT, decision on the use of DDT to control malaria was deferred *(23)*. The World Wildlife Fund and many industrialized countries have proposed a strict ban on all DDT uses by 2007. Alternatives exist for crop application; however, no such alternatives currently exist for malaria control. Each year malaria strikes 500 million people, killing 2.7 million, mostly children. *(24)*.

6

Toxaphene

Khanenia and Zhiravlev demonstrated in 1944 that chlorination of terpenes found in turpentine led to products with increased toxicity to body lice, and the insecticidal properties of this polychloroterpene mixture were reported in 1947. In 1948, Hercules Powder Company marketed this as the insecticide "Hercules 3956®," produced by the chlorination of terpenes such as alpha-pinene and camphene in the presence of light and free radical initiators. The official common name of the mixture is camphechlor, but it is widely referred to by its trade name of toxaphene. Toxaphene is a mixture of polychlorobornanes and camphenes with average of 64-67% chlorine or $C_{10}H_{10}Cl_8$ (25), and consists of more than 300 isomers and congeners, even though there are theoretically 16640 isomers and congeners, most of which are not formed and are chiral (26). Toxaphene is highly toxic and bioaccumulative to fish and birds, and is carcinogenic to rats and mice; as a result, its pesticide registration in the U.S. was cancelled by the U.S. EPA in 1983, and it was banned in the United States in 1986 (26). Many other countries followed the U.S. lead; however, production continued in Nicaragua until the early 1990s. The former Soviet Union was also a significant producer and user of polychloroterpenes and the former East Germany was a major producer and exporter until 1990 (27). Toxaphene use in western Europe was very limited in comparison to the United States (27). Toxaphene had been used extensively in the southern United States on cotton, and high levels in the water, sediment and biota of the Great Lakes were traced to atmospheric transport and deposition (28,29). Indeed, the soils of the cotton growning states continue to be a source to the atmosphere years afterwards (30). Concentrations of toxaphene have declined in lake trout from Lake Ontario but not in trout and smelt from Lake Superior and the northern part of Lake Michigan in 10-15 years. Toxaphene also remains a major OC (organochlorine) contaminant in lake trout from lakes and rivers outside of the Great Lakes in Canada and the United States (31) as well as in marine mammals in the Canadian Arctic and Greenland (32). The possibility of toxaphene-like compounds being produced as by-products of chlorine bleaching at pulp and paper mills has been speculated upon following the detection of mono- and dichlorobornanes in pulp mill effluents. However, recent studies of sediments upstream and downstream of pulp mills in the United States have concluded that this is very unlikely (33). Toxaphene may still be in use in some countries because it is relatively cheap and easy to produce and a potential replacement for banned chemicals such as DDT. However, with the signing of the UNEP agreement toxaphene is likely to be consigned to the history books.

Diene-Organochlorine Insecticides

Many insecticides of this large diene-organochlorine group (including chlordane/heptachlor, dieldrin/endrin/aldrin, mirex and chlordecone) are formed by an initial route leading to a bridged ring system, involving the Diels-Alder reaction between a compound containing a double or triple bond, generally activated by conjugation at the 1,4-positions *(34)*. Clordecone, mirex, aldrin, endrin and chlordane were included in the UNECE POPs Protocol under the LRTAP (Long-Range Transboundary Air Pollution) Convention *(2)*.

Chlordecone and Mirex

The key starting material, hexachlorocyclopentadiene ("hex"), undergoes self-condensation, with further reaction yielding the cage structure chlordecone ($C_{10}Cl_{10}O$), which had been marketed in the United States as Kepone. Pesticide registration of Kepone was cancelled in the United States in 1978 following widespread contamination of the James River and the Chesapeake Bay from factory releases *(35)*. Mirex ($C_{10}Cl_{12}$), another cage structure, is formed by further chlorination of chlordecone with PCl_5, or by self-condensation of "hex" with $AlCl_3$.

Chlordane

Technical grade chlordane is a mixture of at least 120 compounds, with the major constituents being cis- and trans-chlordane, heptachlor, cis- and trans-nonachlor, chlordene, and others *(36)*. Chlordane was released into the environment primarily from its application as a soil insecticide *(37)*. The United States was the major world producer and user of chlordane. There was limited use in western Europe and the former Soviet Union. In the United States, chlordane was used extensively prior to 1983 and from 1983 to 1988 it was registered for termite control. It was cancelled for this use in 1988; however, production continued for export purposes. In 1997, the sole U.S. manufacturer of chlordane voluntarily ceased production at all its national and international facilities.

Aldrin, Dieldrin, Endrin, and Heptachlor

World sales of aldrin, and dieldrin ceased in 1991 when the major manufacturer voluntarily stopped production. Endrin production ceased in the mid-1980's. Dieldrin was mainly used as a soil insecticide. In tropical countries it was used for the control of vectors of diseases, mainly malaria. It is no longer manufactured in Canada and the US, and its use is now restricted for termite control. Manufacture in Europe, especially for export to developing countries, continued until the late 1980s. It is also a degradation product of aldrin. It is extremely persistent in soil (half-life >7 years) and has a long half life in biota *(36)*. It is the most potent carcinogen of the major organochlorine pesticides. Aldrin was produced by the Diels-Alder reaction of norbornadiene with "hex." Epoxidation of Aldrin leads to Dieldrin, while chlorination allylic to this double bond yields Heptachlor. Endrin is also an epoxide, but with opposite stereochemistry to Aldrin and Dieldrin *(9)*.

Non-Insecticidal PBTs

In addition to the chlorinated insecticides, some other industrial chemicals either intentionally manufactured or resulting as by-products of industrial processes or combustion are also of concern as PBTs. These include polychlorinated biphenyls (PCBs), hexachlorobenzene, octachlorostyrene, polychlorinated napthalenes (PCNs), polychlorinated dibenzodioxins (PCDDs), polychlorinated dibenzofurans (PCDFs), polycyclic aromatic hydrocarbons (PAHs), chlorinated paraffins, and brominated flame retardants. The terms chlorinated and polychlorinated tend to be used interchangeably for these substances, and no differences are intended in this discussion.

Polychlorinated Biphenyls (PCBs)

Polychlorinated biphenyls (PCBs) consist of 209 isomers and congeners, with a 10,000 fold range in Kow values between the mono-substituted 2-chlorobiphenyl and the fully-substituted decachlorobiphenyl *(38)*. First produced in 1929, commercial mixtures range from 21% (Arochlor 1221) to 60% chlorine (Arochlor 1260), and are made by the reaction of biphenyl with anhydrous chlorine catalyzed by iron filings or ferric hydroxide. PCBs were manufactured in Austria, China, Czechoslovakia, France, Germany, Italy, Japan, the Russian Federation, Spain, the United Kingdom and the United States and exported for use to virtually every country in the world *(39)*. PCBs have been used for dielectric fluids in transformers and capacitors, as well as flame retardant plasticizers for paints, sealants and plastics, etc. In 1976, the U.S. Congress banned, with limited exceptions, the manufacturing, processing, distribution, and use of PCBs, except in a "totally enclosed manner" and similar action was taken in Japan, Canada and western European countries. This action was based upon their toxicity and widespread environmental distribution as a result of their high degree of persistence and bioaccumulation. Nevertheless, large quantities of PCBs have remained in use especially in closed systems, e.g., in transformers. There have been a number of court decisions and additional regulations in the U.S. clarifying transformer, as well as other enclosed uses and PCB disposal *(40)*. In Germany, Bayer ceased production in 1983, and in 1989 production, marketing, and use of PCBs were prohibited *(41)*. During the recent negotiations for UNECE POPs protocol *(2)* it was revealed that Russia continues to have a PCB manufacturing capacity and that PCBs continue to be used as dielectric fluids in that country *(22)*.

Polychlorinated Dibenzo-*p*-dioxins (PCDDs) and Polychlorinated Dibenzofurans (PCDFs)

In 1872, Mertz and Weitz in Germany reported preparation of the first chlorinated dibenzo-*p*-dioxin, the fully octachloro derivative, although its structure was not determined until 1957, with the report by Sanderman et al. of the first synthesis of 2,3,7,8-tetrachlorodibenzo-*p*-dioxin (TCDD) *(42)*. The parent compound, dibenzo-*p*-dioxin, was prepared in 1866 by Lesimple from triphenyl phosphate and lime, and the structure determined in 1871 by Hoffmeister *(43)*. TCDD and other PCDDs and the corresponding 2,3,7,8-tetrachlorodibenzofuran (TCDF) and PCDFs have never been produced intentionally, but are by-products of various industrial processes including chlorine bleaching in paper making, metal smelting and processing. PCDDs and PCDFs are also produced by combustion processes. Toxicological effects ultimately linked to TCDD were observed in workers following an industrial accident during the production of the herbicide 2,4,5-T (2,4,5-trichlorophenoxyacetic acid) on March 8, 1949 at a Monsanto plant in Nitro, West Virginia *(15)*. The TCDD was later shown to arise from the condensation of two molecules of the 2,4,5-trichlorophenol starting materials under certain conditions. TCDD was subsequently implicated as an impurity in exposure from the herbicide Agent Orange, consisting of a 1:1 mixture of the n-butyl esters of 2,4,5-trichlorophenoxyacetic acid and the 2,4-disubstituted derivative. In 1976, an industrial accident in Seveso, Italy resulted in the release of a large quantity of dioxin by-product from an uncontrolled reaction. Mass balance information regarding the relative importance of industrial by-product sources versus combustion processes is undergoing refinement. Analysis of lake sediment data for North America and western Europe as well as archived soils and plants in the UK *(44)* show that levels of PCDDs and PCDFs were very low until 1920-1940, corresponding to a period of significant increase in the manufacturing of synthetic organic chemicals. This finding is also consistent with low levels in ancient mummies and 100-400 year old frozen Eskimos compared to those in human tissue today in industrialized countries *(45)*.

Hexachlorobenzene

Hexachlorobenzene (HCB) appears to be the most widespread of these chlorinated by-products. HCB (not to be confused with the cyclohexane derivative BHC or benzene hexachloride) was introduced commercially as a fungicide for wheat in 1933 *(37)*. It also had industrial uses in organic syntheses as a raw material for synthetic rubber; a plasticizer for polyvinyl chloride; as a rubber peptizing agent in the manufacture of nitroso and styrene-type rubbers; additive for pyrotechnic compositions for the military; a porosity controller in the manufacture of electrodes; and an intermediate in dye manufacture. HCB use as a fungicide was banned in the US, Canada and Europe in the 1970s *(46)*. The major global sources of HCB are combustion processes such as waste incineration, and aluminium casing manufacture *(47)*, and pesticide use *(48)*. In some processes e.g. combustion of wastes, HCB may be formed concurrently with other chlorinated compounds including chlorinated dioxins and furans, hexachlorobenzene, octachlorostyrene, chlorobiphenyls, and chloronaphthalenes. HCB is an impurity in the pesticides pentachlorophenol, dacthal, atrazine, picloram, pentachloronitrobenzene, chlorthalonil, and lindane *(46)*.

Acknowledgements

The assistance of Elizabeth Tunis, History of Medicine Division, National Library of Medicine; Jack Eckard, Reference Librarian, Rare Books and Special Collections, Countway Library of Medicine, Boston; and Jill Rosenshield, Associate Curator, Special Collections, University of Wisconsin Library is gratefully acknowledged.

References

1. Buccini, J. *Progress in Developing a United Nations Convention on Persistent Organic Pollutants (POPs)*, *Organohalogen Compounds* **1999**, *43*, 459-460.
2. United Nations Economic Commission for Europe (UNECE). Protocol To The 1979 Convention On Long-Range Transboundary Air Pollution On Persistent Organic Pollutants. http://www.unece.org/env/lrtap/
3. Gribble, G.W. *The Natural Production of Organochlorine Compounds*, in *Chlorine and Chlorine Compounds in the Paper Industry*, V. Turoski, Ed., Ann Arbor Press, Chelsea, MI, 1998, pp. 89-108.
4. Gribble, G.W., *The Natural Production of Organobromine Compounds*, *Organohalogen Compounds* **1999**, *41*, 23-26.
5. *Chlorine* in *The New Encyclopaedia Britannica, Micropaedia*, Vol. 3, 15ᵗʰ Edition, Chicago, 1995, p. 250.
6. *Bromine* in *The New Encyclopaedia Britannica, Micropaedia*, Vol. 2 15ᵗʰ Edition, Chicago, 1995, p. 542.
7. Klassen, C.D. (Ed.) *Casarett and Doull's Toxicology: The Basic Science of Poisons*, 5ᵗʰ Edition, McGraw-Hill, New York, 1996, 643.
8. Faraday, M. *XX. On New Compounds of Carbon and Hydrogen, and On Certain Other Products Obtained during the Decomposition of Oil by Heat*, *Philosophical Transctions of the Royal Society of London*, 1825, 440-466.
9. Brooks, G.T. *Chlorinated Insecticides: Retrospect and Prospect*, in *Pesticide Chemistry of the 20ᵗʰ Century*, J.R. Plimmer, Ed., ACS Symposium Series 37, American Chemical Society, Washington, DC, 1977, p.2
10. Li, Y.-F., L.A. Barrie, T.F. Bidleman and L.L. McConnell. *Geophys. Res. Lett.* **1998**, 25, 39-41.
11. Li, Y.-F., D.J. Kai and A. Singh. *Arch. Environ. Contam. Toxicol.* **1998**, 35, 688-697.
12. Klassen, C.D. (Ed.) *Casarett and Doull's Toxicology: The Basic Science of Poisons*, 5ᵗʰ Edition, McGraw-Hill, New York, 1996, 643.
13. Büchel, K. *Chemistry of the Pesticides*, Wiley-Interscience, New York, 1983.
14. Magill, F.N. (Ed.) *Great Events from History II: Ecology and Environment Series. The United States Bans DDT*, Vol. 3, 1966-1973, Salem Press, Pasadena, CA, 1995.

15. Klassen, C.D. (Ed.) *Casarett and Doull's Toxicology: The Basic Science of Poisons*, 5th Edition, McGraw-Hill, New York, 1996, 671-673.
16. Zeidler, O. *Beitrag zur Kentniss der Verbindungen zwischen Aldehyden und aromatischer Kohlenwasserstoffen*, Inaugural Dissertation der Philosophen-Facultat der Universitat-Strasbourg, Wien, 1873.
17. Metcalf, R.L. *A Century of DDT. J. Agricul. Fd. Chem.* **1973**, *21*, 511-519.
18. Carson, R. *Silent Spring*, Houghton Mifflin, Boston, 1962.
19. Connell, D.W., Miller, G.J., Mortimer, M.R., Shaw, G.R., and Anderson, S.M. Occurrence and Behavior of Persistent Lipophilic Contaminants in the Southern Hemisphere, *Organohalogen Compounds* **1999**, *41*, 339-342.
20. *Insect Control Technology* in *Kirk-Othmer Encyclopedia of Chemical Technology*, 3rd Edition, Vol. 13, Wiley-Interscience, New York, 1981.
21. Li, Y.F., D.J. Cai and A. Singh. Adv. Environ Res. 1999, 2, 497-506.
22. Arctic Monitoring and Assessment Program. Multilateral Cooperative Project on Phase-out of PCB Use, and Management of PCB-contaminated Wastes in the Russian Federation. 1999. Oslo, Norway. (http://www.grida.no/amap/amap.htm).
23. Hileman, B. *Paring of Persistent Pollutants Progresses, Chemical and Engineering News*, September 20, 1999, p. 41.
24. Hileman, B. *Dilemma over Malaria, Chemical and Engineering News*, September 20, 1999, p. 9.
25. Brooks, G.T. *Chlorinated Insecticides: Technology and Application*, Vol. I, CRC Press, Cleveland, 1974, pp. 85-183, 205-210.
26. Oehme, M. and Vetter, W. *Toxaphene, a Different Environmental Problem, Organohalogen Compounds* **1999**, *41*, 561-564.
27. Heinisch E.; Kettrup A.; Jumar S.; Wenzel-Klein S.; Stechert J.; Hartmann P.; Schaffer P., *Schadstoff Atlas Ost-Europa*. Chapter 2.9, p39-47. Pub: Ecomet/Landsberg-Lech, Germany.**1994.**
28. Swackhamer, D.L., Hites, R.A.. *Environ. Sci. Technol.* **1988**, 22, 543-548.
29. Swackhamer, D.L., Eisenreich S.J., Long D. T. Atmospheric deposition of toxic contaminants to the Great Lakes: Assessment and importance. Volume II: PCDDs, PCDFs and Toxaphene in sediments of the Great Lakes. Report to US EPA, Great Lakes National Program Office, Chicago, IL. 1996.
30. Hoff, R., Bidleman T.F., Eisenreich, S.J. *Chemosphere*, **1993**, 27, 2047-2055.
31. Kidd, K.A., D.W. Schindler, D.C.G. Muir, W.L. Lockhart and R. H. Hesslein. *Science*, **1995**, 269, 240-242.
32. DeMarch, B., C. DeWit and D. Muir. *Persistent Organic Pollutants*. In: *AMAP Assessment Report. Arctic Pollution Issues*, Chapter 6. Arctic Monitoring and Assessment Program. Oslo. Norway. 1999, pp. 183-372.
33. McDonald, J.G., Shanks, K.E., and Hites, R.A. *Are Pulp and Paper Mills Sources of Toxaphene to Lake Superior and Northern Lake Michigan? Organohalogen Compounds* **1999**, *41*, 581-586.
34. Brooks, G.T. *Chlorinated Insecticides: Technology and Application*, Vol. I, CRC Press, Cleveland, 1974, pp. 84-183.

12

35. Evers, B., Hawkins, S., Ravenscroft, M., Rounsaville, J.F., and Schultz, G. (Eds.), *Organochlorine Insecticides*. In *Ullmann's Encyclopedia of Industrial Chemistry*, 5th Edition, Vol. A14, VCH, Weinheim, 1989, 278-320.

36. Howard, P.C. 1991. *Handbook of Environmental Fate and Exposure Data for Organic Chemicals*. Lewis Publ., Chelsea MI. 684 pp.

37. Extension Toxicology Network (Extoxnet). http://pmep.cce.cornell.edu/profiles/extoxnet/carbaryl-dicrotophos/chlordane-ext.html. 1999.

38. Ramamoorthy, S. and Ramamoorthy, S. *Chlorinated Organic Compounds in the Environment: Regulatory and Monitoring Assessment*, Lewis Publishers, Boca Raton, FL, 1997, 125-127.

39. United Nations Environment Program (UNEP). *Guidelines for the Identification of PCBs and Materials Containing PCBs*. First Issue. 1999. UNEP Chemicals, Geneva Switzerland. 34 pp.

40. U.S. Environmental Protection Agency, Office of Pollution Prevention and Toxics, *PCB Q&A Manual*, 1994 edition, PDF file, 221 pages, http://www.epa.gov/opptintr/chemtype.htm.

41. Schüürmann, G and Markert, B. *Ecotoxicology*, John Wiley and Sons and Spektrum Akademischer Verlag, New York, 1997, p. 343.

42. Rappe, C. *2,3,7,8-Tetrachlorodibenzo-p-dioxin (TCDD) – Introduction*. In *Dioxin: Toxicological and Chemical Aspects*, F. Cattabeni, A. Cavallaro, and G. Galli, Eds., SP Medical and Scientific Books, New York, 1978, pp. 9-11.

43. Evers, B., Hawkins, S., Ravenscroft, M., Rounsaville, J.F., and Schultz, G. (Eds.), In *Ullmann's Encyclopedia of Industrial Chemistry*, 5th Edition, Vol. A12, VCH, Weinheim, 1989, p. 130.

44. Kjeller, L-O., K.C. Jones, A.E. Johnston and C. Rappe. *Environ. Sci. Technol.* **1991**, 25; 1619-1627.

45. Webster, T. and Commoner, B. *Overview: The Dioxin Debate*. In *Dioxins and Health*, A. Schecter, Ed., Plenum Press, New York, 1994, 1-6.

46. Courtney, K.D. *Environ. Res.* **1979**, 20, 225-266.

47. Westberg, H., Selden, A. *Organohalogen Compounds* **1994**, 20, 355-358.

48. Bailey, R. Global Hexachlorobenzene Emissions. Report to the Chlorine Chemistry Council of the Chemical Manufacturers Assoc. 1998. 54pp.

Chapter 2

The 'B' in PBT: Bioaccumulation

D. T. H. M. Sijm

**National Institute of Public Health and the Environment,
P.O. Box 1, NL–3720 BA Bilthoven, The Netherlands**

Bioaccumulation results in concentrations of substances in organisms that are much higher than the concentrations in their surrounding medium or food. Contaminants that accumulate through the foodchain must be (1) taken up, (2) persistent against biotransformation, and (3) excreted at a very slow rate. Most of the existing knowledge on bioaccumulation is based on studies on organic substances in aquatic species, on predators who prey on aquatic species, on terrestrial plants, and on the subsequent foodchain. The physical-chemical parameter which is often used as a surrogate for the bioaccumulation potential of organic substances from water and food is the octanol/water partition coefficient (Kow), while that for the route via air is the octanol/air partition coefficient (Koa). For metals or other non-organic substances, that strongly bioaccumulate, no surrogate physical-chemical parameters are known.

A polar bear does not drink millions of liters of seawater, but ultimately does receive its concentration of PCBs of ca. 1-10 mg/kg in its fat (1-2) from water. If the concentration of PCBs in polar seawater is similar as in the Great Lakes, in which the PCB concentration is approximately 0.5 ng/L (3), the bioaccumulation factor for PCBs in polar bears would be around 20,000,000 L/kg! This bioaccumulation factor would even be two orders of magnitude higher if the aqueous concentration of PCBs of ca. 10 pg/L in the Bering Sea would be chosen (4). In other words, the polar bear

© 2001 American Chemical Society

has to extract an equivalent of millions of liters of water to accumulate the PCBs in his fat to such high levels. Again, the polar bear does not drink the large amounts of water, but receives the PCBs from his food. Its food includes seals exclusively, which in turn prey on fish, which in turn prey on even smaller aquatic species. Whereas the polar bear is not, its food is thus in direct contact with water.

The present chapter illustrates that for contaminants to show up high in the foodchain, they have to meet three important criteria: (1) they must be taken up, (2) they must be persistent against biotransformation processes along the foodchain, and (3) they must be excreted at a very slow rate. Only two types of substances meet these criteria, i.e. persistent, hydrophobic organic substances and some (organo)metals. In order to understand uptake and the slow depuration, exposure as well as depuration routes for different types of organisms will be explained. Two modulating factors, i.e. biotransformation, which reduces persistence and increases depuration, and bioavailability, which affects uptake, will finally be explained.

Definitions

Bioaccumulation can simply be viewed as the process of a chemical moving from an organism's medium (water, sediment, soil or air) or diet into the organism. Uptake by respiratory organs (gills, lungs, skin, etc.) is an important route for aquatic, benthic and terrestrial species. Uptake by leaves exposed to air can be an important route for terrestrial plants. Uptake from the food by the gastro-intestinal tract is an important uptake route for sediment and soil ingesting organisms, such as worms, and for animals higher in the food-chain, such as mammals, e.g. polar bears, or fish-eating birds. Bioaccumulation thus takes into account all possible exposure routes. *Bioconcentration* is the process of a chemical moving from water to an organism, and thus only takes into account the aqueous compartment as exposure medium. *Biotransformation* is the enzyme catalyzed conversion of a contaminant.

Bioaccumulation kinetics are usually described by a simple first-order one-compartment model (Table I). Depending on the route of exposure, the uptake and depuration rates may change (Table I). In the following sections, each route of uptake, and subsequently, the depuration routes are described.

Exposure Routes

Exposure may or may not result in uptake. Contaminants that are highly bound to abiotic particles, may not or slowly be taken up. Contaminants that have a molecular dimension which is too big to allow passage over gill membranes will not be taken up from water, although they may be taken up from the food. As shown in the introduction, polar bears will receive organic contaminants primarily from their food rather than directly from water. Other organisms, however, receive most of their contaminant burden from water, again others from sediment, soil, or air. In the following sections it is shown that primarily physiological and habitat parameters, e.g. body weight and where the organism resides, will determine the primary exposure route. Physical-chemical parameters will then add to the magnitude of bioaccumulation and the rates of exchange.

Table I. Bioaccumulation models for different routes of exposure.
Bioaccumulation is described by: $dC_o/dt = k_{u,i} \cdot C_{ambient} - k_e \cdot C_o$, where C_o is the concentration of the contaminant in the organism (in mg/kg), $k_{u,i}$ is the uptake rate constant, where i indicates the exposure route, k_e is the depuration rate constant, and t is time (in d).

Exposure route	Concentration in The exposure Medium ($C_{ambient}$)	Uptake rate constant ($k_{u,i}$)	Depuration rate constant (k_e)
Water	C_w (in mg/L)	$k_{u,w}$ (in L/kg · d)	k_e (in 1/d)
Sediment	$C_{sediment}$ (in mg/kg)	$k_{u,sed}$ (in kg/kg · d)	k_e (in 1/d)
Soil	C_{soil} (in mol/kg)	$k_{u,soil}$ (in kg/kg · d)	k_e (in 1/d)
Air	C_{air} (in mol/m^3)	$k_{u,air}$ (in m^3/kg · d)	k_e (in 1/d)
Food	C_{fd} (in mol/kg)	$k_{u,fd}$ (in kg/kg · d)	k_e (in 1/d)

In general, fish and phytoplankton take up contaminants from water, terrestrial plants take up substances from air and soil, and mammals take up contaminants from food (Figure 1). The general simplistic view on uptake is depicted in Figure 2. Substances enter the organism from an external medium or via the gastro-intestinal tract, are transported across a stagnant external layer, further across a membrane, and finally enter the organism. It is the fugacity-driven concentration difference between the external medium and the internal body of the species that causes the substance to move from outside to inside, or the reverse (5). Most organic substances are taken up passively, i.e. it is the concentration (or fugacity) difference that causes uptake. Some substances, among them a few metals, can be taken up against a concentration gradient, where active uptake processes, such as ATP-driven ion channels, play a role. Other substances are being adhered to the species, such as extremely hydrophobic organics on air particles, which are scavenged by plants (6).

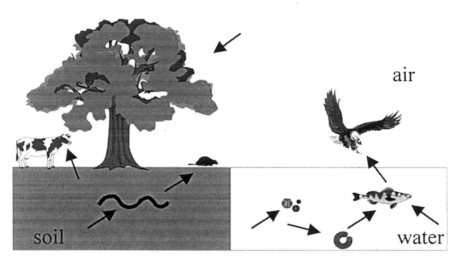

Figure 1. Species in the different environmental compartments will be exposed to substances via various routes, i.e. water, air, soil, sediment, and/or food. Arrows indicate chemical uptake routes.

Water

Most aquatic organisms receive their contaminants directly from water. The main exposure route is via the respiratory organs, such as the skin, the gills, or the cell membranes. Small, unicellular organisms take up or excrete oxygen, carbon dioxide, minerals and nutrients via their cellular membranes. This membrane is meant to selectively pass molecules in order to maintain homeostasis and thus to selectively take up and excrete certain substances. Heavy metals such as Cd^{2+}, Hg^+, etc. can be taken up by membrane systems that were designed to take up Na^+, K^+, Mg^{2+}, Ca^{2+}, etc. Neutral organic substances, such as polychlorinated phenols (PCPs), polychlorinated biphenyls (PCBs), polycyclic aromatic hydrocarbons (PAHs), etc. generally travel passively through the lipid bilayers comprising the membranes.

Most substances are taken up through the respiratory surfaces, such as skin, gills, and membranes. Therefore, the rate of uptake is usually a function of the respiratory surface area. For example, Sijm et al. (7) showed that uptake rates of hydrophobic organics in fish linearly increased with gill surface area. Since the surface area does not linearly relate to aquatic organisms' body weight, but approximately to 1/3 of it, uptake rate constants theoretically relate to body weight to the power -1/3. For body weights ranging 15 orders of magnitude, i.e. for algal species in the low μg range to trout in the kg range, the uptake rate constant was empirically found to relate to the body weight to the power -0.23 (8), which is close to the theoretical exponent. For hydrophobic organics, the uptake rate is thus primarily related to body weight, or respiratory surfaces, among different aquatic species.

For fish, the uptake rate of organic substances from water is thus highly related to the gill ventilation rate. A small fish, such as a 0.5 g fathead minnow ventilates approximately 1 liter of water each day. A large fish, such as a 1 kg trout ventilates approximately 100 liter water each day. Hence, each year, the fathead minnow and the trout ventilate approximately 700 and 36,500 liters, respectively. While ventilating, the fish will extract organic contaminants from water, and if these substances are eliminated slowly, the contaminants will bioaccumulate in the fish.

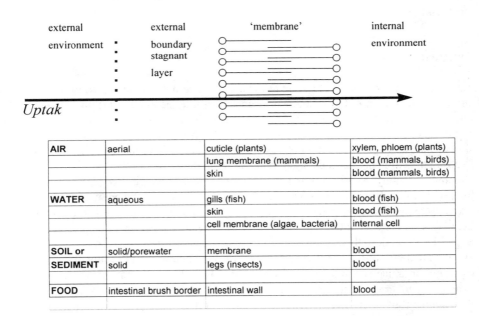

AIR	aerial	cuticle (plants)	xylem, phloem (plants)
		lung membrane (mammals)	blood (mammals, birds)
		skin	blood (mammals, birds)
WATER	aqueous	gills (fish)	blood (fish)
		skin	blood (fish)
		cell membrane (algae, bacteria)	internal cell
SOIL or	solid/porewater	membrane	blood
SEDIMENT	solid	legs (insects)	blood
FOOD	intestinal brush border	intestinal wall	blood

Figure 2. Schematic picture of the most important barriers for uptake for organisms exposed to air, water, soil or sediment, or food.

Within one species, uptake further depends on physical-chemical parameters. This can be explained by taking a further look into the uptake process (Figure 2). For uptake from water, the substance needs to pass the aqueous boundary layer and the membrane itself. Furthermore, the external water is assumed to be well mixed and homogenous. Large aquatic organisms actively ventilate water through their respiratory organs. Smaller species, such as algae and bacteria do not have a ventilation organ, but their motility may drive refreshment of the surrounding water. Inside the organisms, either blood has the distributing function, or in the case of unicellular organism, distribution is mainly by diffusion. It has been shown that for organic substances with a log Kow < 3, the rate limiting step is transfer across the membrane, while for organic substances with a log Kow > 3, the rate limiting step is transfer across the aqueous boundary layer (9). The largest uptake rate constants within a species are found for organic substances with log Kow > 3. Hence, the Kow is an important physico-chemical parameter which affects the uptake rate. Other physico-chemical parameters that affect the uptake rate are molecular weight, molecular size, charge and degree of ionization (10-11). In particular, large molecular dimensions may prevent substances from passing the respiratory surfaces (12), although these large substances may be taken up through the gastro-intestinal tract (13).

For other substances, such as metals, uptake highly depends on the environmental properties, such as salinity and pH, which affects the speciation of the metal and its uptake rate.

Sediment

Benthic species that live in or close to the sediment are exposed to sediment and contaminants in it. In the sediment, substances may be in equilibrium with the interstitial or overlying water. Organisms in the sediment may thus take up these contaminants from either the aqueous phase or directly from the sediment through the skin or the gastro-intestinal tract (GIT). Sediment-ingesting organisms will be exposed to sediment and the contaminants in it via their GIT. In the GIT, contaminants may desorb from the sediment into an aqueous phase and be assimilated by the organism from this aqueous phase. Thus, how much of the contaminants are taken up by benthic organisms directly from the sediment, and how much indirectly, i.e. via the interstitial or overlying water, is still subject of discussion.
Some theories adopt the Equilibrium Partitioning Theory (EQT) which simply assumes that equilibrium will exist between the concentrations in sediment and (benthic) organisms, and the interstitial water (14). It then does not matter how exactly organisms obtain their contaminant, i.e. from the solid or aqueous phase. A few studies show that direct uptake from sediment does occur (15-16). The uptake rate constants from sediment for organic substances vary several orders of magnitude for similar substances (e.g. 17-20).

Soil

Species that live in or on the soil are exposed to soil and contaminants in it. In the soil substances may be in equilibrium with the interstitial water (porewater). Organisms in the soil may thus take up these contaminants from either the aqueous phase or directly from the sediment. Soil-ingesting organisms will be exposed to sediment and the contaminants in it via their GIT. In the GIT, contaminants may desorb from the soil into an aqueous phase and subsequently be taken up. Thus, how much of the contaminants are taken up by organisms directly from the soil, and how much indirectly, i.e. via the porewater, is still subject of discussion. As is the case with with sediment, some theories adopt the EQT which simply assumes that equilibrium will exist between the concentrations in soil and organisms, and the porewater. Several studies showed that uptake and bioaccumulation can be explained by the EQT for hydrophobic and (partly) dissociated organic substances (21). Belfroid et al. (22) showed that the direct uptake from food or from the soil only dominates the porewater route of uptake for very hydrophobic substances. As for sediment, few studies have systematically investigated the relationship between either physiological or physico-chemical parameters on the uptake rate of substances from soil.

Air

Air-breathing organisms, such as mammals, or organisms that are in close contact with air, such as terrestrial plants and trees, are exposed to contaminants in the air. For mammals, the uptake of hydrophobic organics from air is negligible when compared to the uptake via the foodchain. For terrestrial plants, however, air may be the dominating route of exposure for organic substances (6).
The cuticular layer of plants has a strong capacity to extract the organic chemicals from air. There are some interspecies studies relating bioaccumulation to different plant properties, but as yet there is no strong evidence relating the surface area or other plant properties to uptake. One of the physical-chemical properties that relates to bioaccumulation in plants is the octanol-to-air partition coefficient (Koa). BCF increases with increasing Koa to the power a, where the exponent a highly depends on the plant (23).
Organisms that eat plants, such as cattle, take up contaminants along with the plants. Thus, indirectly they take up contaminants from the air, since the contaminant burden in the plants is the result of the plants scavenging the air. Such transfer ratios for contaminants, i.e. the ratio between the concentration in cow's milk or beef and the concentration in air, have been reported occasionally (e.g. 24).

Food

Chemical uptake from food seems to be related mainly to the efficiency by which the chemicals are taken up and to the feeding rate. The efficiency may depend on physico-chemical properties, such as Kow, molecular weight and degree of ionisation, and on the quality of the food (20,25). The feeding rate is the amount of food an organism consumes per unit of body weight per day. For organic chemicals, the uptake efficiency for fish is approximately 50%, but varies greatly. The variation in uptake efficiency probably is explained by differences in type of food and the age of the organism. For earthworms, the contribution of dietary uptake to the concentration in the organism is small, except for hydrophobic chemicals (log Kow > 5) in soils with a high organic matter content (22). Whereas molecular size seems to limit the uptake of contaminants from water, even large molecules such as decabromodiphenylether, which has a molecular weight of 959 are taken up from the food by fish, albeit at an uptake efficiency of less than 1% (13).

Various Routes of Uptake

In general, for aquatic, benthic and terrestrial organisms, uptake of organic substances from water (bulk water, interstitial or overlying water or porewater) seems to be the most relevant route of uptake when organisms are exposed to the substance via multiple routes. Only very hydrophobic organic substances are taken up predominantly from food (22,26-29). For plants, air is the dominating route, for mammals and other air-breathing organisms, food is the major route. The dominant route of uptake for metals may be any route.

Depuration Routes

There are many routes by which an organism can eliminate a substance or by which it can reduce the concentration of that substance (see e.g. 30). A few of these depuration routes are depicted in Figure 3, such as passive and active excretion, growth, reproduction, and biotransformation. Passive depuration is elimination of the substance through the same route as uptake occurs, i.e. via the respiratory surfaces, however, in the opposite direction. Active excretion involves energy-driven processes that eliminate the substance from the organism. Growth of the organism simply results in dilution of the substance, and thus in a lower concentration, although the substance does not necessarily have to leave the species. Excretion by reproduction may be via eggs or sperm, or via lactation. Biotransformation will transform the substance into another one, which by definition eliminates the parent substance.

For organic substances, passive excretion to water generally depends on the hydrophobicity of the substance, usually expressed by Kow and the molecular weight of the substance, and further by the lipid content and the weight of the organism (10). As with chemical uptake, passive depuration processes thus depend on the relative size of the respiratoru surfaces (10). Depuration of hydrophobic organics is so slow because of the preference of the organic substances is much higher for the lipid-rich or organic carbon (31) parts of the organisms than for its surrounding medium. Depuration is either expressed as a depuration rate constant (in 1/d) or as a half-life (in d). The half-life of a substances can vary orders of magnitude, depending on the species. For example, the half-life of a PCB is in the range of hours in phytoplankton (8), whereas it is years in the eel (32).

For metals, which are usually not stored in the fatty tissues of an organism, there is yet no single physico-chemical property that relates to depuration to water or to bioaccumulation in general. Again, depuration of some metals may be slow, because the preference of the metal for body constituents, e.g. proteins or bone tissue, is much higher than for its surrounding medium.

Depuration to soil or sediment depends on biological and/or physical-chemical properties, but how is still poorly understood. Although the same properties relate to depuration to water and depuration to soil or sediment, the organic matter content of the soil seems important for the depuration to soil (33).

All other depuration routes depend on the biological properties of an organism, and further depend on physiology, age, habitat, season, etc. Usually, the depuration routes have not been studied individually. Therefore, it is very difficult either to distinguish or to quantify the individual routes.

Since depuration generally decreases with increasing body size, large organisms will eliminate substances at a slower rate than smaller organisms. Large organisms are thus more vulnerable to biomagnification than smaller ones.

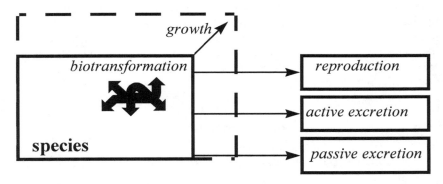

Figure 3. Schematic of the various routes of depuration of a substance from a species.

Foodwebs

Some top predators accumulate high amounts of contaminants through their foodchain. Individuals of the same species, however, may contain different levels of contaminants, depending on the habitat or foodweb they inhabit. For example, Kidd et al. (34) explained the different levels of toxaphene and other hydrophobic contaminants in lake trout and burbot, which are top predators in Canadian lakes, by differences in the foodweb structure. They used isotope ratios ($^{15}N/^{14}N$) to identify the length of the foodchain and noted that foodchain length was postively related to the extent of biomagnification in the two species in different Canadian lakes.

To fully understand the routes through which substances move through the foodchain, several foodweb models can be used. Some examples of foodweb models are that for the herring gull (35), an aquatic foodchain (36), a benthic and pelagic food chain (37), and Contaminants in Aquatic and Terrestrial ecoSystems (CATS) (38).

Bioaccumulation

For each species, the resulting bioaccumulation factor, i.e. the resultant of exposure and depuration, can be distinguished. The BCF (bioconcentration factor) is used when the primary medium of chemical exposure is water. The BMF (biomagnification factor) is used when the primary medium of chemical exposure is food. When air is the primary medium of chemical exposure the bioaccumulation factor (BAF) is used and when soil or sediment is the primary medium of chemical exposure the BSAF (biota-soil(or sediment)-accumulation factor) is used.
Most of the information on bioaccumulation is obtained for neutral organic substances. For those substances, log BCF generally increases with log Kow, until log Kow 5-6, where the BCF may not further increase, or even drops (39-40). BCF values may reach numbers as high as 10^6 L/kg. BMF and BSAF values are generally much lower than 0.1 kg/kg for substances with a log Kow <3, and may reach values up to approximately 10 kg/kg for substances with a log Kow > 4 (26,39,41-43). The main reason why the concentration in the predator may become higher than in its food is that within the GIT, the concentration of the contaminant in the food increases due to food digestion (43). BAF values for exposure via air may reach numbers as high as 10^9 m³/kg (6,24,45).

Modulating Factors

The most important modulating factors on bioaccumulation and exposure are biotransformation and bioavailability. Biotransformation mainly affects the depuration rate, and thus the magnitude of the bioaccumulation or biomagnification factors. Bioavailability does not alter the magnitude of the bioaccumulation factor, but it does affect the actual exposure concentration.

Biotransformation

Biotransformation will transform the substance into (an)other one(s), i.e. the metabolite(s), and by definition eliminates the parent substance. For example, the non-2,3,7,8-substituted chlorinated dibenzo-p-dioxins do not accumulate through the foodchain, whereas the 2,3,7,8-substituted dioxins accumulate strongly (e.g. 46). The metabolite may be either more toxic or less toxic than the parent, and may be more easily excreted or be more slowly eliminated than the parent compound. In general, biotransformation of hydrophobic organic substances leads to more polar substances, which will be faster excreted and have a lower bioaccumulation potential than their parent (e.g. 46). However, there are notorious exceptions. For example, benzo(a)pyrene is biotransformed, but one of its metabolites is very reactive, forms a stable adduct with DNA, and is excreted much more slowly than the parent. Another example is the methylsulphone-PCBs (48), which are metabolites of PCBs and are excreted more slowly than the parent PCBs.

Species differences with regard to biotransformation, have a large effect on the presence of hydrophobic organic substances through the foodchain. For example, some PCB congeners are not found in one top predator but are found in other top predators or lower throphic level organisms (e.g. 49).

Bioavailability

In any environmental compartment, substances that show a strong tendency to bioaccumulate, in general also exhibit a strong tendency to bind to non-polar particles present in the media, e.g. suspended solids in water, to soil or sediment particles, gaseous particles in air or to the feed that is being eaten by the predators. The strength of the bond between substance and particle affects the rate at which the substance can be taken up from the environmental medium by an organism. For example, highly hydrophobic substances in air will bind strongly to gaseous particles. These particles may be scavenged by plants, but the substances will most likely not be taken up by the plants themselves (6). Cattle that feed on the grass, however, will take up the plants, associated particles and substances, and thus be exposed to substances that were formly present in the air. In water, the particles may reduce the truly dissolved concentration of hydrophobic substances, which results in a lower actual exposure concentration to an organism from water (50). The substances, however, may be taken up in the particulate form via the food.

The EPT assumes that when either the concentration in the soil, the pore water, or the organism is known, the concentrations in the other two can be estimated, at equilibrium. Although concentrations of chemical in pore water are rarely reported, some studies have confirmed the expected relationship between the concentrations in organisms and the concentrations in the soil or sediment, respectively (overview in 51). However, in other studies, significant deviations, up to a factor of 30, were observed between the concentration in soil/sediment and that in the organism (51). These deviations may originate from different BCF values caused by (selective) feeding behaviour and additional uptake, biotransformation and bioturbation, and/or different Kp values caused by ageing or sequestration, composition of organic material and the presence of oil (51).

Thus, although the EPT relates the concentrations in pore water to concentrations in soil or sediment and organisms, deviations from EPT are observed that require further attention. Some explanations relate to biological factors, other relate to physical-chemical factors that reduce bioavailability.

24

Final remarks

In the introduction it was stated that a polar bear receives its PCB-burden indirectly from the aqueous compartment, and thus indirectly extracts millions of liters of water. The previous sections provide the necessary information to understand this process. The following back-of-the-envelope calculation further illustrates this analogy. We assume that a polar bear weighs 500 kg, of which 20 % is lipid, and eats 10 kg of seal skin and fat each day. We assume that the seal weighs 50 kg and eats one 5 year-old trout of 1 kg each day. The PCB concentration in the fish results from the extraction of approximately 180,000 L of water (5 yr x 365 d/yr x 100 L/kg/d). A seal thus consumes an equivalent PCB burden of approximately 67,000,000 L of water each year (365 d x 1 kg/d of fish). Subsequently, the polar bear consumes an equivalent PCB burden of approximatley 5,000,000,000 L of water each year (0.2 seal per day x 365 d/yr).

Each year, the bear thus receives an equivalent of approximately 50 mg of PCBs (5,000,000,000 L x 10 pg/L). If this is distributed over the polar bear's lipid, this will result in a concentration of approximately 0.5 mg/kg in the lipid. Each subsequent year, the lipid concentration will increase, because PCBs are depurated slowly and the PCBs are not biotransformed appreciably by polar bears.

Conclusions

The present paper shows that for bioaccumulation through the foodchain to occur, a substance has to meet three conditions, it has to be taken up from its surrounding medium or its food, it needs to be persistent, i.e. it should not be biotransformed, and it should be eliminated at a very slow rate. Biotransformation reduces the persistence of a substance and increases the depuration rate. Bioavailability may affect how much of a substance is available for uptake. Most known substances that bioaccumulate through the foodchain are hydrophobic organic substances, although some (organo)metals are also known.

References

1. Norstrom, R.J.; Simon, M.; Muir, D.C.G.; Schweinsburg, C.E. *Environ. Sci. Technol.* **1988**, *22*, 1063-1071.
2. Muir, D.C.G.; Wagemann, R.; Hargrave, B.T.; Thomas, D.J.; Peakall, D.B.; Norstrom, R..J. *Sci. Total Environ.* **1992**, *122*, 75-134.
3. Baker, J.E.; Eisenreich, S.J. *Environ. Sci. Technol.* **1990**, *24*, 342-354.
4. Iwata, H.; Tanabe, S.; Sakai, N.; Tatsukawa, R. *Environ. Sci. Technol.* **1993**, *27*, 1080-1098.
5. Mackay, D. *Multimedia environmental models. The fugacity approach.* Lewis Publishers, In.c, Chelsea, Michigan , USA.

6. McLachlan, M.S. *Environ. Sci. Technol.* **1999**, *33*, 1799-1804.
7. Sijm, D.T.H.M.; Verberne M.E.; de Jonge, W.J.; Pärt, P.; Opperhuizen, A. *Toxicol. Appl. Pharmacol.* **1995**, *131*, 130-135.
8. Sijm, D.T.H.M.; Broersen, K.W.; de Roode, D.F.; Mayer, P. *Environ. Toxicol. Chem.* **1998**, *17*, 1695-1704.
9. Gobas, F.A.P.C.; Opperhuizen, A.; Hutzinger, O. *Environ Toxicol Chem* **1986**, *5*, 637-646.
10. Sijm, D.T.H.M.; van der Linde, A. *Environ. Sci. Technol.* **1995**, *29*, 2769-2777.
11. Saarikoski, J.; Lindström, R.; Tyrnelä, M.; Viluksela, M. *Ecotoxicol. Environ. Saf.* **1986**, *11*, 158-173.
12. Opperhuizen A. *Bioconcentration of hydrophobic chemicals in fish.* In Poston, T.M.; Purdy, R., Eds, *Aquatic Toxicology and Environmental Fate,* American Society for Testing and Materials, ASTM STP 921, Philadelphia, PA, USA, 1986, pp. 304-315.
13. Kierkegaard, A.; Balk, L.; Tjärnlund, U.; de Wit, C.A.; Jansson, B. *Environ. Sci. Technol.* **1999**, *33*, 1612-1617.
14. DiToro, D.M.; Zabra, C.S.; Hansen, D.J.; Berry, W.J.; Swartz, R.C.; Cowan, C.E.; Pavlou, S.P.; Allen, H.E.; Thomas, N.A.; Paquin, P.R. *Environ. Toxicol. Chem.* **1991**, *10*, 1541-1583.
15. Kukkonen, J.; Landrum, P.F. *Aquatic Toxicol.* **1995**, *32*, 75-92.
16. Leppänen, M.; Kukkonen, J.V.K. *Environ. Toxicol. Chem.* **1998**, *17*, 2196-2202.
17. Oliver, B.G. *Can. J. Fish. Aquat. Sci.* **1984**, *41*, 878-883.
18. Oliver, B.G. *Environ. Sci. Technol.* **1987**, *21*, 785-790.
19. Landrum, P.F. *Aquatic Toxicol.* **1988**, *12*, 245-271.
20. Gobas, F.A.P.C.; Muir, D.C.G.; Mackay, D. *Chemosphere* **1988**, *17*, 943-962.
21. Van Gestel, C.A.M.; Ma, W.C. *Ecotoxicol. Environ. Saf.* **1988**, *15*, 289-297.
22. Belfroid, A.; Seinen, W.; Van Gestel, K.; Hermens, J.; van Leeuwen, K. *Environ. Sci. Pollut. Res.* **1995**, *2*, 5-15.
23. Kömp, P.; McLachlan, M.S. *Environ. Sci. Technol.* **1997**, *31*, 2944-2948.
24. Thomass, G.O.; Sweetman, A.J.; Lohmann, R.; Jones, K.C. *Environ. Sci. Technol.* **1998**, *32*, 3522-3528.
25. Clark, C.E.; Mackay, D. *Environ. Toxicol. Chem.* **1991**, *10*, 1205-1217.
26. Thomann, R.V. *Can. J. Fish. Aquat. Sci.* **1981**, *38*, 280-296.
27. Thomann, R.V.; Connolly, J.P. *Environ. Sci. Technol.* **1984**, *18*, 65-71.
28. Thomann, R.V.; Connolly, J.P.; Parkerton, T.F. *Environ. Toxicol. Chem.* **1992**, *11*, 615-629.
29. Walker, S.L.; Gobas, F.A.P.C. *Environ. Toxicol. Chem.* **1999**, *18*, 1323-1328.
30. Sijm, D.T.H.M.; Seinen, W., Opperhuizen, A. *Environ. Sci. Technol.* **1992**, *26*, 2162-2174.
31. Skoglund, R.S.; Swackhamer, D.L. *Environ. Sci. Technol.* **1999**, *33*, 1516-1519.
32. De Boer, J.; van der Valk, F.; Kerkhoff, M.A.T.; Hagel, P.; Brinkman, U.A.Th. *Environ. Sci. Technol.* **1994**, *28*, 2242-2248.
33. Belfroid, A.C.; Sijm, D.T.H.M. *Chemosphere* **1998**, *37*, 1221-1234.
34. Kidd, K.A.; Schindler, D.W.; Hesslein, R.A.; Muir, D.C.G.; *Sci. Total Environ.* **1995**, *160/161*, 381-390.

35. Norstrom, R.J.; Clark, T.P.; MacDonald, C.R. *The herring gull bioenergetics and pharmacodynamic model.* National Wildlife Research Centre, Canadian Wildlife Service, Environment Canada, Hull, Québec, Canada, 1991.
36. Gobas. F.A.P.C. *Ecological Modelling* **1993,** *69,* 1-17.
37. Morrison, H.A.; Gobas, F.A.P.C.; Lazar, R.; Whittle, D.M.; Haffner, G.D. *Environ. Sci. Technol.* **1997,** *31,* 3267-3273.
38. Traas, T.P.; Stäb, J.A.; Kramer, P.R.; Cofino, W.P.; Aldenberg, T. *Environ. Sci. Technol.* **1996,** *30,* 1227-1237.
39. Opperhuizen, A.; Sijm, D.T.H.M. *Environ. Toxicol. Chem.* **1990,** *9,* 175-186.
40. Nendza, M. *QSARs of bioconcentration: validity assessment of log P_{OW}/log BCF correlations.* In: Nagel, R.; Loskill, R., Eds., *Bioaccumulation in aquatic systems,* VCH Weinheim, Germany, 1991, pp. 43-66.
41. Morrison, H.A.; Gobas, F.A.P.C.; Lazar, R.; Haffner, G.D. *Environ. Sci. Technol.* **1996,** *30,* 3377-3384.
42. Hendriks A.J. *Chemosphere* **1995,** *30,* 265-292.
43. Tracey, G.A.; Hansen, D.J. *Arch. Environ. Contam. Toxicol.* **1996,** *30,* 467-475.
44. Gobas, F.A.P.C.; Zhang, X.; Wells, R. *Environ. Sci. Technol.* **1993,** *27,* 2855-2863.
45. Tolls, J.; McLachlan, M.S. *Environ. Sci. Technol.* **1994,** *28,* 159-168.
46. Sijm, D.T.H.M.; Opperhuizen, A. *Chemosphere* **1988,** *17,* 83-99.
47. De Wolf, W.; de Bruijn, J.H.M.; Seinen, W.; Hermens, J.L.M. *Environ. Sci. Technol.* **1992,** *26,* 1197-1201.
48. Brandt, I.; Bergman, Å. *Chemosphere* **1987,** *16,* 1671-1676.
49. Kannan, N.; Reusch, T.B.H.; Schulz-Bul, D.E.; Petrick, G.; Duinker, J.C. *Environ. Sci. Technol.* **1995,** *29,* 1851-1859.
50. Hamelink, J.L.; Landrum, P.F.; Bergman, H.L.; Benson, W.H. *Bioavailability. Physical, chemical, and biological interactions;* SETAC Special Publication Series, Lewis Publishers, Boca Raton, USA, 1994.
51. Belfroid, A.C.; Sijm, D.T.H.M.; van Gestel, C.A.M. *Environ. Rev.* **1996,** *4,* 276-299.

Fate and Behavior

Chapter 3

Persistent Organic Pollutants in the Coastal Atmosphere of the Mid-Atlantic States of the United States of America

Steven J. Eisenreich, Cari L. Gigliotti, Paul A. Brunciak, Jordi Dachs, Thomas R. Glenn IV, Eric D. Nelson, Lisa A. Totten, and Daryl A. Van Ry

Department of Environmental Sciences, Rutgers, The State University of New Jersey, 14 College Farm Road, New Brunswick, NJ 08901 (E-mail: eisenreich@envsci.rutgers.edu)

Abstract

The concentrations of polychlorinated biphenyls (PCBs; ~ 60 congeners), chlordanes (cis, trans chlordane; cis, trans nonachlor), polycyclic aromatic hydrocarbons (PAHs; 36 compounds), and nonylphenols (NPs; 11 isomers) were measured in the urban, suburban and coastal atmosphere of the NY-NJ area near the lower Hudson River Estuary (HRE). Concentrations exhibited significant seasonal and directional variability and were driven by local and regional emissions from combustion sources (PAHs), surface exchange at the atmosphere-land and atmosphere-water interfaces sub-regionally (PCBs, NPs) and long range transport from areas to the south and west (chlordanes). PCB concentrations in the region are typically higher than observed elsewhere reflecting the mix of historical uses and urban/industrial density plus volatilization from estuarine waters. PAH concentrations are high and reflect the mix of strong combustion sources in the region, especially vehicular and heating emissions. NPs in the atmosphere are reported for the first time and their source is dominated by volatilization from the HRE. Chlordanes exhibit low concentrations except when warm air masses transport volatilized pesticides from the S/SE United States.

I. Introduction

Wet deposition via rain and snow, dry deposition of fine/coarse particles, and gaseous air-water exchange are the major atmospheric pathways for persistent organic pollutant (POP) input to the Great Waters such as the Great Lakes and Chesapeake Bay (1-3). The Integrated Atmospheric Deposition Network (IADN) operating in the Great Lakes (4, 5) and the Chesapeake Bay Atmospheric Deposition Study (CBADS)

© 2001 American Chemical Society

(6) were designed to capture the *regional* atmospheric signal, and thus sites were located in background areas away from local sources. However, many urban/industrial centers are located on or near coastal estuaries (e.g., Hudson River Estuary and NY Bight) and the Great Lakes. Emissions of pollutants into the urban atmosphere are reflected in elevated local and regional pollutant concentrations and localized intense atmospheric deposition that is *not* observed in the regional signal *(4, 5)*. The southern basin of Lake Michigan as one such location is subject to contamination by air pollutants such as PCBs and PAHs, Hg and trace metals *(1-3)* because of its proximity to industrialized and urbanized Chicago, IL and Gary, IN. Concentrations of PCBs and PAHs are significantly elevated in the Chicago and coastal lake offshore as compared to the regional signal *(7-10)*. Higher atmospheric concentrations are ultimately reflected in increased precipitation *(11)* and dry particle fluxes of PCBs and PAHs *(12)* and trace metals *(13, 14)* to the coastal waters as well as enhanced air-water exchange fluxes of PCBs *(15)*. The story is similar in the Chesapeake Bay *(16, 17)*.

The Hudson River Estuary is bordered by the densely urbanized and industrialized area of New York City, Connecticut, and northern New Jersey and in a prevailing transport regime downwind of other large atmospheric emission sources: Philadelphia, PA, Wilmington, DE, the Baltimore-Washington complex, and the Ohio River Valley. Except for Chesapeake Bay *(1)*, there is little information on atmospheric concentrations and deposition and fate of persistent organic pollutants (POPs) in the Mid-Atlantic States. Our goal was to perform long-term spatially targeted measurements of atmospheric POPs such as PCBs, chlorinated pesticides, PAHs, and emerging chemicals near the Hudson River Estuary (HRE). Here we report the first year of atmospheric concentrations including seasonal and spatial trends for PCBs, chlordanes, PAHs and nonylphenols (NPs) as an example of emerging chemicals.

II. Setting and Brief Experimental Description

Atmospheric research and monitoring stations were established at three locations surrounding the Hudson River Estuary (HRE) (Figure 1). The first site was established at the Rutgers Gardens Meteorological Station in August 1997 in a suburban area outside of New Brunswick, New Jersey (NB) about 1.5 km both from US Highway 1 and the NJ Turnpike (suburban site). The second site was established at Sandy Hook, NJ in February 1998 on a sandy spit reef extending into the NY Bight and serving as the boundary between Raritan Bay/Hudson River Estuary and the Atlantic Ocean (marine/coastal site). The third site was established at the Liberty Science Center (LSC) in Jersey City, NJ in July 1998 in the heart of the urban/industrial complex, on the HRE and across the Hudson River from Manhattan and the Statue of Liberty (urban-industrial site). Each site is equipped with a modified organics Hi-volume air sampler with a quartz fibre filter (QFF) and polyurethane foam (PUF) adsorbent operating at a calibrated flow rate of ~0.5 m^3/min. At each site, 24-hour integrated air samples were collected every 6 days for

30

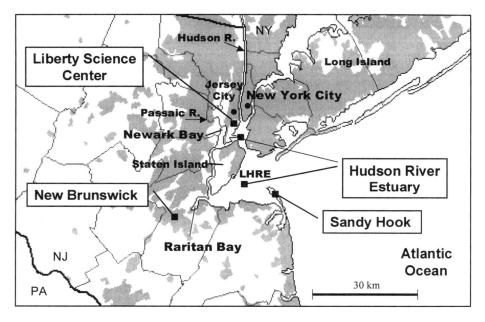

Figure 1. Map of land and water sampling sites as part of the measurement of POPs in the NY-NJ area and the lower Hudson River Estuary.

the first year. The sampling frequency was changed to once in 12 days to match the IADN as well as other long term monitoring programs. Integrated, wet-only precipitation samples were also collected every 12 (or 18) days using MIC collectors with 0.21 m^2 stainless steel collection funnels and attached glass columns filled with XAD2 resin. Additionally, an intensive sampling campaign was undertaken in the summer of 1998 that included air and water sampling from the Raritan Bay and lower HRE.

Samples were extracted with petroleum ether (PUF), dichloromethane (QFF) and acetone/hexane (XAD), and cleaned up and fractionated on 3% H_2O-deactivated alumina columns. The PCB fraction was analyzed on a HP 5890 GC equipped with a 60m high resolution glass capillary column (DB-5) and a Ni-63 electron capture detector. Samples were analyzed for chlordane (cis/trans), MC5, nonachlor (cis/trans), oxychlordane and PAHs using a HP 6890 Gas Chromatograph/5973 Mass Selective Detector in EI mode and selective ion monitoring. Nonylphenols and octylphenol were analyzed in the third fraction on a HP 5890 GC/5972 MS. Correlative measurements of total suspended particulate matter (TSP), particulate matter < 2.5 μm in aerodynamic diameter ($PM_{2.5}$) and particulate organic carbon (OC) and elemental carbon (EC) with supporting meteorological measurements (wind speed and direction, rainfall intensity, temperature) were also made but will not be reported extensively here. Back trajectories were obtained for each sampling interval using the NOAA HYSPLIT model. A complete quality assurance protocol was established and followed on all analytes and media.

III. Atmospheric Concentrations and Behavior

Polychlorinated Biphenyls (PCBs)

Urban/industrial areas are major sources of atmospheric polychlorinated biphenyls (PCBs) to surrounding waters (7, 11, 14, 15). Atmospheric transport from major urban/industrial areas can lead to significant loadings of PCBs to surrounding terrestrial and aquatic ecosystems (11, 18-22). Pathways for atmospheric PCB deposition to environmental surfaces include air-water exchange, air-vegetation exchange, wet deposition, and dry deposition. Once these aquatic and terrestrial surfaces have been loaded with PCBs, they may also act as a source to the regional atmosphere.

Measurements of atmospheric PCBs in the New Jersey/New York area are few although the Chesapeake Bay has been extensively studied (15, 16, 21, 23). Analysis of sediment and water from wastewater treatment systems reveals an extensive impact in the New Jersey/New York area (17, 24-27). For example, Durrell and Lizotte (24) found total PCB influent concentrations in wastewater treatment facilities ranging from 31 to 625 ng L^{-1} during normal flow (110 ng L^{-1} average). The other major source of PCBs to the Hudson River Estuary (HRE) derive from upstream loading of PCBs resulting from historical industrial releases and the continuing release from contaminated sediments (28). With elevated concentrations in water and sediments, and based on the vapor pressures of these compounds, atmospheric transport must play a key role in the cycling of PCBs.

High concentrations of PCBs in water and sediments have been well-documented for several rivers in the New Jersey/New York metropolitan area (17, 25). In contrast, little is known about atmospheric concentrations of PCBs in this area. Atmospheric PCBs derive from historical uses in the urban/industrial setting, surface to air exchange from local waters and terrestrial areas, emissions from landfills, and from sub-regional and long-range transport into the metropolitan area.

Table I shows the atmospheric PCB concentrations at the New Brunswick suburban, Sandy Hook coastal and Liberty Science Center urban/industrial sites. Concentrations of the gas (G) + particle (P) phase in air and in precipitation were highest at the LSC site. In comparison to other studies, the ΣPCB concentrations (~60 congeners) at the Sandy Hook coastal site are approximately 2X those reported at Sturgeon Point, NY, an Integrated Atmospheric Deposition Network (IADN) site (4, 5), but similar to reported values in the Chesapeake Bay, MD (15, 21). Although SH is a marine/coastal site, PCB concentrations are sufficiently high to suggest that "local" sources impact this site. Precipitation volume-weighted mean concentrations are higher at New Brunswick than at both the Sandy Hook site and Chesapeake Bay. The PCB concentrations measured in precipitation at the Sandy Hook site are comparable to those reported in Chesapeake Bay (21).

Figure 2 shows a comparison of ΣPCB concentrations (G+P) at the NB, SH, and LSC sites plotted with a temperature profile from NB. Average concentrations over the sampling period from October 1997 through November 1998 were higher at NB (560 pg m^{-3}) than at Sandy Hook (457 pg m^{-3}) from the commencement of sampling in February 1998 through November 1998. A comparison of the average ΣPCB concentrations over the same sampling period (February through November 1998) showed an even greater difference in concentration between the NB (605 pg m^{-3}) and SH (457 pg m^{-3}) sites. The patterns of ΣPCBs between the two sites are statistically different (p<0.05) indicating a different mix of sources and sinks. ΣPCB concentrations at NB exhibit a large temporal variability ranging from 60 pg m^{-3} to 2300 pg m^{-3} (1 SD = ±410 pg m^{-3}). Such significant variability in temporal signal is unusual with respect to studies in the Great Lakes and Cheaspeake Bay (4, 22, 29, 30). Similarly, the Sandy Hook site also exhibits significant variability with a range fo PCB concentrations of 84 pg m^{-3} to 1,100 pg m^{-3} (1 SD = ±300 pg m^{-3}). Variability of this magnitude generally is indicative of local and regional emission sources.

During a mini-intensive sampling campaign in July 1998, 6 and 12 hour air samples were simultaneously collected at 4 locations including over the water of the lower HRE (see map Figure 1). The concentrations at NB and SH co-vary ($r^2 = 0.96$) indicative strongly suggestive of a common source and atmospheric dispersion (Figure 3A). Concentrations at the LSC located amidst the NJ/NY urban-industrial complex show elevated concentrations in comparison to NB and SH. Over-water gas phase

Table I. Comparison of Recent PCB Measurements in Air and Precipitation.

PCBs Gas + Particle	New Brunswick, NJ (this study)	Sandy Hook, NJ (this study)	Liberty Science Ctr, NJ (this study)	Chesapeake Bay (15)	*Sturgeon Point, NY (4)	Chicago, IL (7)
Air (pg m^{-3})						
18	42.6	33.6	75.5	19.9	20	191
16+32	47.2	30.1	83.8	25.4		204
28	29.2	21.0	57.8	62.9		**432
52+43	32.0	32.4	56.0	15.8	16	95.7
41+71	9.6	10.0	21.8	15.8		111
66+95	45.8	39.5	75.4	19.3		303
101	17.2	14.4	26.5	33.3	10	51.4
87+81	8.9	7.2	12.6	6.8		29.1
110+77	18.9	14.3	25.0	3.7		90.7
149+123+107	6.6	6.0	10.2	7.9		28.7
153+132	6.5	6.0	10.4	7.1		70.9
163+138	7.4	6.2	9.9	10.1		42.8
187+182	2.4	2.5	2.6	4.4		7.8
174	1.3	1.0	1.7	2.3		4.9
180	1.9	1.5	2.1	1.8		44.4
ΣPCBs	560 ± 417	457 ± 322	959 ± 802	510	370	3100
Precipitation (ng L^{-1})						
ΣPCBs	1.3	0.97	5.1	1.6	0.7	29.3
Reference	(this study)	(this study)	(this study)	(23)	(4)	(10)

* gas phase only
** includes #31

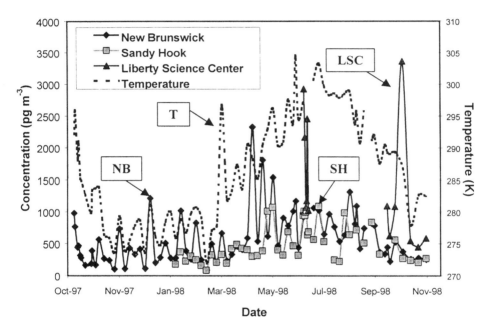

Figure 2. Seasonal ΣPCB concentrations (pg m⁻³) at New Brunswick, Sandy Hook, and the Liberty Science Center sampling sites in New Jersey as compared to daily average temperatures.

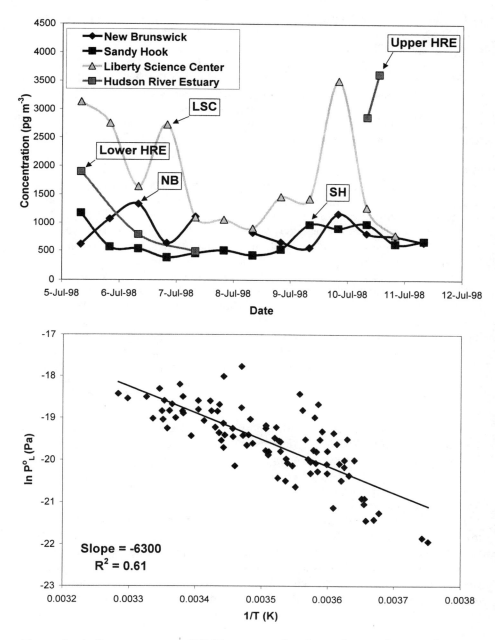

Figure 3. A. Concentrations of ΣPCBs measured in the study area during July 1998 at land and over-water sites (top). B. The relationship of ln P°_L (Pa) of ΣPCBs to 1/T at the NB site (bottom).

PCB concentrations the Hudson River Estuary ranged from 480 to 3,500 pg m^{-3} (mean = 1,800 pg m^{-3}) and were significantly higher than concentrations reported by Nelson et al. *(15)* for the Chesapeake Bay (590 pg m^{-3}). Simcik et al. *(7)* reported values of between 140 to 1130 pg m^{-3} over southern Lake Michigan, and Hornbuckle et al. *(31)* reported values of 670 to 2200 pg m^{-3} over southern Green Bay, Lake Michigan. Both of these latter sites are near emisison sources and waters of elevated PCB concentrations. The highest air concentration was measured over the water of the lower Hudson River opposite Manhattan implicating volatilization of PCBs from the water as a major source of atmosphere PCBs.

Tidal Effects

The dynamic variability of the atmospheric PCB signal is similar, in many respects, to the temproal variation on the estuarine tidal cycle. PCBs are relatively inert to atmospheric degradation processes on the time scale of days *(22, 32)* and their source to the atmosphere is most likely volatilization from water and land surfaces *(7, 33, 34)*. Point sources of PCBs and emission inventories to the atmosphere are unknown, although the availbility of PCBs in the New Jersey/New York area is evident in the sediment record and may represent a source to the atmosphere *(18, 35, 36)*. Volatilization of PCBs from sediments exposed to the air may very well contribute to enhanced atmospheric concentrations; thus a correlation between the tidal cycle and atmospheric PCB concentrations would reflect the importance of this input pathway. This mechanism is supported by studies showing that contaminated sediments release PCBs to a greater extent when wet compared to when dry *(35-37)*. However, regressions against the river flow and the magnitude of the tidal cycle of several rivers in the region showed no significant correlation.

Temperature Effects

The temperature dependence of atmospheric PCB concentrations is now well-documented *(38, 39)*. Specifically increases in atmospheric and therefore presumably land surfaces, results in the emission of PCBs driven both by the influence of T on the surface-air exchange process, and the mass of mobilizable PCBs available. At the NB site, temperature explains 61% of the total variability in ln PCB concentrations based on a Clausius-Clapeyron type plot (Figure 3B). Other researchers have found a "hockey stick effect" *(39)* where concentrations plateau below approximately 0-4°C. Figure 3 shows that ΣPCBs and individual congener concentrations do not exhibit such a plateau in this study. The occurrence of temperatures below 0°C was rare in the winter 1997-8 in the Mid-Atlantic States. The magnitude of the slope of the ln C vs. 1/T plot is indicative of the $\Delta H_{surface-air}$ Proximate PCB sources usually are reflected in a steeper slope *(38, 39)*. In this study, the slope of –6300 is similar to those from the urban areas such as industrial Bloomington, IN area (-7000 to -6000) *(38)*.

Assuming that temperatures at the Sandy Hook site vary in the same manner as temperatures at New Brunswick, a plot of ln [PCB] vs 1/T yields a slope of –4375 and a r^2 of 0.58. This slope is less steep than the slope seen at the New Brunswick site

(-6300) but is within the range of the rural/Great Lakes area (-5000 to –4000) *(38)*. This suggests a stronger contribution of longer-range transport and/or perhaps water-air exchange at the SH site (volatilization from water versus land). Future studies will focus on the magnitude of the air-water vs. air-land exchange of PCBs as a long-term source to the regional atmosphere.

Chlordanes

The atmospheric transport and deposition of organochlorine pesticides is a significant pathway for input of these chemicals to northern latitudes and the arctic *(40-42)*. Atmospheric transport of these chemicals into mid-latitude ecosystems such as the Great Lakes and Chesapeake Bay is also significant *(4, 21, 33)*. To assess the transport and dynamics of chlordanes in the coastal, urban, suburban and rural mid-Atlantic region, measurement of air concentrations provides insight into local, regional and long-range atmospheric transport and impacts to this highly industrial and populous region. Here we present data from 20 months of operation as well as a summer intensive campaign intended to elucidate diurnal variability. Back trajectories for each sampling day are used to explain concentrations that deviate substantially from mean values and to support our hypothesis that most chlordane compounds derive from long range transport.

Continuous measurements of chlordanes in New Jersey and in the lower Hudson River Estuary have yielded data indicative of minimum impact from local sources and major impact from regional and long range atmospheric transport. ΣChlordane concentrations (Σ chlordanes = trans+cis chlordane + trans+cis nonachlor) fall within a latitudinal gradient reported by others (Table II). Average Σchlordane concentrations were similar for both suburban and coastal sites over the twenty months of study at 123 and 93 pg m^{-3}, respectively (Figure 4). Trans/cis chlordane ratios also agree well with the north to south gradient for both sites (1.1 and 1.2 for New Brunswick and Sandy Hook, respectively). The ratio is diagnostic of the "relative" weathering of the trans isomer, which is most likely oxidized to oxychlordane. Particulate chlordane concentrations were a smaller percent of total chlordane in warm weather (<10%) but could contribute >50% of total concentration in cold weather. This gas-particle partitioning behavior is also described by other researchers *(43)*.

Temperature vs. Transport
ΣChlordane concentrations vary with T as observed by others *(44, 45)*. Highest concentrations occurred in July and lowest concentrations occurred in December. A Clausius-Clapeyron type plot applied to the individual chlordane isomers yields r^2 values ranging from 0.23-0.39, which are less than reported in other regions *(33, 44,*

38

Table II. ΣChlordane Concentrations and Cis/Trans Chlordane Ratios in a
Latitudinal Gradient in North America.

Sampling Site	Reference	ΣChlordane (pg m^{-3})	Trans/Cis Chlordane Ratio
Arctic	(41, 65)	4.0 ± 0.1	0.8 ± 0.1
Eagle Harbor, MI	(44)	13 ± 12	1.0 ± 0.2
Sleeping Bear Dunes, MI	(44)	22 ± 20	1.0 ± 0.2
Sturgeon Point, NY	(44)	37 ± 29	0.9 ± 0.3
Egbert, Ontario	(33)	39 ± 35	1.0 ± 0.3
New Brunswick, NJ	this study	123 ± 113	1.1 ± 0.1
Sandy Hook, NJ	this study	93 ± 117	1.2 ± 0.1
Columbia, SC	(45)	300 ± 147	1.6 ± 0.4
College Station, TX	(66)	1050 ± 180	
Midwest Soils	(67)		1.1 ± 0.5
Technical Chlordane	(45)		1.85
Technical Chlordane at equilibrium with air at 20°C (45)			1.30

ΣChlordane = sum of cis, trans chlordane + cis, trans nonachlor

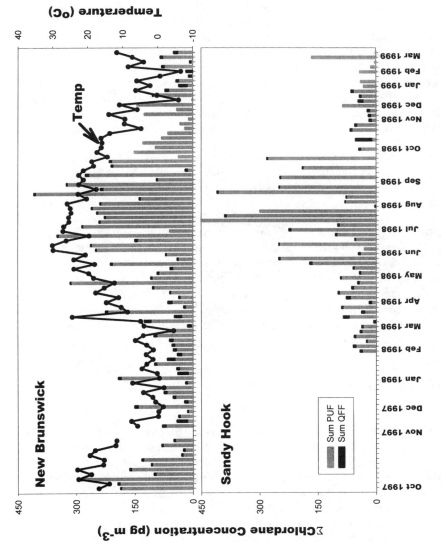

Figure 4. Sum of chlordane concentrations in the atmosphere at the suburban New Brunswick and coastal Sandy Hook sites in New Jersey (pg m⁻³).

45)(Figure 5). Although T explains between 20 and 40% of the variability in the log normalized concentrations, factors other than temperature govern chlordane concentrations in New Jersey air. When ln trans-chlordane concentrations are plotted against 1/T, ~25% of the residuals were greater than one standard deviation (Figure 6). Those residuals greater than one standard deviation were evaluated against three day back trajectories at the beginning, middle and end of the sampling period *(46)*. When observed air concentrations were greater than one SD above the regression line, air masses originated from the southern part of the United States. When air concentrations were less than one standard deviation below the regression line, air originated in central Canada and the Great Lakes region (Figure 7). High residuals rarely occurred when temperatures were close to 0°C. Most of the high residuals (<60%), both positive and negative, occur when air temperatures exceed 17°C. Thus high concentrations were linked closely to air masses and presumably sources from the southern US, and not to local sources.

The trans chlordane isomer is oxidized to oxychlordane by both biotic *(47)* and possibly by abiotic processes. During a summer mini-intensive sampling campaign (12 hour day and night sampling), diurnal variations in trans chlordane were observed that supported atmospheric oxidation. This is evident in the ratios of trans/cis chlordane and trans chlordane/oxychlordane. Trans/cis chlordane ratios are 1.05 during the day and 1.2 at night. Trans chlordane/oxychlordane ratios are higher at night by a factor of almost two, indicating more oxychlordane is present during daytime hours. Diurnal changes in both trans chlordane and oxychlordane suggest a photochemical oxidation process although photo-isomerism occurs as well *(48)*. Trans-chlordane may be converted to "photo trans-chlordane". Daily variation of ΣChlordane concentration is too dependent upon other factors to show any diurnal variability, and the mean concentration is similar for day and night samples at both NB (suburban) and SH (coastal).

Polycyclic Aromatic Hydrocarbons (PAHs)

Atmospheric transport and subsequent deposition delivers PAHs to locations both local and remote from emission source areas. Because PAHs arise from the incomplete combustion of fossil fuels and wood, they are ubiquitous in the atmosphere with particularly high concentrations measured in urban and industrial regions *(5, 7, 16)*. Sources of PAHs include vehicular traffic, oil and gas consumption (particularly used in home heating), petroleum refining, coal burning, and municipal and industrial incinerators. Emissions of PAHs in the urban/industrial areas increase chemical loadings to water bodies downwind of densely populated urban centers and industrial complexes *(7, 14, 16)*.

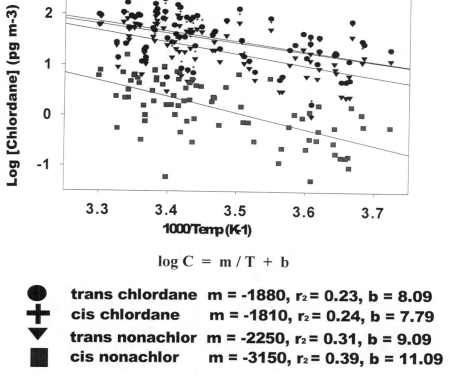

$$\log C \;=\; m \,/\, T \;+\; b$$

⬤ **trans chlordane** $m = -1880$, $r_2 = 0.23$, $b = 8.09$
✚ **cis chlordane** $m = -1810$, $r_2 = 0.24$, $b = 7.79$
▼ **trans nonachlor** $m = -2250$, $r_2 = 0.31$, $b = 9.09$
■ **cis nonachlor** $m = -3150$, $r_2 = 0.39$, $b = 11.09$

Figure 5. Relationship of log concentration of chlordane species with 1/T.

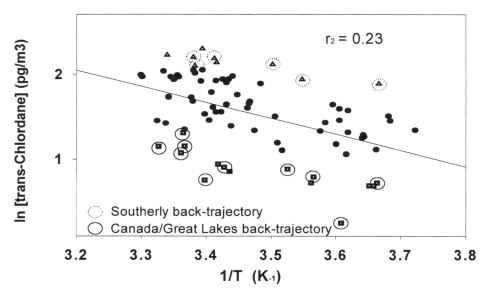

Figure 6. Regression of ln [trans-Chlordane] vs. 1/T; outlying data points (1 sd from mean) are shown as triangles and squares. Outlying points with three day back-trajectories from central Canada and from the southern US are circled.

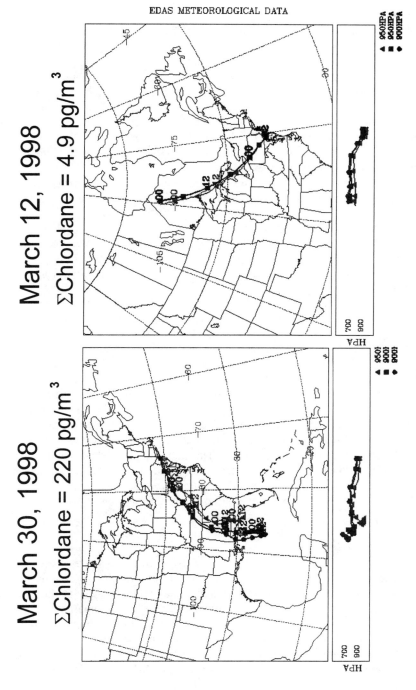

Figure 7. Three day back-trajectories for select outlier samples at New Brunswick.

Spatial Occurrence

ΣPAHs is defined here as the sum of 36 PAHs listed in Figure 8. The suburban NB gas phase ΣPAH concentrations range from 3.5 to 84 ng m^{-3} and were on average 2.4 times greater than at the coastal SH site where concentrations ranged from 2.8 to 42 ng m^{-3}. Particulate phase ΣPAH concentrations were approximately 2.5 times higher at NB where concentrations ranged from 0.38 to 12 ng m^{-3}, than at SH where concentrations ranged from 0.15 to 4.0 ng m^{-3}. The higher PAH concentrations measured at New Brunswick are consistent with the impact of multiple local major traffic arteries and residential areas. The SH site is less impacted than the NB site by PAH pollution due to its location on a peninsula away from the immediate impact of heavy traffic arteries, industry or urbanization.

In comparison to remote sites in the Great Lakes Basin *(4, 5)* the measured PAH concentrations indicate that the coastal Sandy Hook site cannot be classified as rural or remote. Sandy Hook is impacted by local residential, urban, and/or industrial regions in three directions: New York City to the north, the New Jersey urban/industrial complex to the northwest, and the heavily populated New Jersey coastline to the west, south, and southwest. Sandy Hook exhibits a diluted mix of PAH signals from the mainland, because the prevailing local winds derive from the W and SW (See map-Figure 1).

In Table III, gas and particulate phase PAH concentrations are compared to concentrations reported elsewhere. PAH concentrations at SH are comparable to those measured at Sturgeon Point, NY, an Integrated Atmospheric Deposition Network (IADN) site located on the eastern shore of Lake Erie. This site is within 50 miles of Buffalo, NY and Erie, PA, both of which are large steel manufacturing areas. Gas and particulate PAH concentrations at the NB site were not statistically different from those measured during two AEOLOS sampling campaigns in the Chicago – Lake Michigan and Baltimore – Chesapeake Bay systems *(7, 16)*. The New Brunswick, Baltimore, and Lake Michigan sites are considered "impacted" due to proximity to large urban/industrial source areas. The particulate PAH concentrations measured in Newark, NJ during the Airborne Toxic Element and Organic Substances (ATEOS) study in the early 1980s were often an order of magnitude higher than those observed here and likely reflect pollution levels before the effective implementation/enforcement of modern emission control technologies *(49)*.

Temperature/Seasonal Effects

The seasonal PAH distributions (G+P) for New Brunswick are presented in Figures 8 and 9. The gas phase distribution for all seasons is dominated by phenanthrene (PHEN) and methylated phenanthrenes (MePHENs) with the second largest contributions from fluorene and fluoranthene. The most apparent difference between the summer and winter distributions is the relative contributions of MePHENs and PHEN. MePHENs are more important than PHEN in the winter while PHEN is more important in the summer. This is consistent with the increased importance of home heating emissions in the winter when MEPHENs dominate.

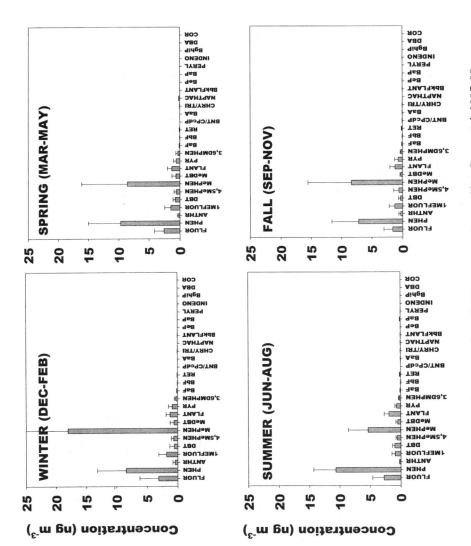

Figure 8. Atmospheric gas phase PAH concentrations at New Brunswick 1997-98.

Table III. Atmospheric PAH Concentrations (ng m^{-3}) from This and Other Studies.

Sampling Site	Reference	Phenanthrene		Pyrene		Benzo[b+k]fluoranthene		Benzo[a]pyrene	
		Gas	Particulate	Gas	Particulate	Gas	Particulate	Gas	Particulate
New Brunswick, NJ	this study	8.9(4.6)	0.16(0.16)	0.69(0.46)	0.14(0.12)	0.012(0.012)	0.32(0.30)	0.037(0.064)	0.088(0.096)
Sandy Hook, NJ	this study	4.8(3.3)	0.083(0.052)	0.41(0.28)	0.070(0.052)	0.0027(0.0019)	0.12(0.12)	0.0023(0.00087)	0.033(0.035)
Newark, NJ	(51)	NA	NA	NA	1.6(1.7)	NA	1.3(0.92)	NA	0.78(0.69)
Sturgeon Point, NY	(4)	4.0(0.068)	0.0060(0.077)	0.51(0.10)	0.074(0.071)	0.019(0.034)	0.074(0.071)	0.013(0.062)	0.044(0.076)
Eagle Harbor, MI	(4)	0.86(0.12)	0.019(0.069)	0.19(0.17)	0.022(0.016)	0.019(0.034)	0.022(0.016)	0.0093(0.023)	0.011(0.047)
Chicago, IL	(7)	64(46)	3.7(7.4)	9.0(8.4)	5.9(11)	0.29(0.38)	6.6(2.4)	0.080(0.082)	3.0(5.9)
Lake Michigan	(7)	9.9(9.6)	0.14(0.15)	1.6(1.8)	0.21(0.17)	0.12(0.23)	0.59(0.74)	0.014(0.030)	0.13(0.14)
Baltimore, MD	Dachs et al (unpublished)	9.8(11)	0.086(0.034)	1.4(1.3)	0.14(0.070)	0.0011(0.0030)	0.16(0.071)	0.00015(0.00055)	0.071(0.041)

(Numbers in parentheses are 1 standard deviation)

47

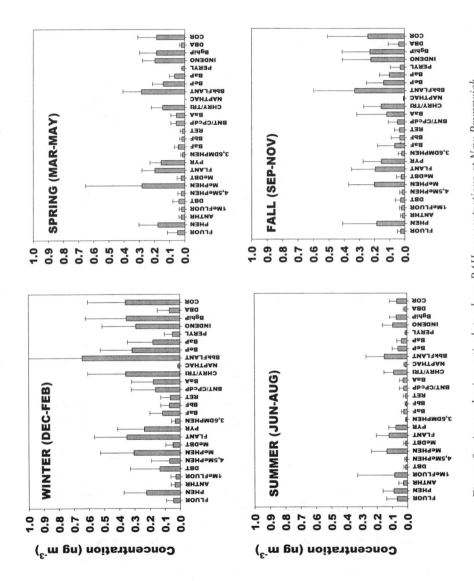

Figure 9. Atmospheric particulate phase PAH concentrations at New Brunswick 1997-98.

The particulate phase PAH distribution is dominated by the higher molecular weight compounds that are typically associated with soot particles *(50)*. The particulate phase concentrations at NB are often more than an order of magnitude less than gas phase concentrations. The particulate PAH distributions are dominated by benzo[b+k]fluoranthene (B[bk]FLANT) in winter and fall. The seasonal profiles show that particulate PAH concentrations are higher overall in the winter than in any other season. Previous studies have shown that increased fossil fuel combustion for home heating coincident with lower atmospheric mixing heights and reduced solar irradiation causes elevated particulate PAH concentrations in the winter *(50, 51)*. However this is not always the case as demonstrated for Chicago - coastal Lake Michigan *(7)*. To assess if winter concentrations derive from human activity or seasonal T controls, an assessment of the influence of temperature is necessary.

At NB, temperature accounts for only 10% (p = 0.005) and 16% (p = 0.001) of the variability in PHEN and MePHENs concentrations, respectively. For the coastal SH gas phase PAH concentrations, temperature accounts for 10% (p = 0.025) and 1% (p = 0.57, not significant) of the variability for PHEN and MePHENs, respectively. The low correlation in log [PAH] vs. 1/T found in this study indicates that gas-phase PAH concentrations are not driven by air-surface exchange as are many other semi-volatile organic compounds like PCBs. Compounds like PAHs are derived from combustion-related emissions. Thus, the lack of T control on atmospheric concentrations is consistent with seasonal differences in emission sources and source strengths.

The seasonal variation in concentrations (Figure 10) is presented for (G+P) PHEN, MePHENs, B[bk]FLANT, and benzo[a]pyrene (BaP) at the NB site in the form of air concentration versus T. The latter two PAHs are predominately found in the particulate phase and as such are normalized by total suspended particulate (TSP) concentrations. The decrease of MEPHENs concentrations with temperature is stronger and opposite of that for PHEN. Unlike PHEN, the majority of the maximum MePHEN concentrations occur during the coldest days of the year. It is unlikely that atmospheric oxidative reactions account for this difference in the seasonal trend between the two compounds *(52)*. MePHENs decrease in concentration with increasing temperature coincident with the high MW and mostly particle-bound B[bk]FLANT and B[a]P. This suggests that the source(s) contributing to the observed increase in MEPHENs in the winter is the same as for the high molecular weight PAHs.

Figure 10 demonstrates that the concentrations of the two predominately particulate phase PAHs increase with colder temperatures. For days with higher temperatures (>10°C) at the NB site, TSP-normalized PAH concentrations demonstrate little variability. In contrast, when the temperatures drop below ~ 10°C, there is a statistically significant (p < 0.001) increase in the variability in PAH concentrations. Correlations of concentrations with temperature are low when regressing one year of data (PHEN/TSP (r^2 = 0.090), MePHENs/TSP (0.081), BbkFLANT/TSP (0.105), and BaP/TSP (0.116)). In the winter, increased fossil fuel consumption for home heating is likely the major source to this particulate PAH burden, rather than temperature-controlled air-surface exchange.

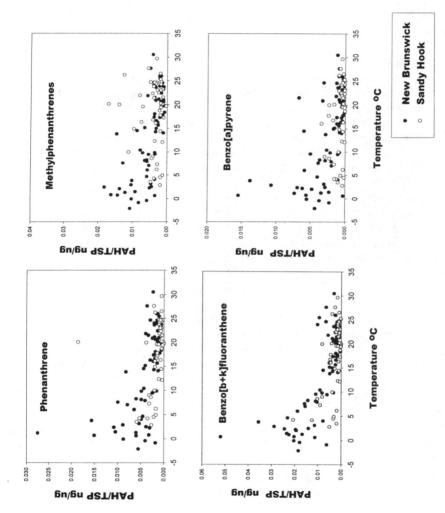

Figure 10. Concentrations of four PAHs as a function of temperature at New Brunswick and Sandy Hook 1997-98. Phenanthrene and methylphenanthrenes are primarily in the gas phase; benzo [b+k] fluoranthene and benzo[a]pyrene are primarily in the particle phase.

At coastal SH, PAH concentrations show no significant seasonal variation. The proposed "winter influence" by home heating is not seen to the same extent at Sandy Hook, most likely due to dilution of the signal during transport. On days at Sandy Hook when high BbkFLANT concentrations were observed, the local winds come directly from heavily populated New York City and Long Island located to the N and NE of Sandy Hook.

Transport Vectors

Because a dependence on sources is not factored into the log $[Conc]_{gas} = a + m/T$ equation, it is likely that the variability in PAH concentration that is not explained by temperature may be attributed to inputs from localized or longer-range source areas. To determine which source areas influence PAH concentrations in New Jersey's coastal atmosphere, 3-day back trajectories were constructed. Back trajectory analysis of those days with "extreme high or low concentrations" ($\pm$ 1 standard deviation from the mean) lead to a determination of which vectors correspond to high and low concentration *(46)*.

The dates of highest gas phase PHEN and MePHENs concentrations occurred when atmospheric transport to the N-NE up the I-95 corridor (from S/SW) through Baltimore, MD, Wilmington, DE, Philadelphia, PA and Camden, NJ dominated. The days of lowest concentration at NB occurred when the back trajectories showed the air masses were derived from the NW and NW, less densely populated areas. When winds derive from the coastal Atlantic Ocean, PAH concentrations were on average not statistically different from the annual average PAH concentration of 15 $\pm$ 9.3 ng m^{-3} at SH ($p<0.05$). When the winds came from the east, the average PAH concentration was 8.3 $\pm$ 5.4 ng m^{-3}. Since other than ship traffic, there are no local sources to the east of Sandy Hook, the PAH concentrations must be attributed to local and regional air circulation patterns. For some sampling days, high PAH levels are consistent with relatively stagnant air masses allowing for concentrations to build to the levels sampled over a few days. The coastal sea-breeze effect, where the previous day's PAHs are transported from the land out to the ocean as the land cools and the water remains warm. The following day, when the land warms again, the PAHs are returned to the land and subsequently are measured. However, a much better understanding is needed to elucidate the role of sea-breeze effect on the fate and transport of POPs.

Nonylphenols (NPs)

Nonylphenol polyethoxylates (NPEOs) have been widely used as surfactants in many industrial, agricultural and household applications. Biodegradation of NPEOs in aquatic environments leads to the formation of nonylphenols (NPs), which are persistent, bioaccumulative and estrogenic *(53-55)*. To date, NPs have been reported only in aquatic environments *(56-58)*. The occurrence of NPs in the atmosphere has recently been demonstrated for the first time and potential sources of NPs to the atmosphere in the lower HRE have been described *(59, 60)*.

Concentrations of NPs in water (dissolved + particulate) in the lower HRE ranged from 15 to 95 ng L^{-1}. These concentrations are more than one order of magnitude lower than NP concentrations reported for other impacted aquatic ecosystems such as the Krka River Estuary (300 to 45000 ng L^{-1}) *(61)* and the Glatt River (20 to 1200 ng L^{-1}) *(56)*. However, these concentrations are 10 to 100x higher than water concentrations of PCBs found in the Lower Hudson River estuary, even though it is known that the Hudson River has been historically impacted by PCBs *(28)*.

The Henry's Law constants for NPs, estimated as the ratio of the subcooled liquid vapor pressures to the aqueous solubilities, are 3 x 10^{-5} to 4 x 10^{-5} atm m^3 mol^{-1} for the different NP isomers *(62, 63)*. These values are sufficient to support water-air exchange of NPs to the atmosphere and lead to the first detection of NPs in atmospheric samples. Gas phase NP concentrations ranged from below detection limits in one sample to 81 ng m^{-3} and from 0.020 to 51 ng m^{-3} for the aerosol phase. Table IV gives total atmospheric (G+P) NP concentrations for samples taken during 1998. These concentrations are comparable to PAH concentrations and two orders of magnitude higher than PCB concentrations reported in the same area. The spatial variability observed (Table IV), with higher concentrations at the SH site and over the bay than at the urban LSC site, is consistent with volatilization from the waters that surround the former two sites. However, high concentrations were also observed at the suburban NB site in summer, which suggests that sources other than volatilization from waters may also be important at land sites. Indeed, NPEOs are used in pesticide formulations *(64)*. Thus, after biological transformation of NPEOs, volatilization of NPs from soil may be a potential source to the atmosphere.

Direct evidence that volatilization from the lower HRE is a source of NPs to the local atmosphere (SH and LSC sites) was obtained by estimating the fugacity of NPs in the dissolved and gas phases: $f_w = C_w H$; $f_g = C_g RT$; where C_w (mol L^{-1}) and C_g (mol L^{-1}) are the NPs concentrations in the dissolved and gas phase, respectively, R is the gas constant (atm L mol^{-1} K^{-1}) and T is the temperature (K). Due to the similarity of the water temperatures (293-295 K) to 298 K, Henry's law constants were not corrected for temperature. The average water/air fugacity ratio (f_w/f_g) was 18 and ranged from 1.3 to 69. Fugacity ratios greater than 1 provide conclusive evidence that net volatilization from the water is a source of NPs to the atmosphere.

Further evidence that air-surface exchange processes drive atmospheric concentrations of NPs is given by the seasonal trends of NP concentrations. Figure 11A shows the gas phase NP concentrations at NB from June to December 1998. Gas phase concentrations exhibit a maximum during the summer months (June through September) while lower concentrations were measured during fall and early winter. Gas phase PCB concentrations exhibit a temperature dependence due to their remobilization from aquatic and terrestrial surfaces during periods of higher temperature. This process is described by an integrated form of the Clausius-Clapeyron equation (log $C_g = a (1/T) + b$) where T is the temperature (K), C_g is the NP concentration in the gas phase, and *a* and *b* are the parameters obtained by least square regression. Figure 11B, showing the linear correlation of log C_g with 1/T

Table IV. Total Atmospheric Concentrations (Gas + Particle) of NPs in lower Hudson River Estuary Region. Includes Range and Number of Samples Analyzed in 1998.

Sampling Site (see Figure 1)	Nonylphenols ($ng\ m^{-3}$)
New Brunswick Suburban	**14** (0.2 - 81.1) n = 27
Sandy Hook Coastal	**12** (0.3 - 108) n = 38
Liberty Science Center Urban/Industrial	**6.4** (1.3 - 24) n = 23
Lower Hudson River Estuary Over-Water	**30** (9.6 - 75) n = 5

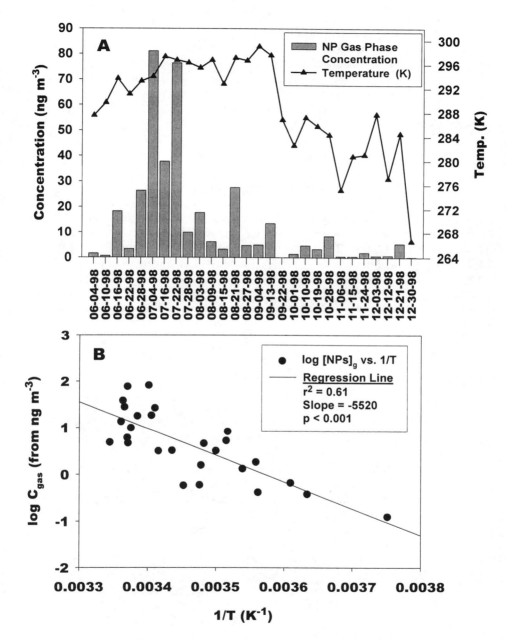

Figure 11. A. Nonylphenol concentrations at New Brunswick (June 6, 1998 - October 1, 1998) (top). B. The relationship of log [NPs] with 1/T (bottom).

54

obtained for the NPs concentrations at NB, indicates that T variation explains 61% of the variability in the gas-phase concentrations.

In summary, this study provides the first evidence of the atmospheric occurrence of NPs and implicates volatilization from aquatic and perhaps terrestrial environments as sources. A corollary to this study is that aquatic and terrestrial environments containing high concentrations of organic pollutants with appropriate Henry's law and/or octanol-air constants will contribute to the contamination of the local and regional atmosphere.

Acknowledgements

This work was funded in part by the Hudson River Foundation (Project Officer: D. Suszkowski), the New Jersey Sea Grant Program (NOAA)(Project Officer: M. Weinstein), the New Jersey Department of Environmental Protection (Project Officer: S. Nagourney), and the New Jersey Agricultural Experiment Station. J. Dachs acknowledges a postdoctoral fellowship from the Spanish Ministry of Education and Culture, and L. Totten acknowledges postdoctoral support from the Camile and Henry Dreyfus Foundation. We acknowledge the assistance of Rainer Lohmann in the summer field intensive.

References

(1)   Baker, J. E. In *Atmospheric Deposition of Contaminants in the Great Lakes and Coastal Waters*; Baker, J. E., Ed.; SETAC Press: Pensacola, FL., 1997, 451 p.
(2)   Eisenreich, S. J.; Baker, J. E.; Zhang, H.; Simcik, M. F.; Offenberg, J. H.; Totten, L. *Environ. Sci. Tech.* 2000, In Review.
(3)   Ondov, J. M.; Caffrey, P. F.; Suarez, A. E.; Borgoui, P. V.; Holsen, T.; Paode, R. D.; Sofuoglu, S. C.; Sivadechathep, J.; Lu, J.; Kelly, J.; Davidson, C. I.; Zufall, M. J.; Keeler, G. J.; Landis, M. S.; Church, T. M.; Scudlark, J. *Environ. Sci. Tech.* 2000, In Review.
(4)   Hoff, R. M.; Strachan, W. M. J.; Sweet, C. W.; Chan, C. H.; Schakleton, M.; Bidleman, T. F.; Brice, K. A.; Burniston, D. A.; S., C.; Gatz, D. F.; Harlin, K.; Schroeder, W. H. *Atmos. Environ.* 1996, *30*, 3505-3527.
(5)   Hillery, B. R.; Simcik, M. F.; Basu, I.; Hoff, R. M.; Strachan, W. M. J.; Burniston, D.; Chan, C. H.; Brice, K. A.; Sweet, C. W.; Hites, R. A. *Environ. Sci. Tech.* 1998, *32*, 2216-2221.
(6)   Cotham, W. E.; Bidleman, T. F. *Environ. Sci. Tech.* 1995, *29*, 2782-2789.
(7)   Simcik, M. F.; Zhang, H.; Franz, T.; Eisenreich, S. J. *Environ. Sci. Tech.* 1997, *31*, 2141-2147.

(8)   Harner, T.; Bidleman, T. F. *Environ. Sci. Tech.* 1998, *32*, 1494-1502.
(9)   Green, M. L.; Depinto, J. V.; Sweet, C. W.; Hornbuckle, K. C. *Environ. Sci. Tech.* 2000, In Review.
(10)   Offenberg, J. H.; Baker, J. E. *Environ. Sci. Tech.* 1997, *31*, 1534-1538.
(11)   Franz, T. P.; Eisenreich, S. J.; Holsen, T. M. *Environ. Sci. Tech.* 1998, *32*, 3681-3688.
(12)   Paode, R. D.; Sofuoglu, S. C.; Sivadechathep, J.; Noll, K. E.; Holsen, T. M.; Keeler, G. J. *Environ. Sci. Tech.* 1998, *32*, 1629-1635.
(13)   Caffrey, P. F.; Ondov, J. M.; Zufall, M. J.; Davidson, C. I. *Environ. Sci. Tech.* 1998, *32*, 1615-1622.
(14)   Zhang, H.; Eisenreich, S. J.; Franz, T. P.; Baker, J. E.; Offenberg, J. H. *Environ. Sci. Tech.* 1999, *33*, 2129-2137.
(15)   Nelson, E. D.; McConnell, L. L.; Baker, J. E. *Environ. Sci. Tech.* 1998, *32*, 912-919.
(16)   Offenberg, J. H.; Beker, J. E. *J. Air & Waste Mngmt. Assoc.* 1999, *49*, 959-965.
(17)   Iannuzzi, T. J.; Huntley, S. L.; Bonnevie, N. L.; Finley, B. L.; Wenning, R. J. *Arch. Env. Contam. Tox.* 1995, *28*, 108-117.
(18)   Bremie, G.; Larsson, P. *Atmos. Environ.* 1998, *27*, 398-403.
(19)   Swackhamer, D. L.; McVeety, B. D.; Hites, R. A. *Environ. Sci. Tech.* 1988, *22*, 644-672.
(20)   Hornbuckle, K. C.; Eisenreich, S. J. *Atmos. Environ.* 1996, *30*, 3935-3945.
(21)   Baker, J. E.; Poster, D. L.; Clark, C. A.; Church, T. M.; Scudlark, J. R.; Ondov, J. M.; Dickhut, R. M.; Cutter, G. In *Atmospheric Deposition of Contaminants in the Great Lakes and Coastal Waters*; Baker, J. E., Ed.; SETAC Press: Penacola, FL., 1997, 171-194.
(22)   Hillery, B. R.; Basu, I.; Sweet, C. W.; Hites, R. A. *Environ. Sci. Tech.* 1997, *31*, 1811-1816.
(23)   Leister, D. L.; Baker, J. E. *Atmos. Environ.* 1994, *28*, 1499-1520.
(24)   Durell, G. S.; Lizotte Jr., R. D. *Environ. Sci. Tech.* 1998, *32*, 1022-1031.
(25)   Huntley, S. L.; Iannuzzi, T. J.; Advantaggio, J. D.; Carlson-Lynch, H.; Schmidt, C. W.; Finley, B. L. *Chemosphere* 1997, *34*, 233-250.
(26)   Bopp, R. F.; Simpson, H. J.; Olsen, C. R.; Kostyk, N. *Environ. Sci. Tech.* 1981, *15*, 210-216.
(27)   Bopp, R. F.; Simpson, H. J.; Olsen, C. R.; Trier, R. M.; Kostyk, N. *Environ. Sci. Tech.* 1982, *16*, 666-676.
(28)   Farley, K. J.; Thomann III, R. V.; Cooney, T. F.; Damiani, D. R.; Wands, J. R. "An Integrated Model of Organic Chemical Fate and Bioaccumulation in the Hudson River Estuary," Report to the Hudson River Foundation, 1999.
(29)   Oehme, M.; Haugen, J.-E.; Schlabach, M. *Environ. Sci. Tech.* 1996, *30*, 2294-2304.
(30)   Lee, R. G. M.; Jones, K. C. *Environ. Sci. Tech.* 1999, *33*, 705-712.
(31)   Hornbuckle, K. C.; Achman, D. R.; Eisenreich, S. J. *Environ. Sci. Tech.* 1993, *27*, 87-98.

56

(32)   Panshin, S. Y.; Hites, R. A. *Environ. Sci. Tech.* 1994, *28*, 2008-2013.
(33)   Hoff, R. M.; Muir, D. C. G.; Grift, N. P. *Environ. Sci. Tech.* 1992, *26*, 276-283.
(34)   Honrath, R. E.; Sweet, C. I.; Plouff, C. J. *Environ. Sci. Tech.* 1997, *31*, 842-852.
(35)   Chiarenzelli, J. R.; Scrudato, R. J.; Wunderlich, M. L.; Oenga, G. N.; Lashko, O. P. *Chemosphere* 1997, *34*, 2429-2436.
(36)   Bushart, S. P.; Bush, B.; Barnard, E. L.; Bott, A. *Environ. Tox. Chem.* 1998, *17*, 1927-1933.
(37)   Ravikrishna, R.; Valsaraj, K. T.; Yost, S.; Price, C. B.; Brannon, J. M. *J. Haz. Materials* 1998, *60*, 89-104.
(38)   Wania, F.; Haugen, J.-E.; Lei, Y. D.; Mackay, D. *Environ. Sci. Tech.* 1998, *32*, 1013-1021.
(39)   Hoff, R. M.; Brice, K. A.; Halsall, C. J. *Environ. Sci. Tech.* 1998, *32*, 1793-1798.
(40)   Bidleman, T. F.; Wideqvist, U.; Hansson, B.; Soederlund, R. *Atmos. Environ.* 1987, *21*, 641-654.
(41)   Bidleman, T. F.; Patton, G. W.; Walla, M. D.; Hargrave, B. T.; Vass, W. P.; Erickson, P.; Fowler, B.; Scott, V.; Gregor, D. L. *Artic* 1989, *42*, 307-313.
(42)   Oehme, M. *Ambio.* 1991, *20*, 293-297.
(43)   Bidleman, T. F.; Billings, W. N.; Foreman, W. T. *Environ. Sci. Tech* 1986, *20*, 1038-1043.
(44)   Cortes, D. R.; Basu, I.; Sweet, C. W.; Brice, K. A.; Hoff, R. M.; Hites, R. A. *Environ. Sci. Tech.* 1998, *32*, 1920-1927.
(45)   Bidleman, T. F.; Alegria, H.; Ngabe, B.; Green, C. *Atmos. Environ.* 1998, *32*, 1849-1856.
(46)   Subhash, S.; Honrath, R. E.; Kahl, J. D. *Environ. Sci. Tech.* 1999, *33*, 1509-1515.
(47)   Dearth, M. A.; Hites, R. A. *Environ. Sci. Tech.* 1991, *25*, 1279-1285.
(48)   Knox, J. R.; Khalifa, S.; Ivie, G. W.; Casida, J. E. *Tetrahedron* 1973, *29*, 3869-3879.
(49)   Greenberg, A.; Darack, F.; R., H.; Lioy, P.; Daisey, J. *Atmos. Environ.* 1985, *19*, 1325-1339.
(50)   Aceves, M.; Grimalt, J. O. *Environ. Sci. Tech.* 1993, *27*, 2896-2908.
(51)   Lioy, P. J.; Daisey, J. M.; Greenberg, A.; Harkov, R. *Atmos. Environ.* 1985, *19*, 429-436.
(52)   Simo, R.; Grimalt, J. O.; J., A. *Environ. Sci. Tech.* 1997, *31*, 2697-2700.
(53)   Giger, W.; Brunner, P. H.; Schaffner, C. *Science* 1984, *225*, 623-625.
(54)   Ahel, M.; Giger, W.; Koch, M. *Water Res.* 1994, *28*, 1131-1142.
(55)   White, R.; Jobling, S.; Hoare, S. A.; Sumpter, J. P.; Parker, M. G. *Endocrinology* 1994, *135*, 175-182.
(56)   Ahel, M.; Giger, W.; Schaffner, C. *Wat. Res.* 1994, *28*, 1143-1152.
(57)   Marcomini, A.; Pavoni, B.; Sfriso, A.; Orio, A. A. *Mar. Chem.* 1990, *29*, 307-323.

(58)   Kannan, N.; Yamashita, N.; Petrick, G.; Duinker, J. C. *Environ. Sci. Tech.*
        1998, *32*, 1747-1753.
(59)   Dachs, J.; Van Ry, D. A.; Eisenreich, S. J. *Environ. Sci. Tech.* 1999, *33*,
        2676-2679.
(60)   Van Ry, D. A.; Dachs, J.; Gigliotti, C. L.; Brunciak, P. A.; Nelson, E. D.;
        Eisenreich, S. J. *Environ. Sci. Tech.* 2000, In Review.
(61)   Kvestak, R.; Terzic, S.; Ahel, M. *Mar. Chem.* 1994, *46*, 89-100.
(62)   Bidleman, T. F.; Renberg, L. *Chemosphere* 1985, *14*, 1475-1481.
(63)   Ahel, M.; Giger, W. *Chemosphere* 1993, *26*, 1461-1470.
(64)   Field, J. A.; Reed, R. L. *Environ. Sci. Tech.* 1996, *30*, 3544-3550.
(65)   Hoff, R.A.; Chan, C.H. *Chemosphere* 1986, *15*, 4452-4499.
(66)   Atlas, E.; Giam, C.S. *Water, Air, Soil Polln.* 1988, *38*, 19-36.
(67)   Aignar, E.J.; Leone, A.D.; Falconer, R.L. *Environ. Sci. Tech.* 1998, *32*,1162-
        1168.

Chapter 4

Modeling Irreversible Sorption of Hydrophobic Organic Contaminants in Natural Sediments

Wei Chen[1], Amy T. Kan[2], and Mason B. Tomson[2]

[1]Brown and Caldwell, 1415 Louisiana, Suite 2500, Houston, TX 77002
[2]Department of Environmental Science and Engineering,
MS 519, Rice University, Houston, TX 77005

The ability to quantitatively predict the resistant release of sediment-associated hydrophobic organic contaminants is extremely important in managing contaminated sediments. Many laboratory and field observations support the irreversible sorption model proposed by the authors, in which irreversible sorption is attributed to the physical binding of organic molecules in sediment organic matter. This kind of physical binding is not affected by a number of physical and chemical changes in the environment, such as competitive sorption, discharge of caustic materials, and external mechanical disturbance. Quantitatively, irreversible sorption of hydrophobic organic contaminants can be well modeled with an irreversible sorption isotherm, which accounts for the contribution of both the reversible and irreversible compartments. Since the parameters of the irreversible isotherm can be readily estimated or measured, it provides a simple yet more accurate approach to predicting the long-term resistant release of contaminants sorbed in sediments. The irreversible sorption model may significantly improve the accuracy of fate-transport modeling and risk assessment. It should also provide guidance to environmental decision making.

Introduction

Sediment contamination has become a serious nation-wide problem. To facilitate sediment management and to improve sediment quality, it requires a thorough understanding of the physical, chemical, and biological processes controlling contaminant-sediment interactions. Among the many processes, sorption and desorption are possibly the most fundamental yet the most poorly understood

© 2001 American Chemical Society

processes. In particular, the release of sorbed contaminants from sediments is highly resistant and often cannot be predicted with the conventional models (e.g., the conventional linear model assuming instantaneous reaction, isotherm linearity, and sorption-desorption reversibility). Nonetheless, the ability to quantitatively model the long-term resistant release of sediment-associated organic contaminants is of central importance to the accuracy of fate and transport modeling, the efficiency of remediation and biodegradation processes, and the effectiveness of environmental regulation and decision making. A series of studies have been conducted by the authors *(1-6)* to understand the nature of irreversible sorption and to seek a mathematical model that can be used to quantify the irreversible sorption of hydrophobic organic contaminants in natural sediments. This chapter summarizes 1) the conceptual and mathematical models of irreversible sorption derived by the authors; 2) some experimental observations supporting these models; 3) factors affecting irreversible sorption; and 4) the significance and impact of irreversible sorption.

Conceptual and Mathematical Models

Numerous laboratory and field observations have shown that chemicals sorbed in soils and sediments typically reside in two compartments -- chemicals in the reversible compartment desorb readily, while those associated with the irreversible compartment exhibit great resistance to desorption (Figure 1). Note that various terms have been used in the literature to describe the same phenomenon *(1,7-10)*. "Irreversible compartment" as used herein is to contrast with "reversible compartment", from which the desorption follows the same linear isotherm as sorption does; it does not imply that no desorption occurs. The term "irreversible" has the same thermodynamic meaning as that defined by Adamson and others *(11-13)*. A unique feature of the irreversible compartment is that compounds with different physical and chemical properties behave similarly. The authors have observed that most compounds exhibited a similar partition coefficient of $10^{5.53 \pm 0.48}$ with regard to the irreversible compartment, even though the K_{ow} values of these compounds vary over several orders of magnitude *(3,5)*. In addition, the irreversible compartment seems to have a distinct maximum sorption capacity, which depends on the specific chemical-sediment combination.

These unique features indicate that desorption may occur in an environment which is different from the one that sorption occurs. Accordingly, the authors have proposed a conceptual model to interpret irreversible sorption *(2)*. This conceptual model is illustrated in Figure 2 -- sorption occurs through hydrophobic partitioning, thus the isotherm is linear; during or after sorption, some conformational change or physical rearrangement of sediment organic matter happens, and a portion of sorbed compounds is entrapped within sediment matrix and becomes unavailable to desorption, and thus, desorption includes a reversible and an irreversible phase. Also, since irreversible sorption is due to physical binding, compounds with different physical-chemical properties may behave similarly.

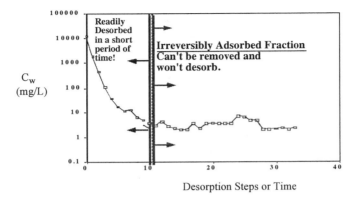

Figure 1: A typical bi-phasic desorption – compounds associated with the reversible fraction desorbed in a few desorption steps, while compounds associated with the irreversible fraction showed great resistance and remained in sediment after extensive desorption. Data are from Kan et al., 1997 *(Ref. 2)*.

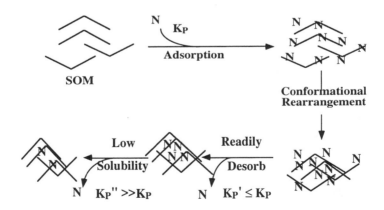

Figure 2: A pictorial representation of the irreversible sorption model (Reproduced from *Ref. 2*).

On the basis of this conceptual model, the authors have derived an irreversible sorption isotherm to quantitatively model the biphasic desorption *(3)*. The isotherm includes the contribution of both the reversible and irreversible compartments to overall desorption and can be expressed as:

$$q = K_{oc} \, f_{oc} \, C + \frac{K_{oc}^{irr} \, f_{oc} \, q_{max}^{irr} \, f \, C}{q_{max}^{irr} \, f + K_{oc}^{irr} \, f_{oc} \, C} \tag{1}$$

where q and C are the solid and solution phase concentrations, respectively; K_{oc} and K_{oc}^{irr} are the partition coefficients of the reversible and irreversible compartments, respectively; f_{oc} is the organic carbon content of sediment; q_{max}^{irr} is the irreversible sorption capacity, i.e., the maximum size of the irreversible compartment; and f is the fraction of q_{max}^{irr} filled during sorption. The first term of the right hand side of Equation 1 accounts for the contribution of the reversible compartment to overall desorption and is represented by the conventional linear isotherm; the second term reflects the contribution of the irreversible compartment. Since the irreversible compartment has a well delineated maximum sorption capacity, the second term takes a Langmurian type of form, and is characterized by two parameters -- the irreversible partition coefficient, K_{oc}^{irr}, and the maximum sorption capacity, q_{max}^{irr}. The parameters in Equation 1 can be readily measured or estimated -- f_{oc} can be measured by a number of methods; K_{oc} can be measured in sorption experiments and can be estimated from K_{ow} (e.g., *Ref. 14)*; K_{oc}^{irr} is fairly constant, about $10^{5.53 \pm 0.48}$ (mL/g), for compounds with K_{ow} from $10^{2.3}$ to $10^{6.4}$ *(3,5)*; q_{max}^{irr} can be estimated from compound K_{ow} value and sediment f_{oc} value *(3)*; and f is typically equal to 1 in natural systems. Therefore, Equation 1 provides a simple yet more accurate approach to quantitatively predicting the resistant release of hydrophobic organic contaminants from natural sediments. Furthermore, Equation 1 can be simplified in most cases. At high concentration, the isotherm is reduced to the conventional linear isotherm:

$$q = K_{oc} \, f_{oc} \, C \tag{2}$$

At low concentration, the isotherm is also reduced to a linear isotherm:

$$q = K_{oc}^{irr} \, f_{oc} \, C \tag{3}$$

Note that in Equation 3, the equilibrium partition coefficient, K_{oc}, is replaced by the irreversible partition coefficient, K_{oc}^{irr}.

Experimental

Sorption and Desorption Experiments. Five experiments (Exp. 1 to 5) were conducted to study sorption and desorption. The detailed experimental procedures were discussed in a previous paper *(5)*. Dickinson sediment (1.5% OC) was used as the sorbent and the chemicals used were 1,2-dichlorobenzene (1,2-DCB), 1,2,4-trichlorobenzene (1,2,4-TCB), 1,2,3,4-tetrachlorobenzene (1,2,3,4-TeCB), hexachlorobenzene (HCB), and 1,4-dichlorobenzene (1,4-DCB), respectively. In each sorption experiment, the sediment was saturated with a chemical by repetitive sorption or continuous sorption. In Exp. 1 and 2, a 42 mL glass vial containing 2 g sediment and about 40 mL 1,2-DCB (~60 mg/L) or 1,2,4-TCB (~20 mg/L) solution was tumbled at 1 rpm for 6 days to initiate sorption. At the end of the sorption, the vial was centrifuged and 90% of the supernate was replaced with freshly prepared 1,2-DCB or 1,2,4-TCB solution to conduct another sorption. Aqueous phase concentrations were measured by extracting supernate with isooctane followed by GC/ECD. The repetitive sorption experiments were continued until there was no significant uptake of the chemicals by sediments. In Exp. 3 to 5, each reaction vial contained 10 g sediment, a dialysis bag containing the chemical, and about 35 mL electrolyte solution. The vials were tumbled for 2 to 4 months to saturate the sediments. Upon the completion of the repetitive or continuous sorption experiments, sediment phase concentrations were measured by extracting sediments with methanol/water solution (15:85 volume) *(5)* followed by GC/ECD.

Desorption in Exp. 1 to 5 was initiated by repetitively equilibrating 2 g sediment with about 42 mL electrolyte solution containing 0.01 M NaCl, CaCl$_2$, and NaN$_3$ (3 to 6 days per step). At the end of each step, about 90 % of the solution was replaced with fresh electrolyte solution. After 3 desorptions with electrolyte solution, 3 to 5 Tenax desorptions were performed to accelerate the removal of the reversible fraction -- about 0.5 g Tenax was added to the vial in each step and was replaced with clean Tenax at the end of each desorption. Chemicals desorbed by Tenax were quantified by extracting Tenax with acetone and analyzing the extracts with GC/ECD. After Tenax desorptions, more desorptions were conducted with electrolyte solution at varied time (0.5 to 25 days) to study the desorption from the irreversible compartment. Upon the completion of each desorption experiment, the solid phase concentration was measured to check mass balance.

Competitive Sorption Experiments. Exp. 6 was designed to study the effect of the competitive sorption of trans-cinnamic acid on the release of 1,4-DCB from the irreversible compartment of the Dickinson sediment. The sediment sample was initially saturated with 1,4-DCB, and then desorbed extensively with electrolyte solution and Tenax. The sediment preparation and desorption were discussed in detail in a previous paper *(5)*. Prior to competitive sorption experiments, the sediment sample was further desorbed with electrolyte solution eleven times to insure that only contaminants in the irreversible compartment retained. Competitive desorptions were done by two successive desorptions with an electrolyte solution containing 100 mg/L of trans-cinnamic acid. The concentrations of 1,4-DCB were monitored at the end of each desorption step.

External Perturbation Experiment. The starting sediment sample in this experiment (Exp. 7) was a portion of the sediment sample prepared in the sorption experiment of Exp. 5. The desorption experiment was set up nearly identically to that of Exp. 5. However, the reaction vial was horizontally shaken rather than tumbled to mix the system. Desorption was conducted with both electrolyte solution and Tenax as in Exp. 5.

Model Prediction and Experimental Observations

The five sorption and desorption experiments (Exp. 1 to 5) conducted with five chlorinated benzenes and Dickinson sediment covered wide solution and solid phase concentration ranges -- up to five orders of magnitude. For each experiment, the aqueous phase concentration at the end of sorption was close to the solubility; the aqueous phase concentrations in the last few desorptions were typically less than 2 ppb.

The uniqueness of irreversible sorption can be clearly illustrated with Figure 3. In Figure 3.a, the partition coefficients in the five sorption experiments are plotted with the K_{ow} values of the five compounds. The observed K_{oc} values are nearly linearly related to the K_{ow} values of these compounds. Since 1,2-DCB and 1,4-DCB have the same K_{ow} values, the K_{oc} values of these two compounds are nearly the same. These observations indicate that sorption was driven by the hydrophobic partitioning of organic molecules into sediment organic matter. Figure 3.b, however, shows that the partition coefficients with regard to the irreversible compartment were completely independent of the K_{ow} values. All of the five chemicals exhibited a similar partition coefficient -- $10^{5.42 \pm 0.17}$ mL/g.

Another unique characteristic of the irreversible compartment is that after a short period, desorption from this compartment is independent of time. The time dependency of the aqueous phase concentration was examined by varying the time applied in the last few desorptions, i.e., after the irreversible compartment was reached. It was observed that the aqueous concentration remained essentially the same after a few days, even when desorption time was extended to 25 days. Thus, desorption from the irreversible compartment had probably reached equilibrium.

In Figure 4, the experimentally observed desorption data are compared with model predictions to test the effectiveness of the irreversible sorption model (Equation 1). The symbols are the desorption data of 1,2-DCB, 1,2,4-TCB, 1,2,3,4-TeCB, and HCB from Dickinson sediment (Exp. 1-4) and the lines are plotted with Equation 1. As shown in the figure, the desorption results can be modeled with the irreversible sorption model, even when aqueous phase concentrations cover a range of more than five orders of magnitude. The plateau in each isotherm is approximately the size of the irreversible compartment, i.e., q_{max}^{irr}. In a previous paper, the desorption of 1,4-DCB from four different sediments (0.27 to 4.1% OC) was studied (5). It was found that the desorption data could be modeled by the irreversible isotherm; while the shapes of the four isotherms were similar, the plateaus (q_{max}^{irr}) shifted vertically. It was also observed that q_{max}^{irr} was linearly proportional to the organic carbon content of these sediments.

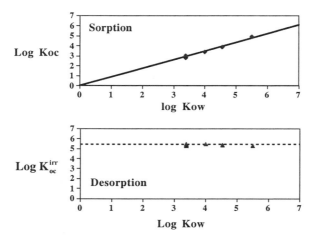

Figure 3: K_{oc} and K_{oc}^{irr} values observed in Exp. 1 to 5. For both plots, data points of 1,2-DCB and 1,4-DCB overlap, because the two compounds have the same K_{ow} value.

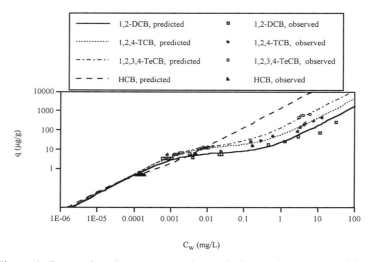

Figure 4: Comparison between experimental observations and model simulations of the desorption of four chlorinated benzenes from Dickinson sediment.

Factors Affecting Irreversible Sorption

Numerous physical and chemical changes in aquatic systems can potentially affect the release of sorbed contaminants from sediments and thus, also affect the effectiveness of the proposed irreversible sorption model. The effect of competitive sorption, caustic treatment of sediment, and external mechanic perturbation on desoption was examined in a recent paper *(6)*.

In Figure 5 is illustrated the effect of competitive sorption of trans-cinnamic acid on the release of 1,4-DCB from the irreversible compartment in Dickinson sediment (Exp. 6). It has been reported that the sorption of 1,3-DCB was suppressed up to 40% when 100 mg/L of trans-cinnamic acid was applied to the sediment *(15)*. In a previous study, the authors also observed that competitive sorption had considerable effect on the desorption from the reversible fraction *(6)* -- the aqueous phase concentrations of aged 1,3-DCB and 1,4-DCB, HCB, and hexachlorobutadiene were increased by a factor of 1.1 to 3.1 when Lake Charles sediment was saturated with naphthalene. However, as in Figure 5, competitive sorption had little effect on the desorption from the irreversible compartment. Before the competitive sorption experiments, the aqueous phase concentration of 1,4-DCB reached a relatively constant value of 1.50 ± 0.47 μg/L in five repetitive desorptions (Step 23-27), with respect to the irreversible compartment. The corresponding partition coefficient, K_{oc}^{irr}, was $10^{5.37}$, and was within the reported range -- $K_{oc}^{irr} = 10^{5.53 \pm 0.48}$ *(3,5)*. The average concentration of 1,4-DCB in the two trans-cinnamic acid sorptions was 1.50 ± 0.84 μg/L, and was almost identical to the concentration before competitive sorption.

These experimental results are in agreement with the irreversible sorption mechanism proposed by the authors -- since irreversible sorption is due to some kind of post-sorption conformational change or physical rearrangement of soil organic matter, unless competitive sorption could alter the physical entrapment, an increase in desorption is unlikely to occur. Other researchers *(16)* observed that while atrazine sorbed to small organic molecules was not retained by a dialysis bag, atrazine sorbed to natural organic molecules was significantly retained. They also suggested that the retention was due to the sorption of atrazine to natural organic matter followed by a physical entrapment.

The effect of caustic treatment of sediment on the desorption from the irreversible compartment was studied in a previous paper by Chen et al. *(6)*. No evidence of enhanced desorption by caustic solution was observed, even when the sediment was treated with 1 M NaOH. It has been reported that organic macromolecules are generally densely coiled at low pH but relatively loose and flexible at neutral or high pHs *(17-19)* . Therefore, applying caustic solution to sediments may either "swell" or dissolve some soil organic matter and result in a release of the sorbed compounds. Thus, these observations might suggest that these irreversibly sorbed molecules were either physically entrapped or sorbed to very insoluble hydrophobic organic materials, e.g., soot or humus in the sediment *(20,21)*. Similar results have been reported in another study, in which caustic treatment had no effect on the release from the resistant fraction *(17)*.

Exp. 7 was conducted to examine the potential impact of external forces on the release of sorbed compounds from the irreversible compartment. The experimental results were compared with those from Exp. 5. Different mixing modes -- tumbling and horizontal shaking were used in Exp. 5 and 7, respectively. While tumbling may simulate excess perturbation due to dredging or storm event, horizontal shaking may better represent the relatively transient conditions in aquatic environments. Results from these two experiments are compared in Figure 6. As seen in the figure, desorption data obtained from tumbling and shaking experiments are very similar, especially in lower concentration range. The q_{max}^{irr} and K_{oc}^{irr} values, as indicated by the plateau and the intercept of the isotherm, are nearly identically for the two isotherms. Thus, it is suggested that chemicals sorbed in the irreversible compartment are relatively stable and not affected by the method of mixing.

Impact of Irreversible Sorption

The existence of irreversible sorption invalidates the assumption that the release of sorbed chemicals can be modeled with the conventional linear reversible isotherm, which is the primary model used to access contaminant fate and transport and to enact environmental quality criteria. The inability of the linear model to predict long-term resistant release is more problematic in natural environments, because contaminants typically exist in lower concentration ranges, within which the resistance to desorption is more significant. The impact of irreversible sorption is two-folded. First, irreversible sorption could greatly reduce the availability of sorbed compounds and therefore could greatly reduce the efficiency of bioremediation and *in-situ* treatment. Previously, the authors evaluated the impact of irreversible sorption on the efficiency of pump and treat systems *(3)*. It was found that when desorption was controlled by the linear-reversible mechanism, only 20 pore volume of water was required to decrease the contamination level from 15 mg/L to 1 µg/L; however, when desorption was controlled by the irreversible sorption mechanism, 3300 pore volume of water was required. On the other hand, since a considerable fraction of sorbed compounds is unavailable to desorption, the same process that reduces bioavailability also reduces the risks of sorbed contaminants. Thus, the current sediment quality criteria could significantly overestimate the potential risk associated with these compounds. As in Figure 7, according to the linear model, a sediment (1.5% OC) containing 1 µg/g 1,2,4-TCB could result in a 20 µg/L concentration in water. However, since the release is much more accurately represented with the irreversible model, the actual concentration in the water would more likely be 0.2 µg/L -- 100 times lower. Thus, if the irreversible sorption model is adopted in environmental decision making, many contaminated sediments might be left alone without any practical concern to the environment or to human health, and the corresponding economical impact would be enormous.

Acknowledgments

This research has been conducted with the support of Hazardous Substance Research Center South and Southwest, the Gulf Coast Hazardous Substance Research

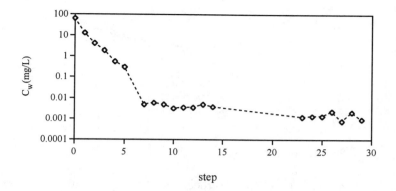

Figure 5: Effect of competitive sorption on the release of 1,4-DCB from the irreversible compartment of Dickinson sediment. The 6th desorption was a Tenax desorption. The last two desorptions were conducted with 100 mg/L trans-cinnamic acid solution.

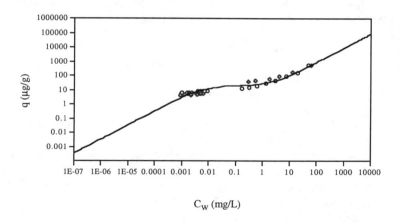

Figure 6: Effect of external mechanical force on the release of 1,4-DCB from Dickinson sediment. Two desorption experiments were conducted using a tumbler and a horizontal shaker, respectively. The diamond symbols are desorption data from tumbling experiment (Exp. 5) and circles are those from shaking experiment (Exp. 7). The line is the model prediction with Equation 1.

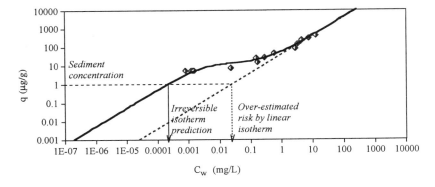

Figure 7: Irreversible sorption model significantly improves the accuracy of risk assessment, while the potential risks of contaminated soils and sediments are often exaggerated by the linear model, especially at relatively low concentrations.

Center, Office of Exploratory Research of the U.S. Environmental Protection Agency and the Defense Treat Reduction Agency. We also thank Dr. C.R. Demas of U.S. Geological Survey Louisiana District for assistance in collecting Lake Charles sediment.

References

1. Kan, A. T.; Fu, G.; Tomson, M. B. *Environmental Science and Technology* **1994**, 28, 859-867.
2. Kan, A. T.; Fu, G.; Hunter, M. A.; Tomson, M. B. *Environmental Science and Technology* **1997**, 31, 2176-2185.
3. Kan, A. T.; Fu, G.; Hunter, M. A.; Chen, W.; Ward, C. H.; Tomson, M. B. *Environmental Science and Technology* **1998**, 32, 892-902.
4. Chen, W.; Kan, A. T.; Fu, G.; Vignona, L. C.; Tomson, M. B. *Environmental Toxicology and Chemistry* **1999**, 18, 1610-1616.
5. Chen, W.; Kan, A. T.; Tomson, M. B. *Environmental Science and Technology* **1999** (submitted).
6. Chen, W.; Kan, A. T.; Tomson, M. B. *Environmental Toxicology and Chemistry* **1999** (submitted).
7. DiToro, D. M.; Horzempa, A. M. *Environmental Science and Technology* **1982**, 16, 594-602.
8. West, C. C.; Waggett, G. G.; Tomson, M. B. In *Proceeding of the* Second International Conference on Ground Water Quality Research, 1985, Stillwater, OK.Second International Conference on Ground Water Quality Research.
9. Vaccari, D. A.; Kaouris, M. *J. Environ. Sci. Health* **1988**, A23, 797-822.

10. Huang, W.; Weber, W. J. Jr. *Environ. Science and Technol.* **1998**, 31, 2562-2569.
11. *Physical Chemistry of Surface*; Adamson, A. W.; J. Wiley & Sons, Inc.; New York, 1990.
12. Burgess, C. G.; Everett, K. H. *Pure and Applied Chemistry* **1989**, 61, 1845.
13. Baily, G. W.; White, J. L.; Rothberg, T. *Soil Science Society of America, Proceedings* **1968**, 32, 222.
14. Karickhoff, S. M.; Brown, D. S.; Scott, T. A. *Water Resources* **1979**, 13, 241-248.
15. Xing, B.; Pignatello, J. J. *Environmental Science and Technology* **1998**, 32, 614-619.
16. Devitt, E. C.; Wiesner, M. R. *Environmental Science and Technology* **1998**, 32, 232-237.
17. Carroll, K. M.; Harkness, M. R.; Bracco, A. A.; Balcarcel, R. R. *Environmental Science and Technology* **1994**, 28, 253-258.
18. Ghosh, K.; Schnitzer, M. *Soil Science* **1980**, 129, 266-276.
19. Cornel, P. K.; Summer, R. S.; Roberts, P. V. *Journal of Colloid and Interface Science* **1986**, 110, 149-164.
20. Chiou, C. T.; Kile, D. E. *Environmental Science and Technology* **1998**, 32, 338-343.
21. Gustafsson, O.; Haghseta, F.; Chan, C.; MacFarlane, J.; Gschwend, P. M. *Environmental Science and Technology* **1997**, 31, 203-209.

Chapter 5

Persistence of a Petroleum Hydrocarbon Mixture in the Soil Profile During Leaching: A Field Experiment

Ishai Dror[1], Zev Gerstl[1], Hilel Rubin[2], Carol Braester[2], and Bruno Yaron[1]

[1]Institute of Soil, Water and Environmental Sciences,
ARO—The Volcani Center, Bet Dagan 50250, Israel
[2]Faculty of Civil Engineering, Technion—Israel Institute
of Technology, Haifa 32000, Israel

The fate and persistence of a petroleum spill under leaching were studied in a field experiment in which the amount of contaminant on the soil surface, the soil moisture status and the leaching pattern were all controlled. Kerosene – a semivolatile mixture of petroleum hyrocarbons - was applied to a sandy loam Mediterranean Red Soil and leached with 500 mm of irrigation and 500 mm of rain water. The field was sampled periodically during 180 days following application and the redistribution and persistence of kerosene components were determined. The major portion of the petroleum product applied to the soils was retained in the upper part of the profile and, despite the amount of leaching water applied, only a slight proportion was transported to the deeper layers. The average residual total concentation of kerosene after 180 days and leaching with 1,000 mm water decreased from 20,000 µg/g soil to 1,900 µg/g soil in the soil upper 20 cm layer. It was shown that soil dissipation and redistribution are the controlling factors affecting the persistence of kerosene components in field soils and that these factors are affected by the properties of the contaminant and the porous medium and the prevailing environmental conditions.

© 2001 American Chemical Society

Introduction

The contamination of soils and of the unsaturated zone from the land surface down to the groundwater with petroleum products is one of the major hazards in the industrial world. The presence of petroleum products in the soil may become a long-term source of contamination of the porous media and the subsurface environment. Unfortunately such cases are common because of storage tank and piping leaks, spills on land surfaces and improper disposal practices (e.g. ,*1,2,3*).

Many cases of petroleum spills on land surfaces have been reported (e.g. ,*1,2,4*), and investigations have been carried out on the redistribution and dissipation of the contaminants over time (e.g., *4,5,6,7,8*). For example, a recent survey of the 25-year-old Nipisi oil spill showed the presence of oil residues at the polluted site with degradation of the residues still occurring at the soil surface and weathered oil still present in the subsurface sediments (*2*). The published data on post-spill behavior of oil in the unsaturated zone are from field surveys under uncontrolled natural conditions and little information on the impact of various factors affecting the fate of petroleum products in the soil environment can be obtained from these studies.

Our experiment, which followed a wide-ranging series of laboratory studies (e.g., *9,10,11,12,13*) was designed to determine, under field conditions, the combined effect of leaching by irrigation and rainwater on the redistribution, dissipation, persistence and composition of a petroleum hydrocarbon mixture. Kerosene - a petroleum product containing more than 100 semi-volatile hydrocarbons - was selected for our experiment. The fate and persistence of the petroleum spill were studied in the field with both the amount of contaminant on the soil surface and the leaching pattern being controlled. The present paper is devoted to reporting the experimental results. A companion publication on validation of a model of kerosene hydrocarbon redistribution under leaching in field soils is in preparation.

Experimental

Materials: The experiment was performed on a sandy loam Mediterranean Red Soil (Rhodoxeralf) at the Volcani Center, located in the coastal area of Israel. Relevant properties of a selected soil profile are presented in Table 1.

Kerosene – an industrial petroleum product - was chosen as the volatile petroleum hydrocarbon mixture (VPHM) for our studies. Figure 1 shows the chromatogram of a kerosene sample and selected identified components. The relevant properties of these components are given in Table 2.

Experimental: Twelve double-ring sampling plots of 0.385 m^2 each were randomly selected within the experimental field of 120 m^2 and kerosene was applied to them at a rate of 5.2 l/m^2 (2.0 l per plot) – amount aquvalent to kerosene residual capacity in the upper 15 cm layer of the soil profile. The experimental field was leached by sprinkler irrigation with 500 mm water during the dry summer season and by 500 mm of rain as the experiment continued during the winter season.

A field experimental set up are presented elsewhere (17).The treatments applied consisted of two leaching rates, pulses of 50 and 250 mm, and two initial soil moisture contents prior to kerosene application, air-dry and field capacity. A non-irrigated plot where the kerosene was added to the air-dry soil was also included in the experiment. The plots were sampled to a depth of 60 cm in increments of 10 cm followed by 20 cm increments to depth of 1 m, and the soil was analyzed for kerosene content. The soil samples were extracted in the field imidiatly upon sampling to reduce lose of volatile hydrocarbons. Replicate samples were taken per double ring.

Table 1: Properties of a selected soil profile from the experimental field

Depth (cm)	Silt %	Sand %	Clay %	Organic carbon %	pH
0-10	6.3	77.5	16.2	1.2	7.7
10-30	5.0	82.5	12.5	1.2	7.5
30-50	3.8	80.0	16.2	0.8	7.8
50-70	8.1	63.1	28.8	0.6	7.9
70-90	7.5	64.4	28.1	0.7	8.0
90-110	12.5	52.5	35.0	0.6	8.1
110-130	13.8	56.2	30.0	0.7	8.2

A companion outdoor experiment was carried out in order to establish the volatilization rate of kerosene in soil upper layer under no leaching and minute biodegradation conditions. Kerosene in the same rate (5.2 l/m^2) as in field experiment was applied to to 5 cm depth layer of air dry sandy loam Mediterranean Red Soil (Rhodoxeralf) which was previously crushed to pass 2 mm screen. The soil was kept dry throughout the experiment. Duplicate samples were taken after 2,6,9,14,24,44,85,and 194 day from application and kerosene content was determined using the same method of sampling and extraction described for the field experiment.

Analytical procedure: Extraction of residual kerosene from the soil was performed in the field immediately after sampling using a solvent mixture of 8 ml water and 6 ml dichloromethane. The soil-solvent mixture was mechanically shaken overnight and the soil was separated from the solvents by centrifugation (10 minutes - 2500 rpm). Phenanthrene was added to the organic phase extract prior to gas chromatography analysis as an internal standard and the spiked sample was analyzed. A Varian gas chromatograph, Model 3400 equipped with a FID detector and DB 1 capillary column (30 m, 0.25 mm, 0.25 μm) was used for chromatographic separation and a Finnigen MAT Magnum GCMS was used to identify the compounds. In both instruments the separation program was 40^{0}C for 2 min, a ramp of 3^{0}C/min to 180^{0}C, hold for 2 min and another ramp of 20^{0}C/min to 250^{0}C.

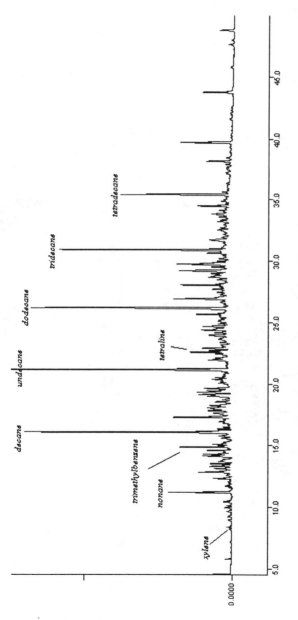

Fig. 1 Gas chromatogram of kerosene

Table 2: Properties of selected kerosene components

Compound	Solubility M (15,16)	M.W g/mol	Density g/ml	b.p. ^{0}C	Vapor pressure Pa (14)
p. xylene	2.02exp-3	107	0.861	137	1498
trimethyl benzene	7.66exp-4	120	0.894	176	408
n-nonane	3.69exp-6	128	0.718	150	825
n-decane	1.00exp-6	142	0.730	173	318
n-undecane	3.23exp-7	156	0.740	195	133
tetralin	1.59exp-7	132	0.973	207	88
n-dodecane	8.76exp-8	170	0.748	215	31
n-tridecane	4.15exp-8	184	0.755	235	1.3
n-tetradecane	1.18exp-8	198	0.762	253	0.4
n-pentadecane	3.37exp-9	212	0.769	270	0.1

Results and Discussion

Persistence of kerosene during leaching should be considered in relation to its redistribution and dissipation in the soil profile, as all processes occurr simultaneously. We present and discuss our results in light of this consideration.

The dynamics of kerosene redistribution in the soil profiles under leaching by 500 mm of irrigation (10 pulses of 50 mm each) followed by 500 mm of rain is depicted in Figure 2. The residual kerosene in the soil decreased as the cumulative amount of irrigation increased (i.e. with time). The applied kerosene, which was calculated to contaminate the soil to a depth of 10-15 cm, was leached in all cases to depths of 30-40 cm. During the first month (up to 300 mm of irrigation) the average kerosene concentration ranged from 2,300 to 4,500 µg/g soil in the upper 40 cm. When leaching and time had increased to 1,000 mm (180 days) the average residual concentration of kerosene hydrocarbons in the same soil layer was less than 500 µg/g soil. The deeper layers, 40 to 100 cm, were characterized by an average residual kerosene content ranging from 20 to 150 µg/g soil. The decrease in concentration can not be explained only by the redistribution of kerosene in depth, but by a combined dissipation-redistribution process, in which dissipation - enhanced by biological, chemical and/or physical factors - is predominant.

The effects of various treatments on the presence and redistribution of kerosene in the soil profile 180 days after its application and following leaching with 1,000 mm of rain and irrigation water, is illustrated in Figure 3. The redistribution with time of the kerosene in the nonirrigated plot, leached only by rainwater, is also included. After 180 days and 1,000 mm of leaching, the kerosene hydrocarbon concentration in the upper soil layer of the two 50 mm pulse irrigation treatments was over 1,000 µg/g soil. We can see that redistribution of kerosene following the various leaching treatments was restricted to the 0-40 cm layer. This trend suggests that kerosene is retained in the upper 40 cm and reaches the subsurface by volatilization and transport in the gas phase and/or by dissolution in the soil solution and transport in the liquid phase. Losses of kerosene components from the initially wet soil were smaller than from the initially dry soil. This behavior may be explained by the fact that the wet soil contains less free pore space, through which volatilization occurs, than the dry soil. The lowest concentration in the upper soil layer was observed in the non irrigated field in which the soil was leached only by 500 mm rain. The low concentration in the upper 10 cm is due to the uninhibited volatilization in the dry soil during the period between the kerosene application and the beginning of the rainy season.

The effects of water regime on kerosene redistribution and persistence are shown by the results of the irrigations with 50 and 250 mm pulses. The dissipation of kerosene was greater in the plots where 250 mm pulse irrigation was applied than in those irrigated by pulses of 50 mm, most probably because of the longer volatilization

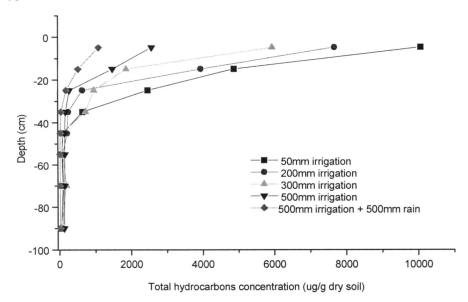

Fig. 2. Kerosene redistribution in the soil profile during leaching with 50 mm pulse irrigation and rain water

periods before and between the irrigations. The difference in the sizes of the irrigation pulses did not significantly affect kerosene redistribution in the soil profile beneath 50 cm. One hundred and eighty days after contaminant application, the redistribution process had ceased in all treatments, and only the persistent fraction of the kerosene remained in the 50-100 cm layer, and the mean residual concentration ranged from 10 to 25 µg/g soil.

The effect of soil moisture content prior to kerosene application (2% in air-dry soil and 17% in soil at field capacity) may also be observed in Figure 3. In this case also, the factor controlling the kerosene concentration in the upper soil layer is the ratio between free and water-saturated pores. When the soil was partially saturated with water (field capacity) the residual concentration was much higher: 3,250 µg/g soil in the initially wet soil compared with 1,000 µg/g in the initially dry soil. This difference decreasesd with increasing depth and below 40 cm the differences in kerosene concentration between these two treatments became insignificant.

It was observed (Figures 2 and 3) that most of the kerosene was retained in the upper soil profile (0-20 cm) during leaching and only small amounts were transported to the deeper layers. Figure 4 presents the time variation of kerosene concentration resulting from the dissipation and persistence of kerosene in the entire soil profile when the field was leached by 10 irrigations of 50 mm each and then by 500 mm of rain. The dissipation of kerosene in the upper layer of the soil profile, depicted

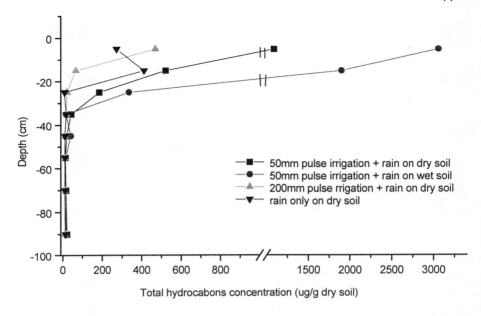

Fig. 3. Redistribution of kerosene in the field soil 180 days after application and leaching with 1000 mm (500 mm irrigation + 500 mm rain) and 500 mm (rain only)

together with that from the entire soil profile (0-100 cm) shows that the dissipation process occurs primarily in the upper soil layer and is only little affected by leaching and redistribution in depth down the soil profile. One hundred and eighty days after the kerosene spill, the residual concentration had decreased from an average of 20,000 to 1,900 µg/g dry soil in the upper layer and to an average of 1,600 µg/g dry soil in the soil profile from 0 to 100 cm depth. This dissipation pattern can be divided into two parts: the first from application to 50 days enhanced volatilization - and the second part from 50 days onward - slow dissipation with a persistent constant residual hydrocarbon concentration of 1,900 µg/g dry soil

Kerosene, a mixture of more then a 100 components with differing physical and chemical properties (Table 2), changes in composition during redistribution and dissipation in soils. The chemical and physical properties of each compound provide the driving force for the partitioning of individual components among the various phases in the soil matrix (NAPL, liquid, gas, and solid). Table 3 shows, as an example, the concentrations of selected components of the kerosene mixture (25% of the total hydrocarbons in the neat kerosene) found in the 0-10 and 60-80 cm soil layers 40 and 180 days after kerosene application to the field and leaching with 500

Table 3: Persistence of selected components of kerosene determined in soil 40 and 180 days after application (μg/g dry soil)

Components	At application (calculated)	After 40 days				After 180 days			
		500 mm irrigation only		0 Leaching with		1000 mm irrigation and rain		500 mm rain only	
Soil depth (cm)	0-10	0-10	60-80	0-10	60-80	0-10	60-80	0-10	60-80
xylene	132	10.9	0	0	0	0.1	0	0	0
n-nonane	485	47.3	1.0	0.8	0.4	0.3	0	0	0
trimethylbenzene	146	37.4	0.4	2.2	0	0.3	0	0	0
decane	1623	282.4	5.8	36.4	0	1.6	0	0.1	0
undecane	1630	527.9	10.1	130.2	1.7	0.7	0	0.9	0
tetraline	332	72.9	0.6	6.7	0	1.2	0	0.2	0
dodecane	1271	606.4	9.8	155.6	3.4	6.9	0.1	2.4	0.1
tridecane	1134	591.7	9.4	187.0	4.7	11.6	0.1	4.2	0.1
tetradecane	687	469.0	6.0	209.7	4.3	7.3	0.1	3.8	0.1
total	7440	2645.9	43.1	728.6	14.5	44.5	0.3	11.6	0.3

and 1000 mm water accordingly. A decrease in concentration compared with the neat kerosene occurred for all the components. The residual component concentrations were greater in the irrigated than in the dry field, and in the upper layer than in the deeper one. In general, we observe that 40 days after applications the residual concentrations of the selected kerosene components in the upper 10 cm layer of the soil profile ranged between 11 and 606 μg/g in the leached soil and between 0 and 210μg/g in the non leached, dry field. The concentrations in the 60-80 cm soil layer at the same time were much lower generally: less than 10μg/g in the leached soil and less than 4.7 μg/g in the dry soil. Six months after application all the components were present in only minute concentrations in all the treatments. Greater residual values were determined in the upper soil layer than in the 60-80 cm layer where the concentrations were below the detection limit.

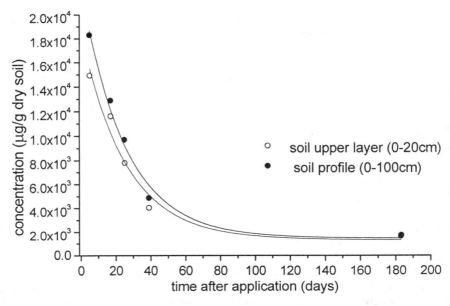

Fig. 4. Dissipation of kerosene in soil upper layer (white dots) and in the whole soil profile (black dots)

The importance of the vapor pressure in determining the behavior of individual components can be seen especially for xylene, nonane and trimethylbenzene, as well as for all the components with vapor pressure greater then 133 Pa at 300C (Table 2). For these compounds we find 75 -92% disappearance from the upper layer of the leached soil after 40 days while for the less volatile compounds we found a residual concentration range of 30-70%. In the 60-80 cm layer the concentration was reduced even more, so such that only 0.3% of the volatile compounds and a little less then 1% of the heavier compounds remained. In the non-irrigated field, 40 days after application, the concentrations of xylene, nonane, trimethylbenzene and tetraline in the 0-10 cm layer were reduced to 0-0.2% of the amount applied and those of less volatile compounds to 8-30%. One hundred and eighty days from the beginning of the experiment, the residual concentrations of all hydrocarbons were very low and only trace amounts of all the studied compounds could be found in the upper 0-10 cm layer of the field that had been leached with 1,000 mm. In the 60-80 cm layer most of the selected compounds had disappeared after 180 days, and the concentrations of the remaining compounds were the same for both leaching treatments (with and without irrigation).

The effect of volatilization from the soil upper layer under dry conditions on the kerosene dissipation from the soil profile was determined in a companion experiment with no leaching and the biodegradation close to zero. The cumulative volatilization of total kerosene and of light (up to C_{11}) and heavy constituents (C_{11} to C_{15}) are shown in Fig. 5. The light fractions represent 27% from the total kerosene and the heavy represent 70%. The experimental data show a first order decay pattern with calculated curves expressed by the equation:

$$y = 100 - Ae^{\frac{-x}{t}}$$

where: y is the concentration (as % of the applied concentration) and x is the time elapsed from application A (as % of the applied concentration) and t [time] are constants. The constant t is related to the rate of kerosene dissipation, with a small t the exponential factor becomes more dominant and the slope becomes steeper causing the function to approach complete evaporation faster. The t values obtained in our experiment were in the descending order: C_{11}-C_{15} fraction (17.5±1.6) > total kerosene (14.1±1.5) > up to C_{11}. (5.7±0.5). The empirical equation provides an acceptable fit with experimental data and quantifies the effect of kerosene components on the volatilization process. The light fraction is the first to evaporate and more than 90% of it was volatilized after 24 days from application compared to 70% for the heavier fraction and 75% for the total kerosene mixture. The pattern and rate of the evaporation from the surface without the effect of leaching and minute biodegradation was similar to the behavior under field condition indicating that volatilization was the major process affecting the persistence of the kerosene mixture. The importance of the vapor pressure of the various compounds of kerosene could be seen from the difference in their volatilization rate.

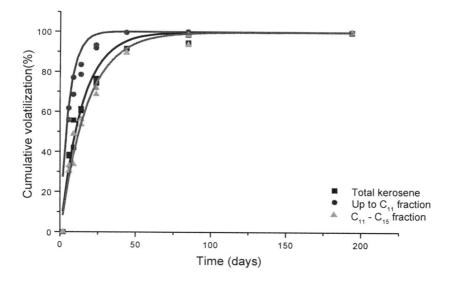

Fig. 5 Cumulative volatilization of kerosene and two groups of kerosene components (up to C_{11} and C_{11} to C_{15}) from the soil upper layer

The redistribution of four different representative compounds in the soil profile during leaching is shown in Figure 6. All the compounds exhibited trends of decreasing average total concentration with increasing depth for the first three irrigations (25 days). The rates of concentration decrease of kerosene residuals decreased in the order: n-nonane = trimethylbenzene > n-dodecane > n-tetradecane. For the two lightest compounds, n-nonane and trimethylbenzene there were large decreases between the first irrigation (5 days) and the 200 mm irrigation (17 days), with no change in the average kerosene concentration in the following 8 days (to 300 mm of irrigation), and transport to a deeper center of mass with continuous reduction in the component concentration is shown after 500 mm of irrigation (39 days). For the heavier compounds, n-dodecane and n-tetradecane the decrease in the residual concentrations were more moderate and in accordance with the lengths of the carbon chains. After 500 mm of irrigation n-nonane, trimethylbenzene, and n-dodecane exhibited the highest concentrations in the deeper layers of the soil profile and not in the 0-10 cm layer. Nonane, the compound with the highest vapor pressure, showed a large concentration decrease in the top soil layer accompanied by an increase in concentration in the 10-30 cm layer compared with the distribution after 300 mm of leaching. For trimethylbenzene, the concentration in the upper 30 cm layers was almost constant. The behavior of n-nonane, n-dodecane and n-tetradecane were alike and the differences in their redistribution and dissipation were due to volatilization and solubility differences, n-nonane volatilized faster than the other n-alkanes and therefore its concentration decreased more strongly It is also more soluble in water and therefore it was transported as a solute more than the other n-alkanes. These difference caused the center of mass of these compounds to move deeper and compounds to disappear more readily as their chain lengths were shorter.

Summary and Conclusions

The present paper considers the effect of leaching by irrigation and rainwater on the persistence of entrapped kerosene under field conditions in a contaminated site. The major proportion of a petroleum product applied to the soil was retained in the upper layers of the field soil profile and, despite the amount of leaching water applied, only a small proportion was transported to the deeper layers. Dissipation and redistribution are the controlling factors affecting petroleum hydrocarbon persistence in field soils and these factors are affected by the properties of the contaminant, and the porous medium and the prevailing environmental conditions. A petroleum product mixture can not be considered as a homogeneous substance but rather as a mixture of hydrocarbons which are redistributed and dissipated in the soil in accordance with their individual properties. The pollution hazard to the unsaturated zone arises mainly within a short time after contaminant application, but a long-term source of contamination may remain. The field and outdoor experiments of the present study with kerosene contaminant have so far indicated that the volatilization in the upper layer was the major process affecting its persistence in the soil profile during leaching. Because of irrigation and rain, depending on the rate and quantity of precipitation, significant quantities of the entrapped contaminant may be advected to deep layers, where the effect of evaporation is diminished.

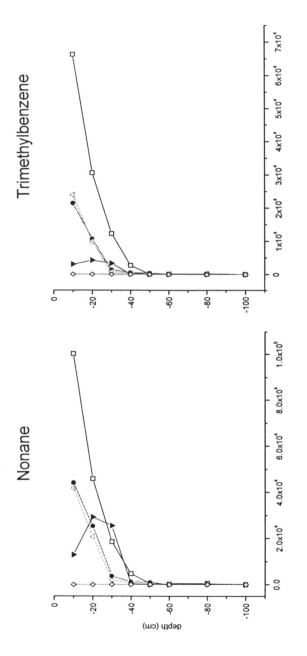

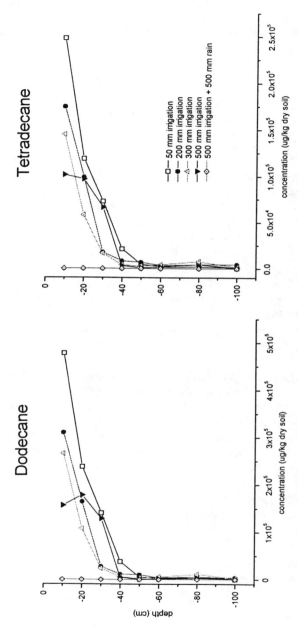

Fig. 6 Redistribution of selected hydrocarbons of the kerosene mixture in the field soil leached with 1000 mm (500 mm irrigation + 500 mm rain) during 180 days following application

84

Acknowledgements

The research was supported by grants from the Israeli Ministry of the Environment, the Water Research Institute of the Technion, and the Belfer Foundation for Energy Research. Thanks are due to Dr V. Glezer for GC-MS determinations.

References

1. Essaid, H. I.; Herkelrath, W. N.; Hess, K. M. *Water Resour. Res.* 1993 29:1753-1770.
2. Wang, Z.; Fingas, M.; Blenkinsopp, S.; Sergy, G.; Landriault, M.; Sigouin, L., Lambert, P. *Environ. Sci. Technol.* 1998 32:2222-2232.
3. Thorn, K. A., Aiken, G. R. *Org. Geochem.* 1998 29:909-931.
4. Eiceman, G. A; McConnon, J. T.; Zaman, M.; Shuey, C.; Earp, D. Ital. *J. Envir. Anal. Chem.* 1986 24:143-162.
5. Durnford, D.; Brookman, J.; Billica, J.; Milligan, D. L. *Ground Water Monitoring Review* 1991 11:115-122.
6. Ostendorf, D.W. *Water Sci. and Technol.* 1990 22:37-44.
7. Rubin, H.; Braxein, A.; Daniels, H.; Rouve, G. *Environ. Sci. Pollut. Control Ser.* 1994 11:355-375.
8. Johnson, R. L.; Perrott, M. *J. Contam. Hydrol* 1991 8:317-334.
9. Yaron, B.; Sutherland, P.; Galin, Ts.; Acher, J. *J. Contam. Hydrol.* 1989 4:347-358.
10. Galin, Ts.; Gerstl, Z.;Yaron, B. *J. Contam. Hydrol.* 1990 5:37 5-385.
11. Gerstl, Z.; Galin, Ts.; Yaron, B. *J. Environ. Qual.* 1994 23:487-493.
12. Yaron, B.; Clavet, R.; Prost, R. *"Soil Pollution Processes and Dynamics"* 1996 Springer –Verlag, Berlin, Germany 312 pp.
13. Jarsjo, J.; Destouni, G.;Yaron, B. *J. Contam. Hydrol.* 1997 25:113-127.
14. *Handbook of Chemistry and Physics*; Weast, R. C. (Ed.) CRC Press, Cleveland, Ohio, 54th edition 1974.
15. Kan, A. T.; Thomson, M. B. *Environ. Sci. Technol* 1996 30, 1369-1376.
16. Mott, V. H. *GWMR* 1995. 157-167.
17. Dror, I., Gerstl, Z. and Yaron, B. *J. Contam. Hydrol.* 1999 (submitted)

Chapter 6

Polychlorinated Biphenyls in Poland: History, Fate, and Occurrence

Janina Lulek

Department of Inorganic and Analytical Chemistry, Karol Marcinkowski University of Medical Sciences, Poznań, Poland

The review presents the polychlorinated biphenyls (PCBs) regulatory history and sources of their emission into Polish environment. The levels of these xenobiotics determined in different environmental matrices from Poland (air, water, sediments, soil and biota) are compared with data obtained in other regions of world. The concentrations of PCBs detected in several classes of Polish food are discussed.

Polychlorinated biphenyls (PCBs) are a class of 209 individual chemical compounds in which 1 to 10 chlorine atoms are substituted onto a biphenyl nucleus.

Since 1929, PCBs were commercially produced as complex mixtures with different trade names (Aroclors – USA; Clophens – Germany; Kanechlors –Japan etc.). These industrial formulations have been widely marketed for a variety of uses, including dielectric fluids in capacitors and transformers because of their chemical stability and high resistance to thermal breakdown (*1,2*).

However, a mass food poisoning which occurred in western Japan in 1968 and the discovery of PCBs in environmental matrices in late 1960s, prompted some industrial countries to take legislative measures for controlling the flux of these pollutants into the environment (*3,4*).

The paper concerns the PCB problem in Poland. It presents the regulatory measures adapted in Polish legislation as well as gives a brief discussion on the environmental, food and human levels of PCBs detected in Polish matrices.

© 2001 American Chemical Society

Regulatory History

In Poland, the problem of PCBs was mentioned for the first time in the mid 1970s, when Falandysz (5) commented on their solubility in lipids and their presence in food products. Nearly at the same time Strojny (6) reported on toxicological activity of PCBs mixtures used in Poland in C type electrical capacitors.

Since 1982, the use of Baltic cod liver oils for medical purposes in Poland has been restricted, in response to the high total PCBs concentrations (9.1 – 18.0 μg/g fat basis), determined by Falandysz (7) and Kannan et al. (8) in these matrices, collected during 1971-80.

In 1983, as a consequence of high PCBs and DDT levels detected in fresh cod livers, the Minister of Health and Social Welfare (9) forbade their transformation to food products.

Recently, the problem of regulatory measures concerning PCBs has become more urgent because of the necessity to conform Polish environmental legislation to the requirements of the European Union. Table I summarizes the recent PCBs regulations defined in Polish governmental acts.

Table I. PCBs Regulations

Deadline of institute	Legal Acts	PCBs Regulations
31.03.1993	Dz.U.R.P. 22/223 –1993 Ordinance of the Minister of the Health and Social Welfare	Maximal Acceptable Concentration of PCBs in cod liver – 5.0 μg/g fat
2.09.1993	Dz.U.R.P. 76/362 – 1993 Ordinance of the Minister of Environmental Protection, Natural Resources and Forestry	PCBs are included in the register of hazardous waste
1.01.1996	Dz.U.R.P. 153/775 –1995 Ordinance of the Cabinet	PCB s are included in I class of pollutants, air emission is punished by fine
27.09.1998	Dz.U. R.P. 79/513 – 1998 Ordinance of the Minister of Work and Social Policy	Workplace threshold limit Values of PCBs – 1mg/m^3 [Aroclor 1242]
1999	Project of the Ordinance of the Minister of the Health and Social Welfare	Maximal Acceptable Concentration of PCBs in drinking water – 0.5 μg/l

Table II reviews standard methods of PCBs determination in different matrices. These methods will be instituted in Poland in 2000 or later. The majority of them are based on European Standard Methods and include the determination of some individual PCBs in cleaned up extracts by capillary gas chromatography.

Table II. PCBs Standards in Poland

No of standard	Subject	Matrix	Deadline of institute
PN-IEC 997:1998	Determination of polychlorinated biphenyls (PCBs) by gas chromatography method	Electroinsulating oils	Instituted
PrPN-C-04579-1	Investigation of content of selected polychlorinated biphenyls (PCB). Determination of PCB No 28,52,101, 118,138,153 and 180 in water by gas chromatography	Water and wastewater	Project
EN 1528-1:1996	Determination of pesticides and polychlorinated biphenyls (PCB). General requirements	Fat food	2000
EN 1528-2:1996	Determination of pesticides and polychlorinated biphenyls (PCB). Extraction of lipids, pesticides and polychlorinated biphenyls and determination of lipids	Fat food	2000
EN 1528-3:1996	Determination of pesticides and polychlorinated biphenyls (PCB). Clean up methods	Fat food	2000
EN 1528-4:1996	Determination of pesticides and polychlorinated biphenyls (PCB). Determination and confirmation tests	Fat food.	2000
Pr EN 12766-1	Determination of polychlorinated biphenyls. Separation and determination of selected PCB congeners by gas chromatography	Petroleum products and used oils	2001

Production, Use and Sources

Although the production of PCBs in Poland was not of major significance on the world market, however Poland produced two original PCB formulations: Chlorofen and Tarnol. Moreover, Poland imported commercial PCBs mixtures from the former Czechoslovakia and USSR (*10*). No data are available on the quantity of production, manufacture time and usage of PCB formulations in Poland.

Determination of PCB isomer and congener composition in technical Chlorofen, using the high resolution gas chromatography-electron capture detector (HRGC-ECD) and HRGC- low resolution mass spectrometry (LRMS) techniques, was performed by Falandysz *et al.* (*11*). A comparison of PCBs group congeners in Chlorofen, Clophen A60 and Aroclor 1262 is shown in Fig.1.

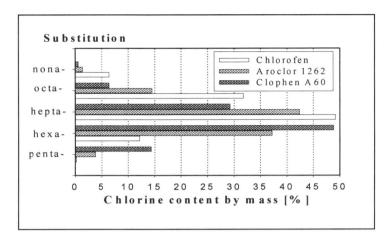

Figure 1. Composition of Chlorofen, Clophen A 60 and Aroclor 1262. (Data from references 11 and 12.)

The presented data indicate that Chlorofen contains larger proportion of nona- and octa-chorobiphenyls and less of penta- and hexa-substituted congeners than Aroclor 1242 and Clophen A60 formulations. According to the results of analyses of individual congeners, Chlorofen also incorporates the largest percent contribution of PCB 180 (2,2',3,4,4',5,5'-Heptachloro biphenyl) (*11*).

Since 1993, the PCBs have been included in the list of hazardous substances (Table 1), but until now the problem of the disposal and/or destruction of waste containing these xenobiotics has not been regulated. The available data (*12*) indicate that in national power plant installations about 1400 t of transformer and capacitor oils are being used.

In 1995-1997, a detailed stock-taking of technical installations which could be using products containing PCBs was made in the southwestern Poland. The results proved that the high-voltage transformers, with few exceptions, are filled with Poland-made oils which do not contain PCBs (13).

The results of the inventory concerning the contents of capacitors and low-voltage transformers are given in Table III.

Table III. Inventory of PCBs Equipment in Southwestern Regions of Poland

Type of Equipment	No of controlled installations	No of equipment containing PCBs	% detected
Low voltage transformers	24 674	69	0.28
Capacitors	34 055	11 004	32.3

NOTE: Data are from Reference 13.

The results indicate condensers as a possible source of PCB emissions to the environment. Unfortunately, no data are available on the contents of PCBs in electroinsulating oils used in many installations not mentioned above (e.g. hydraulic and heat transfer systems, vacuum pumps, etc). The PCB levels determined by Lulek (*14*) and Cęckiewicz (*15*) in randomly collected samples of used oils (Table IV) do not confirmed the data presented by Gurgacz (*16*), that Polish spent oils do not contain the PCBs.

Table IV. Level of PCBs in Some Used Oils

Matrix	Year of collection	No of samples	% detected	Concentration range [µg/g]
Motor oils	1995	13	46	2.88 – 53.42
Transformer oils	1995	13	46	2.33 – 30.67
Mineral oil	N.A.	1	-	96

NOTE: Data are from References 14 and 15. N.A. Not data available.

However it is worth to note that the PCB concentrations determined in examined oils did not exceed 50 µg/g oil (except two samples) which is a generally accepted limit for the definition of special waste in USA and most of countries of Western Europe (*17*).

Environmental Occurrence

PCBs are one of several "global" environmental pollutants, including mercury, lead and certain pesticides. Since the early 1970s, they have been found consistently in many environmental matrices (*1,2, 18-22*).

The first determinati0on of PCBs in environmental matrices in Poland were performed in the mid-1970s by Falandysz (cod liver oil) (5). In the 1980s the same author reported data on PCB levels in certain species of wild birds, marine fish and wild animals (*23*). However, the presence of PCBs in environmental matrices in Poland has attracted real interest since the mid-1990s, which was evidenced among others, by undertaking a monitoring of seven PCBs congeners (28, 52, 101, 118, 138, 153 and 180) in water of selected Polish rivers. Unfortunately, at present the results of this monitoring are not available.

Table V summarizes selected PCB levels determined by different authors in the samples of air, water, river sediments and soils collected in Poland.

The atmosphere is considered an efficient pathway of transportation and dispersion of semivolatile organohalogenated pollutants, including PCBs, in the environment at the global scale (*1*).

The only available data concerning the content of PCBs in the air in Poland, were provided by Falandysz *et al.* (Table V). The mean concentration of total PCBs determined in the ambient air of the coastal city Gdańsk, was 0.36 ± 0.28 ng/m^3 and

Table V. PCB occurrence in Polish Environment

Matrix	Year of collection	Location or soil type	No of samples	Range of PCB content	Ref.
Air	1991-92	Gdańsk	11	0.12 – 1.10	24
Water	1991-92	Wisła River	12	0.12 - 0.30	25
	1997-98	Odra River	23	<0.05 - 14	26
Sediment	1993-94	Northern Poland	20	1.7 - 630	27
		Southern Poland	3	46 - 1300	
	1997-98	Post-flooded of Odra River	13	< 0.05-420	26
Soil	1992	Postmilitary ground	N.A.	N.A.	28
	1990-94	Agricultural and forest	22	< 0.01 - 28	29
		Urban	31	15 – 530	
	1994	Postmilitary ground	7	32 - 3400	30
	1997	Urban	14	< 0.5 - 288	31
	1998	Postmilitary ground	6	9.2 - 288	32

NOTE: Units for content of PCBs in: air – ng/m^3, water- ng/l, sediment and soil – ng/g dry weight, N.A.- no data available.

was almost twice higher than that determined in the lower troposphere over the North and South Atlantic ocean (0.19 ± 0.12 ng/m^3) (18).

The detected PCB level was also higher than that determined on air filters from two UK rural sites in the years 1972-1992 (0.04 –0.81 ng/m^3) (33). The temperature dependent air concentration of PCBs, established by Falandysz et al. (24), may indicate revolatisation of these chemicals from the surface matrix around the city of Gdańsk.

Due to the hydrophobic character of PCBs only trace levels will be found in the dissolved phase. The PCBs adhere to particulate matter and most of this bound PCBs will deposit in sediment or bioconcentrate through one of the aquatic food webs (1).

Although, neither of the current European directives nor World health organization (WHO) recommendations on drinking water quality does state the maximum admissible PCB concentration, the above mentioned project of the Ministry of Health and Social Welfare recommends this value to be 0.5 µg/l (Table I).

The values of PCBs determined in surface waters of selected rivers in Poland are much lower than 0.5 µg/l and do not exceed several ng/m^3 (Table V).

The level of PCBs detected in river bottom deposits of a few selected rivers has confirmed the hypothesis about the adsorption of this group of pollutants on particulate matter. The PCB concentrations were apparently low in sediments collected from the area without evidence of direct impact of urbanization or industrial activity (1.7-2.2 ng/g dry weight) (27). The sediments collected at the sites under direct anthropogenic influence contained PCBs in concentrations between 46 and 630 ng/g dry weight. These values well correspond to the results obtained by Teil et al. (19) for the Seine river bottom sediments collected in Paris (80-365 ng/g dry weight).

For a few years a pilot study on the concentration of PCBs in different types of soils have been performed in Poland. Sampling sites of different types of soils collected between 1992-98 are presented in Fig. 2.

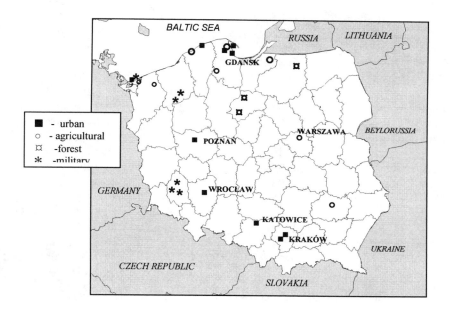

Figure 2. Sampling sites of different type of soils in Poland. (Data from references 28-32.)

The mean PCBs concentrations determined in the agriculture and forest soils do not exceed 20 ng/g dry weight (*29*) and are comparable with the results obtained for agriculture soils from the Jordan river valley (3.3-15 ng/g dry weight) (*34*). The PCB concentrations in the soil samples collected near their potential sources (transformer stations, spill sites, waste dumps) were by one order of magnitude higher than in the agriculture soils (Table V). This observation confirms that the uncontrolled and improper circulation of used oils may be one of the sources of PCBs in the environment in Poland (*14*). The highest PCBs levels, reaching up to 3400 ng/g dry weight have been determined by Falandysz *et al.*(*30*) and Lulek *et al.*(*31*) in randomly selected soil samples from the areas of the former Soviet army bases garrisoned between 1945 and 1993 in the northwestern and southwestern regions of Poland. In view of the above evidence it seems necessary to monitor the content of PCBs in soils, particularly in industrial areas and all former Soviet army bases.

Tables VI and VII present the PCB levels determined over the last several years in the fauna of the Polish coastal area of the Baltic Sea, as well as in some tissues of birds and game animals.

Table VI. Total PCB levels in Fauna from Polish coastal zones of Baltic Sea

Matrix	Year of collection	No of samples	Detected Level	Ref.
Plankton	1991-93	N.A.	210	36
Fish Herring	1981	18	200 – 1100*	37
	1991-93	3	1300	36
	1995	61	709 – 8416	38
	1996	57	802 – 8062	39
Trout rainbow	1982	6	190-530*	37
Pike-perch	1983	3	2 – 140*	37
	1992	1(3)	11000	40
Flounder	1992	3	910 – 9400	41
Perch	1992	2	3000 – 6400	41
Lamprey	1992	2	1000 – 1700	41
	1996	4	1281 – 3131	39
Cod	1992	1(3)	1400	40
liver	1995	9	229 –3954	38
	1996	72	1124 – 14070	39
Sprat	1995	24	571 – 2838	38
	1996	41	705 – 3769	39
Three-spined stickleback	1992	30	2700 – 4200	42
Harbour porpoises	1989–90	3	23000– 42000*	43
Blubber	1991 – 93	4	5700 – 16000	35

NOTE: Units are ng/g lipids or * ng/g wet weight. N.A. – No data available.

Table VII. PCB residues in Wild Birds and Game Animals

Animal	Matrix	Year of collection	No of samples	PCBs Content	Ref.
White tailed eagle	Adipose fat	1984	1	1600	44
	Eggs		2	4.6 – 39	
Lesser spotted eagle	Eggs	1980-84	11	0.27 – 29	44
Black cormorant	Liver	1992	3	34 ± 17	45
Wild boars	Adipose fat	1988	1(50)	0.039	35
		1990 –91	134	0 - 0.165	46
		1995	18	0 – 0.095	38
		1996	52	0.043 – 0.115	39
Roe deer	Adipose fat	1988	1(50)	0.032	35
		1990 – 91	108	0 - 0.054	46
		1995	25	0.050	38
		1996	48	0.015 – 0.022	39
Red deer	Adipose fat	1990 - 91	105	0 - 0.079	36

NOTE: Units are µg/g lipids for adipose fat and liver and µg/g wet for eggs.

PCB Levels in Food and Human

The food chain is widely recognized as the primary pathway of human exposure to PCBs. Table VIII gives the results of determinations of total PCB concentrations in different food products, performed between 1971 and 1996.

Table. VIII. Residues of PCBs in Food in Poland

Matrix	Year of collection	No of samples	% detected	Total PCBs Levels Mean	Range	Ref.
Swine fat	1987-88	3(150)	-	0.003	0.0025 – 0.064	35
	1991	484	46.7	0.090 ± 0.052	0 – 0.888	47
	1994	50	40.0	0.002	0 – 0.019	48
	1995	299	36.8	0.011	0 – 0.037	38
	1996	302	32.4	0.004	N.A.	39
Cows fat	1987	2(100)	-	0.007	0.0065 – 0.0076	35
	1994	470	100	0.032 ± 0.052	0.09 – 0.173	47
	1995	283	94.4	0.016	0 – 0.027	38
	1994	50	76.0	0.005	0 – 0.098	48
	1996	300	97.0	0.019	0 – 0.031	39
Pork ham	1993	168	57.5	0.015 ± 0.018	0 – 0.116	49
Hen fat	1992-93	810	95.2	0.044 ± 0.065	0 – 0.704	47
Hen eggs	1994	455	82.5	0.027 ± 0.031	0 – 0.338	50
Bovine milk	1994	285	96.8	0.017 ± 0.008	0 – 0.051	51
	1995	200	90.5	0.016	0 – 0.100	38
	1996	200	97.0	0.13	0 – 0.019	39
Milk products	1995	50	N.A.	0.025 ± 0.038	0 – 0.258	52
	1996	159	100	0.183 ± 0.062	0.010 ± 0.460	39
Cottage cheese	1995	112	92.6	0.013 ± 0.007	0 – 0.048	53
fermented		124	96.0	0.012 ± 0.04	0 – 0.029	
Cooking oils	1996	79	100	0.035 ± 0.036	0.010 – 0.190	39
Margarine	1996	71	100	0.099 ± 0.076	0.010 – 0.200	39
Vegetable-meat products	1995	63	N.A.	0.004 ± 0.003	0 – 0.010	52
Canned fish	1995	96	100	1.371± 0.710	0.107 – 3.855	38
	1996	33	100	0.813 ± 0.552	0.121 – 1.945	39
liver	1971-83	27	100	6.1 ± 2.4*	3.9 – 10*	9
	1995	24	100	0.503 ± 0.449	Max. 1.578	52

NOTE: Units are µg/g lipid weigh or * µg /g wet weight. N.A. – no data available.

The studies on the concentration of PCBs in fish and fish products initiated by Falandysz (7,9) in the 1970s and continued by the State Fishing Institute in Gdynia

(*38*), were in the 1980s extended to other kinds of food products (meet, meet products) analyzed at the State Veterinary Institute in Puławy (*38*).

Since 1995 PCBs have been included in the list of the most important food pollutants determined within the national monitoring program, undertaken among others to assess the quality of Polish food products (*38, 39*). At present in Poland the admissible levels of PCBs in food and food products are not legally established, except in cod liver, therefore the above presented results can only be referred to the admissible levels determined in legal systems of other countries varying from 0.2 to 0.5 µg/g depending on the product and the country (*54*). The mean values of PCB concentrations determined in all analyzed matrices, except certain species of fish and fish products, are significantly lower than those stated in food control acts in the USA and European Union countries.

The concentrations of PCBs in human tissues determined in Poland, presented in Fig. 3, are comparable with those reported from the USA and other industrialized countries (*55*).

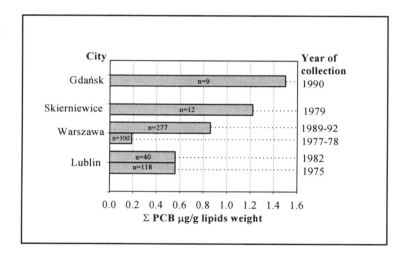

Figure 3. Mean concentrations of total PCBs in human fat from different cities in Poland. (Data from references 55,56.)

Mean concentrations of PCBs detected in human breast milk were relatively high (76 ± 41 µg/l) and did not differ from those found in human milk in other European countries (*57, 58*).

The elevated PCB levels determined in human milk and human tissues do not correspond with the relatively low concentrations of these xenobiotics in the analyzed food products, which may indicate that the compounds get into the human organisms by ways other than through the alimentary chain.

Conclusions

As the main aim of this review was to present the fate and occurrence as well as a historical aspect of PCBs in Poland, the problems related to the analytical procedures used in the determination of these xenobiotics have been disregarded. The results obtained by different authors in different laboratories have been converted to total PCB concentrations, and the results concerning individual congeners, including planar ones, have been omitted.

It should be mentioned that over the almost 20 years of continuous determinations of PCBs in environmental samples in Poland, the analytical procedures used have been changed according to the world trends in the analysis of trace substances. Taking into consideration the requirements of current PCB analysis standards, the laboratories determining these compounds continuously have validated the analytical procedures employed either by using the reference materials or joining the interlaboratory quality control programs both national and international (59).

The main conclusions following from this brief review of data on the environmental occurrence of PCBs in Poland, taking into regard their elevated levels in human milk and tissues and insufficient data on their level in different environmental matrices, are the necessity to complete the relevant legislation, the need of full recognition of the sources of PCBs emission and extension of monitoring of these compounds in the environment.

Acknowledgements

This work was supported partly by Polish State Committee of Scientific Research under Grant 754/PO5/99/16.

References

1. De Voogt, P.; Wells, D.E.; Reutergardh, L.; Brinkman, K.A. *Intern. J. Environ. Anal. Chem.* **1990,** 40, 1-46.
2. Lang, V. *J. Chromatogr.* **1992,** 595, 1-43.
3. De Voogt, P. *Chemosphere* **1991,** 23, 901-914.
4. Reding, R. *J. Chromatogr. Sci.* **1987,** 25, 338-344.
5. Falandysz, J. *Farm. Pol.* **1975,** 26, 197-204.
6. Strojny, J. *Capacitors in Network of Industrial Plants* Wydawnictwo Naukowo-Techniczne, Warszawa, Poland, 1976, pp.180-181.
7. Falandysz, J. *Farm. Pol.* **1977,** 33, 351-356.
8. Kannan, K.; Falandysz, J.; Yamashita, N.; Tanabe, S.; Tatasukawa, R. *Mar. Pollut. Bull.* **1992,** 24, 358-363.
9. Falandysz, J. *Roczn. PZH* **1986,** 37, 9-16.

96

10. Falandysz, J. *Polychlorinated Biphenyls (PCBs) in Environment: Chemistry, Analysis, Toxicology, Levels and Risk Assessment;* ISBN 83-86230-52-5; Fundacja Rozwoju Uniwersytetu Gdańskiego, Gdańsk, Poland, 1999; pp.21.
11. Falandysz, J.; Yamashita, N.; Tanabe, S.; Tasukawa, R. *Intern. J. Environ. Anal. Chem.* **1992,** 47, 129-136.
12. Erickson, M.D. *Analytical Chemistry of PCBs, Second Edition;* ISBN0-87371-923-9; CRC, Lewis Publishers, Boca Raton, New York, 1997; p. 39.
13. Rutkowski, M.; Beran, E.; Gryglewicz, S.; Stolarski, M. *in Management of Hazardous Waste (Gospodarka Odpadami Niebezpiecznymi);* BSE Kancelaria Sejmu: Warszawa, Poland, 1997; pp.42-45.
14. Lulek, J. *Chemosphere* **1998,** 37, 2021-2030.
15. Cęckiewicz, S.; Dettloff, R. in *Proceedings of the Third International Dioxins-Human-Environment Symposium,* September 17-18, 1998, Kraków, pp.1-7.
16. Gurgacz, W. in *Proceedings of First Polish Dioxins-Human-Environment Symposium,* September 22-24, 1994, Krakow, pp. 1-13.
17. Erickson, M.D. *Analytical Chemistry of PCBs, Second Edition;* ISBN0-87371-923-9; CRC, Lewis Publishers, Boca Raton, New York, 1997; p.15.
18. Schreitmüller, J.; Ballschmiter, K. *Fresenius J. Anal. Chem.* **1994,** 348, 226-239.
19. Teil, M.J.; Blanchard, M.; Carru, A. M.; Chesterikoff, A.; Chevreuil, M. *Sci. Total Environ..* **1996,** 181, 111-123.
20. Lead, W. A.; Steinnes, E.; Bacon, J. R., Jones K.C. *Sci. Total Environ.* **1997,** 193, 229-236.
21. Kannan, N.; Tanabe, S.; Okamoto, T.; Tatsukawa, R.; Philips D.J.H. *Environ. Pollut.* **1989,** 62, 223-235.
22. Ockenden, W.A.; Meijer, S.N.; Jones, K.C. in *American Chemical Society, Division of Environmental Chemistry, Preprints of Extended Abstracts Presented at the 217 th ACS National Meeting, Anaheim,CA;* ISSN#1520-0507, Anaheim, CA, March 21-25, 1999, vol. 39, No.1, pp.82-84.
23. Falandysz, J. *Polychlorinated Biphenyls (PCBs) in Environment: Chemistry, Analysis, Toxicology, Levels and Risk Assessment;* ISBN 83-86230-52-5; Fundacja Rozwoju Uniwersytetu Gdańskiego, Gdańsk, Poland, 1999; pp.237-238.
24. Falandysz, J.; Brudnowska, B.; Iwata, H.; Tanabe, S. *Organohalogen Compounds* **1998,** 39, 219-222.
25. Falandysz, J.; Brudnowska, B.; Iwata, H.; Tanabe, S. *Organohalogen Compounds* **1998,** 39, 215-218.
26. Galer, K.; Zygmunt, B.; Wolska, L.; Namieśnik, J. *Chem. Anal. (Warsaw)* **1999** *in press.*
27. Masahide, K.; Falandysz, J.; Brudnowska, B.; Wakimato, T. *Organohalogen Compounds* **1998,** 39, 331-335.
28. Spychała, A. *Biotechnologia* **1994,** 25, 42-49.
29. Masahide, K.; Falandysz, J.; Brudnowska, B.; Wakimato, T. *Organohalogen Compounds* **1998,** 39, 337-342.
30. Falandysz, J.; Kawano, M.; Wakimoto, T. *Organohalogen Compounds* **1997,** 32, 172-176.

31. Lulek, J.; Szafran, B. *Organohalogen Compounds* **1998,** 39, 315-318.
32. Lulek, J.; Szafran, B.; Lasecka, E. in *American Chemical Society, Division of Environmental Chemistry, Preprints of Extended Abstracts Presented at the 217 th ACS National Meeting, Anaheim,CA;* ISSN#1520-0507, Anaheim, CA, March 21-25, 1999, vol. 39, No.1, pp.160-163.
33. Jones, K.C.; Duarte-Davidson, R.; Cawse, P.A. *Environ. Sci. Technol.* **1995,** 29, 272-275.
34. Badawy, M.J., Wahaab, R.A. *Intern. J. Environ. Health Research* **1997,** 7, 161-170.
35. Falandysz, J; Kannan, K. *Z. Lebensm. Unters. Forch.*1992, 195, 17-21.
36. Falandysz, J. Dembowska, A. Strandberg, L. Strandberg, B. Berqvist, P.A. Rappe, C. *Organohalogen Compounds* **1998,** 39, 53-57.
37. Falandysz, J. *Roczn. PZH* **1985,** 36, 447-454.
38. *Rapport from Monitoring of Soils, Plants, Agricultural and Food Products Quality in 1995;* Michny, W., Ed.; ISBN 83-87166-15-4; National Inspection of Environmental Protection, Warszawa, Poland, 1996; pp. 3-159.
39. *Rapport from Monitoring of Soils, Plants, Agricultural and Food Products Quality in 1996;* Michny, W., Ed.; ISBN 83-87166-71-5; National Inspection of Environmental Protection: Warszawa, Poland, 1997; pp. 3-136.
40. Falandysz, J. Dembowska, A. Strandberg, L. Strandberg, B. Berqvist, P.A. Rappe, C. *Organohalogen Compounds* **1997,** 32, 358-363.
41. Falandysz, J. Dembowska, A. Strandberg, L. P.A. Rappe, C. *Organohalogen Compounds* **1998,** 39, 223-228.
42. Falandysz, J. Dembowska, A. Strandberg, L. Strandberg, B. Rappe, C. *Organohalogen Compounds* **1998,** 39, 229-235.
43. Kannan, K. Falandysz, J. Tanabe, S. Tatsukawa, R. *Mar. Pollut. Bull.* **1993,** 26, 162-165.
44. Falandysz, J. Król, W. Jakuczun, B. *Bromat. Chem. Toksykol.* **1987,** 20, 233-239.
45. Falandysz, J. Dembowska, A. Strandberg, L. Strandberg, B. Berqvist, P.A. Rappe, C. *Organohalogen Compounds* **1998,** 39, 47-51.
46. Przybycin, J. Juszkiewicz, T. *Medycyna Wet.* **1993,** 47, 318-319.
47. Niewiadowska, A. Żmudzki, J. *Rocz.PZH* **1996,** 47, 59-64.
48. Zasadowski, A. *Polish J. Environ. Studies* **1995,** 4, 65-67.
49. Żmudzki, J. Niewiadowska, A. Kowalski, B. Szkoda, J. Semeniuk, S. *Medycyna Wet.* **1994,** 50, 623-625.
50. Niewiadowska, A. Żmudzki, J. Semeniuk, S. *Bromatol. Chem. Toksykol.* **1996,** 39, 79-83.
51. Niewiadowska, A. Żmudzki, J. Semeniuk, S. *Roczn.PZH* **1995,** 46, 113-117.
52. Góralczyk, K. Ludwicki, J.K. Czaja, K. Struciński, P. *Roczn. PZH* **1998,** 49. 331-339.
53. Niewiadowska, A. Semeniuk, S. Żmudzki, J. *Roczn.PZH* **1996,** 47, 371-376.
54. Erickson, M.D. *Analytical Chemistry of PCBs, Second Edition;* ISBN0-87371-923-9; CRC, Lewis Publishers, Boca Raton, New York, 1997; p. 54.
55. Tanabe, S. Falandysz, J. Higaki, T. Kannan, K. Tatsukawa R. *Environ. Pollut.* **1993,** 79, 45-49.

56. Ludwicki, J.K. Góralczyk, K. *Bull. Environ. Contam. Toxicol.* **1994,** 52, 400-403.
57. Czaja, K.; Ludwicki, J.K.; Góralczyk, K.; Struciński, P. *Bull. Environ. Contam. Toxicol.* **1997,** 59, 407-413.
58. Czaja, K.; Ludwicki, J.K.; Robson, M.G.; Góralczyk, K.; Struciński, P., Buckley, B. in *American Chemical Society, Division of Environmental Chemistry, Preprints of Extended Abstracts Presented at the 217 th ACS National Meeting, Anaheim,CA;* ISSN#1520-0507, Anaheim, CA, March 21-25, 1999, vol. 39, No.1, pp.150-152.
59. Żmudzki, J. Niewiadowska, A.; Semeniuk, S.; Szkoda, J. in *Quality Problems in Trace Analysis in Environmental Studies;* Kabata-Pendias, A. Szteke, B., Eds.; ISBN 83-86961-12-0; Wydawnictwo Edukacyjne ŻAK: Warszawa, Poland, 1998 pp.265-279.

Chapter 7

Plant Uptake of LAS and DEHP
from Sludge Amended Soil

C. Grøn[1,4], F. Laturnus[1], G. K. Mortensen[1], H. Egsgaard[1],
L. Samsøe-Petersen[2], P. Ambus[1], and E. S. Jensen[3]

[1]Risø National Laboratory, P.O. Box 49, DK–4000 Roskilde, Denmark
[2]VKI, Agern Allé 11, DK–2970 Hørsholm, Denmark
[3]Royal Agricultural University, Agrovej 10, DK–2630 Taastrup, Denmark
[4]Present address: VK1, Agern Allé 11, DK–2970 Hørsholm, Denmark
(e-mail: chg@vki.dk)

Uptake in crop plants of LAS and DEHP from sewage sludge was
studied in controlled greenhouse experiments. Both LAS and DEHP
could be measured in the green parts of barley (highest
concentrations 0.5 µg total LAS/g dw and 0.6 µg DEHP/g dw) and
DEHP also in the green parts of carrot (mean 1.2±0.59 µg DEHP/g
dw). The LAS and DEHP found in the green parts of the plants
probably originated from the air or from dust, and not from the soil
or sludge. LAS from added sludge were found in barley roots at a
maximum concentration of about 20 µg/g dw, whereas
adsorption/uptake of sludge borne DEHP was not observed. LAS
could not be measured (<0.5 µg/g dw) in carrot peel. A low
concentration of DEHP was found in the carrot peel (mean value
3.2±1.6 µg/g dw) that probably did not come from the sludge, but
from the experimental soil used in the experiments. It is concluded,
that atmospheric deposition was the dominating source of LAS and
DEHP in the green parts of the studied crop plants, and that small
amounts could be transferred from the sludge or from the soil to the
roots of the plants.

Background

Sewage sludge is applied to agricultural land partly as waste disposal and partly
as recirculation of nutrients and organic matter (fertilization). With the sludge, a

© 2001 American Chemical Society

variety of organic contaminants are added to the agricultural soils. Linear alkylbenzenesulfonates (LAS, a commonly used tenside) and di-(2-ethylhexyl) phthalate (DEHP, a plasticizer) are among the compounds found in sewage sludges to high concentrations, up to 15 g/kg dry weight (dw) and 0.1 g/kg dw, respectively (1).

Whether a risk of human exposure and damage to the agro-ecosystem by the organic contaminants exists is not yet agreed upon. Still, authorities in several European countries (e.g. Denmark, Sweden, Germany, Austria) have enforced regulations of organic contaminants in sludge spread on farmland. One important issue in evaluating human exposure risk is uptake of the contaminants in crop plants.

In the present study, the uptake of LAS and DEHP in two crop plants, barley and carrot, from sludge amended soil was investigated in controlled greenhouse studies using analytical methods specific for the two contaminants.

Plant uptake of LAS

Alkylbenzenesulfonates (ABS, branched-chain) were reported to be taken up in plants from soil free water cultures with added ^{14}C-labeled ABS (2). Concentrations in the range of 0.1-2 μg/g dw were measured in sunflower stems and leaves and 4-90 μg/g dw in roots, with similar results for barley and with severe inhibition of plant growth by increasing ABS concentrations. With sunflowers grown in soil added labeled ABS, 30-90 μg/g dw were found in the stem and leaves. The ABS found in stem and leaves of barley grown in soil with added ^{14}C-labeled ABS ranged from below the detection limit to 7 μg/g dw. A clear growth inhibition was observed for sunflowers and less for barley (2). The study did not distinguish direct uptake of intact, labeled ABS in the plants from uptake of ABS metabolites or of labeled CO_2 formed by mineralization of added ABS.

Laboratory experiments demonstrated total uptake of LAS, metabolites or CO_2 in ryegrass from soil based upon application of ^{14}C-labeled LAS directly to the soil and yielding concentrations in the range of 0.1-20 μg/g dw (3). In the same study, LAS were reported to be taken up by ryegrass after field application of LAS solution yielding concentrations in the range of 100-1000 μg/g dw in the plants. Here, the LAS contents of the plants were determined using a non-specific colorimetric method for LAS analysis of the plants, and the plant leaves were probably exposed to LAS during application of LAS solution by irrigation. A low uptake from soil into soybean and corn was attributed to uptake of either ^{14}C-labeled LAS or metabolites or of ^{14}C-labeled CO_2 from mineralization of the added LAS (4).

Uptake of LAS, metabolites or CO_2 from sewage sludge spiked with ^{14}C-labeled LAS in grass, radishes, beans and potatoes in the range of 0.04-0.2 μg/g fresh weight (fw) has also been reported (5).

Overall, the reported studies suggest that uptake of LAS in crop plants is possible. However, reported plant concentrations cover four orders of magnitude from 0.1 μg/g dw to 1000 μg/g dw. No studies have been found reporting the uptake of LAS from sludge amended soils using analytical methods specific for LAS, such as HPLC or LC-MS.

Plant uptake of DEHP

In laboratory experiments, up to 1% of the ^{14}C activity added to soil as ^{14}C-labeled DEHP has been found in barley, and up to 0.08 μg/g dw in potatoes in a field experiment, but neither the parent compound DEHP, nor the primary degradation products mono(2-ethylhexyl)phthalate and phthalic acid could be identified in the plants (6).

Uptake of $^{14}CO_2$ formed by mineralization of ^{14}C-labeled DEHP was suggested as the pathway for uptake of the added ^{14}C label from soils (7). Still, root uptake of DEHP partially degraded in the soil to e.g. phthalic acid can not be excluded, as uptake of this compound in wheat and corn has been demonstrated (8).

Uptake of ^{14}C activity from labeled DEHP added with sewage sludge into lettuce, chilli pepper, carrots and tall fescue was demonstrated in laboratory experiments, where intact DEHP could also not be found in the plants (9). Higher concentrations of ^{14}C activity from DEHP in carrot tops than in the roots in this study was explained by root uptake of DEHP or of a polar metabolite followed by translocation to the tops.

After field application of sewage sludge containing approximately 100 μg/g dw DEHP, a five fold increase of DEHP in barley grain to 0.5 μg/g dw was found, as compared to control plots with no sludge added (10). DEHP concentrations above 50 μg/g dw were found in carrots, cabbage and corn grain grown in sludge amended fields (11).

Foliar exposure experiments have indicated leave surface uptake of DEHP of milfoil, and increased concentrations (up to 0.9 μg/g dw) of DEHP was found in milfoil growing near a plasticizer production plant which was known to be emitting DEHP to the air (12). Near phthalate emitting industries, up to 4 μg/g dw DEHP was found in corn leaves, whereas DEHP could not be found (<0.8 μg/g dw) in plants grown on fields amended with DEHP containing sewage sludge, but out of range of the DEHP emitting industries (13).

From studies using addition of radiolabeled DEHP to soil it can be concluded that root uptake of DEHP from soil may be of minor importance, as stated also in a recent, comprehensive review on the environmental fate of phthalate esters (14). The studies on foliar uptake of airborne DEHP demonstrated uptake to low concentrations (0.6-4 μg/g dw) in plants, whereas the two published field studies on plant uptake from sludge amended soil gave plant concentrations differing by two orders of magnitude (0.5-50 μg/g dw).

Conclusive studies on the uptake of DEHP from sludge amended soils using analytical methods specific for DEHP, such as GC-MS, have not been found.

Experimental

Sludge was sampled for the experiments on two occasions from a sewage treatment plant close to Copenhagen, Denmark, treating domestic and industrial sewage (Table I). The sludge was mixed with a sieved (5 mm aperture) sandy soil

(Table II) corresponding to six levels of amendment increasing from 0 to 90 t dw/ha and with four replicates and one plant free control for each treatment. In order to avoid deficiency of N, P, K, Mg, Cu and Mn, these elements were added in excess to the soil before the experiments. Soil and sludge were mixed thoroughly to mixtures with only a few small (<2 mm x 5 mm) particles. The homogeneity of the mixtures was controlled by sampling during distribution of the mixtures among the cultivation pots, followed by subsequent analysis for LAS. For a sludge addition yielding a mean concentration of 69 µg LAS/g dw, the relative standard deviation was 15%, and the analytical relative standard deviation (analyzing freeze dried, grinded and homogenized subsamples of one sample) was 7-16% (Table III). Additional treatments included a heterogeneous "sandwich set up" with a sludge layer corresponding to 90 t dw/ha between two soil layers (barley only). Furthermore, a "spiked set up" was established with additions of LAS and DEHP in solution to soil without sludge, yielding soil LAS and DEHP concentrations corresponding to that obtained with the 90 t dw/ha treatments.

Table I. Sludge characterization

Parameter	Range	Parameter	Range
Water content, %	29.3-29.5	Ash, % of dw	51-52
Total nitrogen, g/kg dw	27.4-28.7	LAS, µg/g dw	3700-5100
Total phosphorus, g/kg dw	29.7-30.7	DEHP, µg/g dw	39-120

Table II. Soil characterization

Parameter	Range	Parameter	Range
Clay, %	3.7	Total nitrogen, %	0.10
Silt, %	3.1	Total phosphorus, g/kg dw	0.38
Sand, %	90.9	LAS, µg/g dw	<0.05-<0.2
Humus, %	2.3	DEHP, µg/g dw	0.13-0.18
Total carbon, %	1.3		

Barley (*Hordeum vulgare* L. cv. Apex) and carrots (*Daucus carota* L.) were sown and cultivated in a greenhouse to young, green plants (barley, 3 weeks) and to maturity (carrots, 3 months), respectively. Barley was grown in 3.2 L glass cultivation pots wrapped with aluminum foil to protect the soil and roots against light. Each pot took 3.5 kg of soil. Carrots were grown in 20 L stainless steel pots, each taking 19-20 kg of soil. Water was supplied to maintain 60-75% of water holding capacity. The light was on for 16-24 hours per day (100-155 µE/m^2/s). The temperature was maintained at 12-18°C (for carrots up to 25°C in the last, hot part of the growth period), and the air humidity varied between 35% and 40% (for carrots in the hot period: 10-33%). The plants were protected against direct dry deposition with a glass dust cover 40 cm and 100 cm above the pots for barley and carrots, respectively, allowing free air circulation around the plants.

Samples of soil, sludge and soil-sludge mixtures were taken after thorough mixing at the set up of each experiment and at the time of harvest. Green plants were cut 1 cm above the soil surface, while roots were cleaned of soil and rinsed with pure water (Milli-Q System). Carrots were separated into green tops, peel and cores before analysis. All sample handling and storage was with stainless steel equipment, glass or aluminum foil. Samples were stored at -18°C until analyzed. All samples for LAS analysis and soil and sludge samples for DEHP analysis were freeze-dried and further homogenized by grinding before analysis. For DEHP, plant samples were analyzed without pretreatment.

General chemical analyses were done using standard methods. Lipid contents of plant material, soil and sludge were determined by extraction with chloroform/methanol and gravimetry, modified according to Folch (15). The performance characteristics obtained for the two dedicated methods used for LAS and DEHP are given in Table III. Note that the analytical detection limits (estimated from the standard deviation obtained by full procedure analysis of blank samples) varied depending upon the amount of sample available and the analytical performance at the time of analysis.

LAS was analyzed using a method (16) for soil, and modified for use with plant material. In the method, two successive Soxhlet extractions with pure water and methanol were followed by clean-up on C8 (reverse phase) and SAX (anion exchange) solid phase extraction columns, concentration by partial evaporation and analysis by HPLC with fluorescence detection (emission 225 nm, excitation 295 nm). Results were calculated as total LAS from quantification of the different LAS homologues.

Table III. Performance of the applied LAS and DEHP analytical methods

Parameter	Analytical principle	Limit of detection	Coefficient of variation	Recovery
LAS	HPLC	0.05-0.5 µg/g dw	7-16%	77-81%
DEHP	GC-MS-MS	0.1-0.5 µg/g dw	15-34%	70-98%

DEHP analysis (modified after ref. 17) was done by extraction with hexane/acetone (1:1), clean up on a Florisil solid phase column (plant samples only), concentration by partial evaporation and analysis by GC-MS-MS (quantification using the sum of ions with m/z=65 and m/z=121 fragmented from the ion with m/z=149).

Results

The sludge amendments increased the growth by up to 22% and 43-44% for dry weight yield of barley (green parts) and carrots (tops and roots), respectively, except for the highest amendment (90 t dw/ha) where the growth decreased. The addition of nutrients with the sludge seemed to outweigh the effects of potentially inhibiting components in the sludge. The decrease observed for the 90 t dw/ha sludge amendment was probably not caused by inhibition by specific components of the sludge, but rather by partially anaerobic conditions developing in the soil-sludge mixtures due to microbial respiration during degradation of the added sludge organic matter. The partially anaerobic conditions for the highest sludge addition were reflected by a high fraction of ammonium nitrogen in this soil-sludge mixture after the growth period. Of the inorganic nitrogen pool, NH_4^+-N constituted 55% for the highest amendment as compared to 2.4-5.5% for the lower sludge amendments (data not shown).

LAS and DEHP in the plants

In the green parts of young barley plants, both LAS and DEHP concentrations were close to the detection limit, in all cases <0.5 µg total LAS/g dw and <0.6 µg DEHP/g dw (Table IV). No trend could be distinguished with increasing sludge additions to the soil. Plants grown in soil with the free compounds added in solution (spiked) and in soil with the plant roots penetrating a solid sludge layer (sandwich)

Table IV. Concentrations of LAS and DEHP in barley

| Sludge addition | LAS | | DEHP | |
t sludge dw/ha	Green parts	Roots	Green parts	Roots
0	<0.3/0.32	<1	0.33/<0.3	0.3
0.4	0.24/0.26	1.1	0.34/0.44	0.4
1.5	<0.2/<0.2	<2	<0.3/0.28	1.0
6	0.26/0.15	2.3	0.43/0.43	0.5
23	0.2/<0.2	19	<0.3/0.35	1.3
90	<0.3/<0.2	13	0.56/0.45	1.3
"90", spiked	<0.5/<0.5	-	<0.5	-
"90", sandwich	0.22	-	0.41	-

NOTE: Units are µg/g dw
NOTE: Data given for separate analyses of two samples, each mixed of material from two pots (green parts), or for analysis of one sample mixed of material from four pots (roots).

exhibited the same low concentrations of LAS and DEHP. Higher concentrations of LAS and less evident also of DEHP were observed for barley roots. An increase in root concentration with increasing sludge addition was observed for LAS.

With carrot peel, core and tops, all LAS concentrations (data not shown) were below the analytical detection limit of 0.5 µg total LAS/g dw, except for a few scattered values measured as 0.5 and 0.6 µg total LAS/g dw. Due to the higher analytical limit of detection, air deposition of LAS on carrot tops as suggested for barley leaves, could not be observed.

For DEHP, the concentrations were highest (mean 3.2±1.6 µg DEHP/g dw for all treatments) for carrot peel, lower for carrot tops (mean 1.2±0.59 µg DEHP/g dw for all treatments) and again lowest for carrot core (mean 0.80±0.42 µg DEHP/g dw for all treatments). The DEHP concentrations in the carrots did not vary with the sludge additions (Table V).

Table V. Concentrations of DEHP in carrots

| Sludge addition | Carrots | | |
T sludge dw/ha	Peel	Core	Tops
0	6.3/2.4	0.7/0.3	1.7/1.3
0.4	3.0/2.7	0.3/0.4	2.3/1.1
1.5	1.9/4.2	0.9/1.3	0.4/0.2
6	5.2/5.1	0.8/1.0	1.2/0.7
23	2.3/2.0	1.0/0.6	1.0/0.9
90	3.7/4.0	1.4/1.6	1.2/1.9
"90", spiked	0.8/1.5	0.4/0.5	1.7/1.7

NOTE: Units are µg/g dw
NOTE: Data given for separate analyses of two samples, each mixed of material from two pots.

Normalized to the lipid content of the plant material (data not shown), carrot core and tops exhibited similar DEHP concentrations, 31±17 µg DEHP/g lipid and 32±14 µg DEHP/g lipid, respectively, whereas the lipid normalized concentration in carrot peel was 156±69 µg DEHP/g lipid. The DEHP concentration in the soil used for the carrot experiments was 0.18 µg/g dw. With a lipid content of the soil of 0.079%, the lipid normalized DEHP content of the soil without added sludge was 228 µg DEHP/g lipid. Similarly, a normalization of the DEHP concentration to the organic matter content of the carrot peel (set equal to the loss on ignition) and the soil (set equal to the humus content) yields normalized values of 3.8 and 7.8 µg DEHP/g organic matter for carrot peel and soil, respectively.

LAS and DEHP in soil

For the low sludge additions, both LAS and DEHP were degraded to below the analytical detection limits after the growth periods of 3 weeks for barley and 3 months for carrots. Still, for the higher sludge additions, up to 86 μg total LAS/g dw and 3.1 μg DEHP/g dw soil-sludge mixture remained in the soil.

Discussion

In contrast to previous reports based upon added ^{14}C-labeled ABS and LAS (*2,3,4,5*), no uptake via roots and translocation to the green parts of barley of intact LAS added to soil as part of sewage sludge was observed. Any presence of LAS here was below or in the range of the contribution from background atmospheric deposition (0.15-0.32 μg total LAS/g dw) or below the analytical detection limit (<0.2-<0.5 μg total LAS/g dw). Atmospheric deposition of LAS can be explained by reported findings of LAS (up to 1500 μg total LAS/g dw) in indoor dust samples (*18*).

In the fine roots of barley, adsorption and/or uptake of LAS added with sludge increased the LAS concentration up to 19 μg total LAS/g dw. This is close to the highest values reported in plant material using ^{14}C-labeled LAS (20 μg/g dw LAS, or metabolites and CO_2 correspondingly), but far from the 1000 μg total LAS/g dw found with non-specific LAS analysis using colorimetry (*3*).

In carrots, LAS concentrations were below the analytical limit of detection (<0.5 μg total LAS/g dw) in the green parts (tops), in the peel and in the core of the taproot. Considering the data for barley roots, it is surprising that LAS with the hydrophobic alkylbenzene structure could not be detected in carrot peel and core. Carrots are considered highly capable of accumulation of hydrophobic contaminants due to a high lipid content (*19*). One reason for the lack of LAS detection can be that the fine root network and not the taproot of carrots is the active part in taking up water and solutes from the soil (*20*). These fine roots are consequently the part of the carrots most exposed to contaminants soluble in the soil water and, in contrast to the barley root samples, they were not included in carrot samples. Following this logic, it is not surprising that LAS was not detectable in the peel and core of the carrots, as any LAS in these parts, as well as in the tops, would need to be translocated to these parts.

The concentrations of DEHP in the green parts of barley and carrots (<0.3-2.3 μg/g dw) were in the same range as concentrations reported in the literature (*12, 13*) resulting from foliar uptake (0.6-4 μg/g dw) at fields near DEHP emitting industries. The DEHP concentrations found in the green plant parts were at the low end of the range (0.5-50 μg/g dw) reported as taken up from soil in field studies (*10, 11*). In the carrot core, the mean DEHP concentration normalized to plant lipid content was equal to that in carrot tops, suggesting that the DEHP concentrations in these two plant compartments were in equilibrium (cf. ref. *20*) and most likely were caused by foliar uptake from the atmosphere. In the carrot peel, the DEHP concentrations found were higher, both on dry weight and on lipid basis suggesting another origin or transport route. Still, the DEHP concentrations in the carrot peel did not increase with the amount of sludge added. Consequently, the sludge is hardly the source of the increased DEHP concentrations in the peel.

The peel concentrations normalized to both lipid and organic matter were below the concentration (normalized correspondingly) found in the unamended soil. This suggests that the DEHP concentration in the soil before addition of DEHP with sludge or as spike was sufficient to cause the measured DEHP concentration in carrot peel by simple partitioning between the organic matter or the lipid phases of soil and carrot peel. A more probable source of carrot peel DEHP may therefore be the low DEHP concentration found as background in the unamended soil. Still, the higher lipid normalized DEHP concentration in carrot peel than in carrot core suggests that the DEHP in the peel can not be translocated across the outer plant membranes to the interior of the plant root.

The low concentrations of LAS and DEHP observed in the plant samples could be due to a rapid degradation of the compounds in the soil causing a virtual "no exposure scenario". This is not the case, at the least for the higher sludge additions, as can be seen from Figures 1 and 2 giving the LAS and DEHP concentrations in the soil before and after the 3 months growth period used for carrots. Evidently, LAS and DEHP remained in the soil-sludge mixtures after the long carrot growth period for the higher sludge amendments. Conversely, lower concentrations of both LAS and DEHP remained, if the compounds were added as spike. The results suggest that the degradation of both LAS and DEHP was inhibited by the partially anaerobic conditions at high sludge amendments, and probably also by a reduced bioavailability due to binding to organic matter in the sludge particles.

The longer exposure period for carrot (3 months) than for barley (3 weeks) may be part of the explanation for the higher concentrations of DEHP observed in carrots than in barley. Here, it should be emphasized as well that the DEHP concentrations in the sludge amended soil were increased by a factor of less than 100 above the analytical limit of detection, whereas the LAS concentrations were increased by more than a factor of 1000 for the highest sludge additions.

The different pattern of LAS and DEHP root uptake/adsorption can alternatively be explained considering the very different properties of the two compound groups. The water soluble (410 mg/L without formation of micelles, ref. *21*) LAS with low binding affinity to the organic matter in the sludge particles ($log(K_{ow})$=1.2-2.5 for C_{10}-C_{13} LAS, ref. *20*) can be transported out of the sludge particles and with the soil solution to those parts of the plant roots that are active in uptake of water and solutes (increased LAS concentrations observed in barley roots with increasing sludge addition). Conversely, DEHP with low water solubility (0.003 mg/L, ref. *14*) and high affinity for binding to the organic matter in the sludge particles ($log(K_{ow})$=7.5, ref. *14*) is trapped in the particles and not available to the plant roots (no or only slight increase in concentrations observed in barley roots). The low concentration of DEHP found in the unamended soil (0.13-0.18 µg DEHP/g dw) is less immobile due to the much lower organic matter content of the soil, compared to that of the sludge particles (Tables I and II). The soil DEHP can slowly equilibrate with the surfaces of

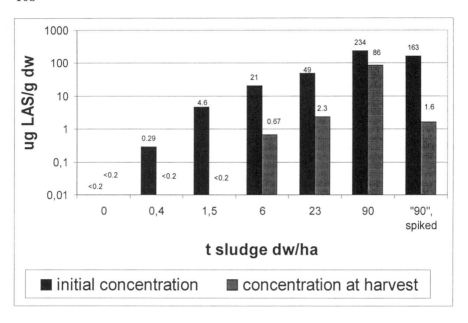

Figure 1. LAS in soil samples at set up and after 3 months of carrot growth.

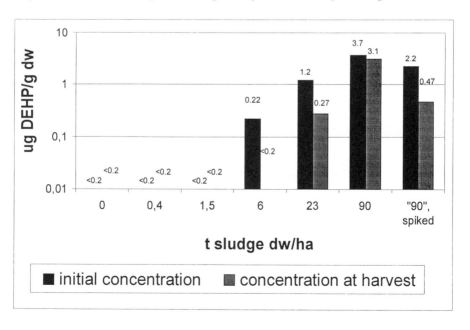

Figure 2. DEHP in soil samples at set up and after 3 months of carrot growth.

plant roots in contact with the bulk of the soil, whether the root is active in water uptake or not (low concentrations of DEHP found in carrot peels but with no relation to the amount of sludge added). For LAS, the inherent concentration in the unamended soil was always less than the analytical detection limit, and with the smaller affinity of LAS for adsorption to the organic matter of the root surfaces, the redistribution between soil and plant roots would not be observed for LAS (no LAS found in barley roots without sludge added, and no LAS found in carrot peel with no active uptake of water and solutes).

The fate of the compounds in the soil-sludge-plant root system is further complicated by the fact that both LAS and DEHP were degraded less when added with sludge particles (63% and 16%, respectively) than when added directly to the soil (99% and 79%, respectively, Figures 1 and 2). There was a more pronounced reduction in biodegradation and therefore probably also in bioavailability for the less soluble and more adsorbable DEHP than for LAS. Also, the occurrence of partially reducing conditions observed in the soil-sludge mixtures at the highest amendment are expected to cause a slower degradation of both LAS and DEHP (see Figures 1 and 2) due to lower biodegradability of the compounds in the absence of oxygen (ref.s *22* and *14*, respectively). Within the interior of the sludge particles, degradation of LAS and DEHP will accordingly be slow due to the anaerobic conditions here. The exposure of plant roots (and other soil organisms) to organic contaminants from sludge added to agricultural soil, thus, depends upon a delicate balance between release from the sludge particles, transport to the root and degradation in the soil.

One additional, complicating factor in evaluating the risk of plant contamination from sludge is the windborne dust and splashes bringing sludge remains to the aerial parts of the plants. In the present study, this aspect was carefully avoided and, therefore, not considered.

Conclusions

The uptake of LAS and DEHP in the green parts of barley and carrot grown in sewage sludge amended soil was predominantly from the air or from dust. The concentrations of both compound groups were in the range from <0.2 µg/g dw to 2 µg/g dw. For barley roots, adsorption or uptake of LAS from added sludge could be observed yielding a maximum concentration of about 20 µg/g dw, whereas adsorption/uptake of sludge borne DEHP could not be observed. In carrot peel, no LAS could be measured (<0.5 µg/g dw), but a low concentration of DEHP (mean value 3.2 µg/g dw) originating from the soil itself was found. A qualitative summary of the uptake pathways for LAS and DEHP in the studied plants is given in Table VI.

The results indicate that even in a worst case scenario with very high sludge amendments, atmospheric deposition and not uptake by plant roots is the dominating source of LAS and DEHP in the green parts of crop plants. Still, low concentrations of DEHP and LAS in root crops can be due to root uptake from soil or sludge.

Table VI. Dominating plant uptake pathways for LAS and DEHP

| | Barley | | Carrot | | |
	Green parts	Roots	Green parts	Carrot core	Carrot peel
LAS	Deposition from dust	Uptake from sludge	-	No uptake found	No uptake found
DEHP	Deposition from air or dust	Uptake from soil	Deposition from air or dust	Translocation from green parts	Uptake from soil

Acknowledgment

The present study was supported by the Danish Environmental Protection Agency and by the Centre for Sustainable Land Use and Management of Contaminants, Carbon and Nitrogen under the Danish Environmental Research Programme 1997-2000.

References

1. Smith, S.R. *Agricultural Recycling of Sewage Sludge and the Environment*; CAB International: Wallingford, UK, 1996.
2. Klein, S.A., Jenkins, D., McGauhey, P.H. *J Water Pollution Control Federation* **1963**, *35*, 636-654.
3. Litz, N., Doering, H.W., Thiele, M., Blume, H.-P. *Ecotoxicol Environ Safety* **1987**, *14*, 103-116.
4. Knaebel, D.B., Vestal, J.R. *Can J Microbiol* **1992**, *38*, 643-653.
5. Figge, K., Schöberl, P. *Tenside Surfactants Detergents* **1989**, *26*, 122-128.
6. Schmitzer, J.L., Scheunert, I., Korte, F. *J Agric Food Chem* **1988**, *36*, 210-215.
7. Topp, E., Scheunert, I, Attar, A., Korte, F. *Ecotoxicol Environ Safety* **1986**, *11*, 219-228.
8. Dorney, J.R., Weber, J.B., Overcash, M.R., Strek, H.J. *J Agric Food Chem* **1985**, *33*, 398-403.
9. Aranda, J.M., O'Connor, G.A., Eiceman, G.A. *J Environ Qual* **1989**, *18*, 45-50.
10. Kirchmann, H., Tengsved, A. *Swedish J Agric Res* **1991**, *21*, 115-119.
11. Webber, M.D., Pietz, R.I., Granato, T.C., Svoboda, M.L. *J Environ Qual* **1994**, *23*, 1019-1026.
12. Løkke, H. Rasmussen, L. *Environmental Pollution (Series A)* **1983**, *32*, 179-199.
13. Müller, J, Kördel, W. *Sci Tot Environ* **1993**, *sup*, 431-437.
14. Staples, C.A., Peterson, D.R., Parkerton, T.F., Adams, W.J. *Chemosphere* **1997**, *35*, 667-749.
15. Folch, J., Lees, M., Stanley, G.H.S. *J. Biol. Chem.* **1957**, *226*, 497-509.
16. Matthijs, E., De Henau, H. *Tenside Surfactants Detergents* **1987**, *24*, 193-199.

17. Holadová, K., Hajslová, J. *Int J Environ Anal Chem* **1995**, *59*, 43-57.
18. Vejrup, K.V., Wolkoff, P., Madsen, J.,Ø., *Proceedings of The 8th International Conference on Indoor Air Quality and Climate*, Indoor Air' 99, 8-13 August 1999, Edinburgh, Scotland **1999**, Vol. 4, pp 244-249.
19. Wang, M.-J., Jones, K.C. *Environ Sci Technol* **1994**, *28*, 1260-1267.
20. Trapp, S., Mc Farlane, J.C. *Plant Contamination*, Lewis Publishers: Boca Raton, FL, 1995.
21. Haigh, S.D., *Sci Tot Environ* **1996**, *185*, 161-170.
22. Hand, V.C., Williams, G.K. *Environ Sci Technol* **1987**, *21*, 370-373.
23. Krueger, C.J., Radakovich, K.M., Sawyer, T.E., Barber, L.B., Smith, R.L., Field, J.A. *Environ Sci Technol* **1998**, *32*, 3954-3961.

Chapter 8

Resistant Desorption Kinetics of Chlorinated Organic Compounds from Contaminated Soil and Sediment

Amy T. Kan[1], Wei Chen[2], and Mason B. Towson[1]

[1]Department of Environmental Science and Engineering,
MS–519, Rice University, Houston, TX 77005
[2]Brown and Caldwell, 1415 Louisiana, Suite 2500, Houston, TX 77002

The chemical release rates from laboratory and field contaminated sediments were studied. Contrary to reports that sorption rates are inversely related to K_{OW}, the slow desorption rates were found to be similar for different compounds. The data were modeled by a two compartment irreversible adsorption and radial diffusion model. Desorption kinetics from the first irreversible compartment can be modeled by radial diffusion and assume an irreversible adsorption constant and soil tortuosity of 4.3. The desorption half life is approximately 2-7 days. Desorption from the second irreversible compartment is very slow, (half-life of approximately 0.32 –8.62 years) presumably caused by entrapment in soil organic matter that increases the constrictivity of the solid phase to chemical diffusion. From the kinetic data, it is deduced that the diffusion pore diameter of the second irreversible compartment is approximately equal to the critical molecular diameter. The mass of chemicals in this highly constrictive irreversible compartment is approximately one fourth of the maximum irreversible, or resistant, compartment. The slow kinetics observed in this study add additional support to the notion that the irreversibly sorbed chemicals are "benign" to the environment.

Particles, or sediments, are the natural depository (and later sources) for contaminants of all kinds including metals, non-metals and neutral organic hydrocarbons. The manner in which contaminants interact with and desorb from particles in natural waters has enormous consequences to all inhabitants of the aquatic

© 2001 American Chemical Society

environment. When hydrophobic chemicals such as petroleum products, industrial wastes and municipal discharges are released into the coastal waterways, the primary fate in every case will generally be to attach to particles in the water column. The manner in which hydrophobic contaminants interact with particles can be either beneficial or detrimental depending upon the specifics of the equilibrium, the kinetics and the mass transport of the adsorption and desorption processes.

Mechanistic interpretation of sorption phenomenon is crucial for prior prediction of contaminant fate in the environment. Even though a large amount of mechanistic laboratory research has been directed toward understanding and quantifying the fate of hydrophobic chemicals in contact with naturally occurring particles in the water column, the ability to predict fate and clean-up protocols is still quite poor. The problem is partly due to the complexity of the environment. Different types of soils, chemicals, contamination level, and environmental chemistry could affect the fate of contaminants significantly.

In this study, the primary interest is the sorption mechanism of trace level neutral organic hydrocarbons to soil containing a moderate amount of organic matter. Several conceptual models have been proposed in the literature to explain the desorption resistance: (I) The existence of a condensed, glassy, organic polymeric matter as adsorbent. The adsorption to the condensed organic matter phase could be kinetically slow, site specific and non-linear (1,2); (II) The presence of high surface area carbonaceous sorbent, e.g. soots. The adsorption to these high surface area carbonaceous materials could be non-linear, site specific, and limited in capacity (3,4); and (III) The sorbed chemicals are irreversibly trapped in the humic organic matrix following sorption (irreversible adsorption) (5). The term "irreversible" is used to imply that desorption takes place from a molecular environment that is different from the adsorption environment. The irreversible compartment has been found to contain a finite maximum capacity (6). Neutral hydrophobic organic compounds desorb at a similar limit, regardless of the physical/chemical properties of the chemicals. The authors in 1985 (7) first proposed the notion of irreversible sorption as a new mechanism important for hydrocarbon transport.

Numerous researchers have shown that the adsorbent surface layer can undergo rearrangement upon adsorption and that this physical-chemical rearrangement can be the cause of non-reversible adsorption (8). These ideas have been tested with a varieties of organic contaminants (5,6,9,10). Recently, Huang and Weber (11) concluded that the entrapment of sorbing molecules within condensed soil organic matter matrices contributes significantly to sorption-desorption hysteresis or sorption irreversibility. Schlebaum et al. (12), presented compelling evidence to show that the anomalous sorption phenomena cannot be explained by the non-linear sorption to a condensed organic phase. Instead, the observations may be attributed to a change in conformation of the humic acid after adsorption, in agreement with the irreversible adsorption model proposed earlier (5). Chiou and Kile (4) proposed that the specific–interaction to high surface area carbonaceous material could reconcile the outstanding features of the nonlinear and the competitive sorption. Interestingly, the data reported by Chiou and Kile (4) can be interpreted as having an apparent saturation capacity of similar magnitude to the irreversible capacity of Kan et al (5,6).

Kan et al. (5) proposed that the sorption and desorption of organic chemicals to and from soil consisted of two compartments: (1) a labile desorption compartment, where the chemical can be readily and reversibly desorbed; (2) an entrapped, or irreversible compartment, where the desorption is hindered by soil organic matter due to the conformational changes in the soil organic matter following adsorption. The term "irreversible" used herein is in the same manner as is commonly used in the physical-chemical literature, e.g., Adamson (*8*), Bailey et al. (*13*), and Burgess et al. (*14*). A minimum thermodynamic requirement for adsorption and desorption to be irreversible is that there be a physical-chemical rearrangement in the solid phase after adsorption occurs, i.e., the desorption takes place from a different molecular environment than that in which adsorption occurs (*8*).

Desorption from both the labile and irreversible compartments are limited by retarded intraparticle radial diffusion. Chemical desorption from the irreversible compartment shows a constant and different partition coefficient, $K_{OC}^{irr} \cong 10^{5.53}$ L/Kg even for chemicals of widely different hydrophobicities. It is proposed that the chemical in the irreversible compartment form an organic "complex" with the organic colloids in the soil organic matter. Therefore, the physical/chemical nature of the adsorbate would be masked by the "complex". A similar conceptualization was proposed by Schulten and Schnitzer (*15*) based upon molecular modeling. Both the kinetic and thermodynamic aspects of desorption, regardless of the chemical hydrophobicity, are expected to be similar. The possible existence of such a complex has been observed previously (*5,16*). In this study, the long term desorption kinetics of neutral organic hydrocarbons from field-contaminated sediment (Lake Charles, LA) were evaluated to further elucidate the mechanisms that control the resistant desorption.

Materials and Methods

Sediments

The Lake Charles sediment was collected near the confluence of Bayou d'Inde and the lower Calcasium River at Lake Charles, Louisiana, by personnel at U.S. Geological Survey in 1995. The sediments were stored in clean plastic zipper bags and chilled to 4°C during shipment. Upon receipt, sediment samples were combined, mixed, and centrifuged at a low temperature to separate the pore water from the sediments, and the sediment was stored in a refrigerator. The sediment contains a large fraction of fine particles. The inorganic fraction of the sediment is mainly quartz, and the organic carbon content is 4.1%. Wet sediment, containing 31.2% moisture, was used exclusively in the experiments. In the following, the reported sediment weight is on dry weight basis. Tenax TA beads (polymeric adsorbent beads, 20/35 mesh, Alltech Associates, Inc., Deerfield IL) were used in desorption experiments. Tenax beads were cleaned before use by successive Soxhlet-extraction with acetone, acetone/hexane (1:1 by volume), methanol, and acetone, respectively, 48 hours for each extraction step. After extensive cleaning, 1 g Tenax beads was extracted with 10-ml acetone. The acetone extract was analyzed by GC/ECD to confirm the cleanliness. Clean Tenax was then baked at 200 C overnight and stored in methanol. Used Tenax was cleaned and regenerated by the same procedure mentioned above. Prior to use, Tenax beads were removed from methanol, baked at 200 C for at least four hours in a clean glass vial. The vial was then sealed and allowed to cool to room temperature. Glassware was cleaned in heated 2% detergent solution (RBS 35 from Pierce Chemical Co., Rockford, IL) overnight at 60 C. Then the glassware was brushed, rinsed with tap water, deionized water, acetone, and oven dried at 200 C.

Desorption experiments

Five experiments were used to study the release of hydrocarbons for the Lake Charles sediments with the presence of Tenax beads as a sink for the contaminant to desorb. The Tenax desorption procedure is similar to that used in the literature (*17,18*). In a typical desorption experiment, 0.7 g sediment was suspended in a vial containing 40 ml electrolyte solution (0.1 M NaCl, 0.01 M NaN$_3$) and 0.5 g Tenax. The vial was capped with a Teflon septum and tumbled at 1 rpm for a designated period of time. Tenax was replaced with clean Tenax at 2, 10, 20, and 30 day intervals. A total of 151 days desorption time was monitored. At the end of each desorption period, the reactor vials were centrifuged for 30 minutes at 1000 g (IEC Centra MP4 Centrifuge, International Equipment Company, MA). The Tenax beads were then carefully taken out with clean stainless steel spatula and transferred to a preweighed 25 ml clear glass vial. Another 0.5 g clean Tenax beads were added into the vial to continue desorption.

The contaminated Tenax was extracted with 6 ml acetone in four successive extractions. The acetone extraction was done on a horizontal shaker (Yamato, Model 1290) overnight. ^{14}C-labeled PCB (2,2'5,5' tetrachlorobiphenyl) or 1,4-dibromobenzene were added to Tenax prior to acetone extraction to determine the extraction efficiency. The contaminant concentration in the acetone extract was analyzed by GC/ECD. Over three repetitive extractions, the cumulative recovery of PCB from Tenax was 98%, even though this might not represent the extraction efficiency of the sample since the contact time between sediments and Tenax is much longer than that between the surrogate and Tenax.

Soxhlet extraction

The initial solid phase concentrations of selected chlorinated hydrocarbons on the sediments were determined with the Soxhlet extraction method followed by GC/MS and GC/ECD analysis. Approximately 10 g of wet sediment were mixed with 10 g anhydrous sodium sulfate (baked at 460 °C for four hours), and were refluxed with dichloromethane for 48 hours. 1,2-dibromobenzene, used as a surrogate standard, was added to the sediment before Soxhlet extraction. After extraction, the extract was passed through a 20-mm ID drying column containing about 10 cm of anhydrous sodium sulfate. The extraction flask and the sodium sulfate column were washed with 100 ml methylene chloride to complete the quantitative transfer of chlorinated hydrocarbons and the surrogate standard. The effluent was collected in a Kuderna-Danish concentrator and concentrated to 2 ml. The concentrated solution was cleaned with a Supelclean™ LC-Florisil SPE tubes (Supelco Inc., Belefont, PA) and a Nylon Acrodisc 13 syringe filter (0.45 μm pore size, Gelman Sciences, Ann Arbor, MI). A portion of the final extract was further concentrated and analyzed by GC/MS to identify the compounds in the sediments. The rest of the extract was diluted with isooctane and quantified with GC/ECD (HP Ultra 2 fused silica capillary column, 25 m l x 0.32 mm internal diameter, 0.52 μm film thickness). The injector temperature was set at 275 C and the detector temperature was set at 390 C. The temperature program was set at an initial temperature of 50 C, for 4 minutes, and increased by 8 C per minute to 280 C and hold at the final temperature for 4 minutes. Equal concentration of 2,5 dibromotoluene was added prior to GC analysis as an internal standard.

Results and Discussion

Since Lake Charles sediments were contaminated for over 30 years, the dissolution of the chlorinated compounds should have ceased long ago. Yet, four compounds, i.e., 1,3 dichlorobenzene, 1,4 dichlorobenzene, hexachlorobutadiene, and hexachlorobenzene still exist in significant and measurable quantity. Figure 1 shows the fractional desorption of the four chlorinated hydrocarbons from the sediments by Tenax desorption over 151 days. Even after 151 days desorption, significant fractions of the contaminants in Lake Charles sediments resist desorption. However, the desorption resistant fractions are extractable by Soxhlet extraction. The plot of fraction remaining vs. time showed a progressive decrease in slope, indicating greater and greater resistance to desorption. The dotted lines in Figure 1 were obtained by exponential curve fitting to an empirical two compartment first order kinetic model (Eq 1) (18).

$$S_t / S_0 = F_1 e^{-k_1 t} + F_2 e^{-k_2 t} \tag{1}$$

Where S_t and S_0 are the mass of sorbed compounds on sediment at time t(d) and at the start of the experiment, respectively; F_1 and F_2 $(=1-F_1)$ are the fractions of contaminant presented in the two slow desorbing sediment compartments, respspectively; k_1 and k_2 (d^{-1}) are the rate constants of the two slow desorbing compartments. For the infinite bath condition, the first order rate constant is related to retarded intraparticle radial diffusion coefficient (D_e) by Eq. 2 (19,20).

$$k \cong \frac{\ln 2}{t_{50\%}} \cong \frac{\ln 2}{0.03 R^2 / D_e} \cong \frac{0.693 \cdot \phi \cdot D_w / \chi_e}{0.03 \cdot R^2 \cdot \rho_s \cdot (1-\phi) \cdot OC \cdot K_{OC}} \tag{2}$$

Where D_e is the effective diffusivity, D_w (cm^2/sec) is the molecular diffusivity of the molecule, χ_e is the effective tortuosity, ρ_s is the solid density (g/cm), k is the first order rate constant (sec^{-1}), ϕ is the intraparticle porosity and R is the particle radius (cm), OC is the fractional organic carbon content, and K_{OC} (cm^3/g) is the organic carbon based partition coefficient. The effective tortuosity ($\chi_e = \chi / K_r$) is the ratio of soil tortuosity factor (χ) and a constrictivity factor (K_r)(19). Theoretically and experimentally proposed tortuosity in unconsolidated material generally ranges between 1.3-3 (19). Constrictivity is a measure of the relative size of the molecule to the pore size. Satterfield et al. (21) proposed the following correlation (Eq. 3).

$$K_r = 0.98 \exp(-4.6\lambda) \tag{3}$$

where λ is the ratio of critical molecular diameter to pore diameter.

As shown in Table 1, the first kinetic rate constants (k_1) are between 0.093 – 0.309 d^{-1} for the four compounds in Lake Charles sediments. The fractional mass of chemicals in the first compartment (F_1) ranges from 0.7-37% for the two

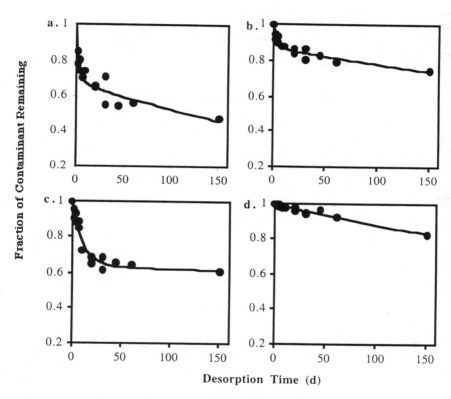

Figure 1. Plots of the fractional mass of four contaminants on Lake Charles sediment versus desorption time (days) following desorption by Tenax beads. The four predominant chemicals in Lake Charles sediment are(a) 1,3-dichlorobenzene, (b) 1,4-dichlorobenzene,(c) hexachlorobutadiene, and (d) hexachlorobenzene.

dichlorobenzenes and hexachlorobutadiene, but is negligible for hexachlorobenzene. The second kinetic rate constants (k_2) are between 0.00022-0.0026 d^{-1} for the four compounds in Lake Charles sediments. The two kinetic rate constants for the three chlorinated benzenes are very similar regardless of the fact that their aqueous solubilities differ by four orders of magnitude. The desorption rates for hexachlorobutadiene are lower by a factor of $2 - 10$ than that of the chlorinated benzenes. The significance of these differences between the desorption of the three chlorinated benzenes and hexachlorobutadienes is not known at this time. The reported two-compartment desorption rates are consistent with the previously reported irreversible adsorption data (10). Fu et al. (10) reported a desorption rate of 0.113 d^{-1} and 0.0016 d^{-1} for naphthalene desorption into water from a laboratory contaminated irreversibly adsorbed sediment.

It is proposed that there existed two kinetically different regimes in the irreversible compartment. Previous work ($5,22-24$) demonstrated that the adsorption and desorption to the labile compartment can be modeled by Eq. 1 and K_{OC}/K_{OW} relation. Since the Lake Charles sediment has been weathered and aged for over 30 years, it is proposed that the kinetic rates correspond to the rates of chemical release from the entrapped and irreversible compartment only, which will be further justified below. Kan et al. (5) observed that most irreversibly adsorbed organic compounds had similar sorption and desorption equilibrium properties ($K_{OC}^{irr} \cong 10^{5.53}$ ml / g). The effective tortuosity (χ_e) can be calculated from Eq. 1 with the experimentally measured rate (k_1) and the same soil parameters ($\rho_s =2.65$, $\phi = 0.05$, R = 50 μm, $K_{Oc} \approx 10^{5.53}$). The calculated χ_e are between 2.5 –8.6 with a geometric mean of 4.3. The effective tortuosity is similar to the soil tortuosity estimated previously for soil and other unconsolidated materials ($19,23$). Therefore, it is proposed that chemical desorption from the first irreversible compartment can be modeled with simple intraparticle radial diffusion of an irreversibly sorbed compound, i.e., the constrictivity factor, $K_r \approx 1.0$. The desorption half life from the first irreversible compartment is between 2 and 7 days.

The desorption rate (k_2) from the second irreversible compartment is two to three orders of magnitude slower than the first irreversible compartment desorption rate (k_1). In Table 1, k_2 values of this study are compared with five other literature reported desorption rate data($10,18,25-27$). The chemicals listed in Table 1 include both polar (atrazine and metolachlor) and nonpolar compounds; the range of K_{OW} values varies by four orders of magnitude. Soil organic carbon contents range from 0.26-7.02%. However, the reported desorption rate constants in Table 1 varied by less than a factor of 3 (except for hexachlorobutadiene). It is proposed that a fraction of the irreversibly adsorbed chemical is trapped in the soil organic matter and that the diffusion from the entrapped compartment is limited by a constrictivity factor. In Table 1 are listed the effective tortuosities (χ_e) calculated from the desorption rate (k_2) and the irreversible isotherm $K_{OC}^{irr} \cong 10^{5.53}$. The differences in the geosorbents may contribute to the large variation in χ_e. The geometric mean of the effective tortuosities (χ_e) is 458 for the eleven compounds, which is within the range of effective tortuosity reported by Ball and Roberts for Borden soil (19). Using the soil tortuosity from compartment 1 data, $\chi = 4.3$the mean effective tortuosity of the second compartment (χ_e, Table 1, Column 8) corresponds to a constrictivity factor, $K_r \approx 0.01$. According to Eq. 2, a constrictivity factor $K_r \approx 0.01$ corresponds to $\lambda \approx 1$, i.e.

Table 1. Comparison of reported long term slow desorption rate constants.

Contaminant	Soil/Sediment OC content	Experiment	K_{OW}	Slow Compartment 1		Slow Compartment 2		Ref.
				Rate constant, k_1 (d^{-1})[a]	χ_e[ab]	Rate constant, k_2 (d^{-1})[a]	χ_e[b]	
1,3 Dichlorobenzene	Lake Charles 4.1% OC	Field contaminated, Tenax desorption	$10^{3.38}$	0.284 (0.32)	3.51	0.0026 (0.68)	383	This study
1,4 Dichlorobenzene			$10^{3.38}$	0.229 (0.14)	4.35	0.001 (0.87)	997	
Hexachlorobutadiene			$10^{4.90}$	0.093 (0.37)	8.60	0.00022 (0.63)	3634	
Hexachlorobenzene			$10^{5.5}$	0.309 (0.01)	2.49	0.0021 (0.99)	641	
Aroclor 1242	Lake Carnegie 2.6% OC	Laboratory contaminated sediment Batch water desorption	$10^{5.3-6.4}$	NA	NA	0.003	351	25
Atrazine	Merrimac find sandy loam 2.6% OC	Field contaminated, Column water desorption	$10^{2.56}$	0.040	NA	0.006	194	26
Metolachlor			-	0.038		0.0019	615	
2,2',3,4',5'-PCB	Lake Ketelmeer 7.02% OC	Field contaminated, Tenax desorption	$10^{6.25}$	NA	NA	0.005	75	18
2,2',4,4',5'-PCB			$10^{6.35}$			0.003	125	
PCBs	Hudson River 1-5% OC	Field contaminated, XAD-4 desorption	$10^{5.3-6.4}$	0.101	NA	0.005	549	27
Naphthalene	Lula fine sand 0.27% OC	Laboratory contaminated sediment Batch water desorption	$10^{3.36}$	0.113	NA	0.0016	2382	10
Geometric mean						0.0021	458	

[a] Number is parenthases are the fractions of contaminant presented in the slow desorbing sediment compartments (F_1 and F_2 of Eq 1).

[b] Calculated from Eq. 1 with the first order rate constant (to the right) and $K_{OC} = K_{OC}^{irr} = 10^{5.53}$ ml/g-OC, $\phi = 0.05$, R= 50 μm.

the diffusion pore diameter is similar in size to the critical molecular diameter. The desorption half lives from the second irreversible compartment ranges from 0.32 to 8.62 years. Therefore, the second compartment represents a fraction that may persist in the environment for a long time, especially because the subsurface environment has much less turbulance than laboratory desorption simulations.

In conclusion, three different desorption regimes (a labile and two irreversible compartments) are proposed to model the desorption of chemicals from soil. First, a fraction of the adsorbed chemicals is labile and readily desorbed. A second fraction of the adsorbed chemicals entrapped by the soil organic matter, as proposed by the irreversible adsorption model. Two kinetic regimes are assumed within the irreversible compartment. A fraction of the entrapped and irreversibly adsorbed chemical desorbs kinetically with a half life of approximately 2-7 days, which is predictable by the radial diffusion and $K_{OC}^{irr} \cong 10^{5.53}$ and a nominal soil tortuosity = 4.3. A second fraction of the entrapped and irreversibly adsorbed chemical desorbs with a half-life of 0.32-8.62 years, which is predictable by a radial diffusion and assumes an additional constrictivity factor of 0.01.

Regardless the mechanism that governs the rate of chemical release, a thermodynamic equilibrium model provides an upper bound prediction for desorption. An irreversible adsorption equation (Eq. 5, (5)) has been proposed to model contaminant desorption and has been tested extensively with reported field data:

$$q = K_{OC} \cdot OC \cdot C + \frac{K_{OC}^{irr} \cdot OC \cdot q_{max}^{irr} \cdot f \cdot C}{q_{max}^{irr} \cdot f + K_{OC}^{irr} \cdot OC \cdot C} \tag{5}$$

where q_{max}^{irr} ($\mu g/g$) is the maximum irreversible capacity, K_{OC}^{irr} is the irreversible compartment organic-carbon based partition coefficient, OC is the organic carbon content, C is the aqueous phase concentration ($\mu g/ml$), f ($0<f<1$) is the fraction of the irreversible compartment that is filled at the time of exposure, which is typically assumed to be equal to 1. The isotherm equation consists of two terms, the first term is a linear term to represent reversible sorption and the second term is a Langmuir-type term to represent irreversible sorption. With this equation and a value of OC, no additional site-specific or chemical-specific information is required to predict the contaminant fate. The isotherm predicts that a significant fraction of irreversibly sorbed chemicals does not desorb into the environment, as expected. A recent paper has shown evidence that the nondesorbing chemicals are not biodegradable (28). If it can be shown that these irreversibly adsorbed chemicals are not bioavailable, one may assume that the irreversibly sorbed chemicals may be "safe" in the environment. The additional kinetic results shown in this study would suggest that approximately a quarter of the irreversibly sorbed compounds is tightly trapped and that desorption is extremely slow. This kinetic effect will add an additional "safety factor" to support the notion that no substantial risk to health or the environment will be added if trace level tightly bound contaminants are left in the soil.

Acknowledgments

This research has been conducted with the support of Hazardous Substance Research Center South and Southwest, the Gulf Coast Hazardous Substance Research Center, Office of Exploratory Research of the U.S. Environmental Protection Agency and the Defense Threat Reduction Agency. We also thank Dr. C. R. Demas of U.S. Geological Survey Louisiana District for assistance in collecting Lake Charles sediments.

References

1 Xing, B.; Pignatello, J. J.; Gigliotti, B. Environ. Sci. Technol. **1996**, 30, 2432-2440.
2 Weber, W. J., Jr.; Huang, W. Environ. Sci. Technol. **1996**, 30, 881-888.
3 McGroddy, S. E.; Farrington, J. W.; Gschwend, P. M. Environ. Sci. Technol. **1996**, 30, 172-177.
4 Chiou, C. T.; Kile, D. E. Environ. Sci. Technol. **1998**, 32, 338-343.
5 Kan, A. T.; Fu, G.; Hunter, M.; Chen, W.; Ward, C. H.; Tomson, M. B. Environ. Sci. Technol. **1998**, 32, 892-902.
6 Kan, A. T.; Fu, G.; Hunter, M. A.; Tomson, M. B. Environ. Sci. Technol. **1997**, 31, 2176-2185.
7 West, C. C.; Waggett, G. G.; Tomson, M. B. In Proceeding of the Second International Conference on Ground Water Quality Research, 1986, Stillwater, OK.Second International Conference on Ground Water Quality Research
8 Adamson, A. W.Physical Chemistry of Surfaces , 5th ed.; John Wiley & Sons: New York, N.Y., USA, 1990.
9 Kan, A. T.; Fu, G.; Tomson, M. B. Environ. Science and Technol. **1994**, 28, 859-867.
10 Fu, G.; Kan, A. T.; Tomson, M. B. Environ. Chem. Toxicol. **1994**, 13, 1559-1567.
11 Huang, W.; Weber Jr., W. J. Environ. Sci. Technol. **1997**, 31, 2562-2569.
12 Schlebaum, W.; Badora, A.; Schraa, G.; Riemsdijk, W. H. V. Environ. Sci. Technol **1998**, 32, 2273-2277.
13 Bailey, A.; Cadenhead, D. A.; Davies, D. H.; Everett, D. H.; Miles, A. J. Trans. Faraday Soc. **1971**, 67, 231.
14 Burgess, C. G. V.; Everett, D. H.; Nuttall, S. Pure and Appl. Chem. **1989**, 61, 1845-1852.
15 Schulten, H.-R.; Schnitzer, M. Soil Science **1997**, 162, 115-130.
16 Devitt, E. C.; Wiesner, M. R. Environ. Sci. Technol. **1998**, 32, 232-237.
17 Pignatello, J. J. Environ. Toxicol. Chem. **1990**, 9, 1117 - 1126.
18 Cornelissen, G.; Noort, P. C. M. V.; Parsons, J. R.; Govers, H. A. J. Environ. Sci. Technol. **1997**, 31, 454-460.
19 Ball, W. P.; Roberts, P. V. Environ. Sci. Technol. **1991**, 25, 1237-1249.
20 Schwarzenbach, R. P.; Gschwend, P. M.; Imboden, D. M.Environmental Organic Chemistry ; John Wiley & Sons: New York, NY, USA, 1993.
21 Satterfield, C. N.; Colton, C. K.; Pitcher, W. H. AIChE J. **1973**, 19, 628-635.
22 Karickhoff, S. W.; Morris, K. R. Environ. Sci. Technol. **1985**, 19, 51-56.

23 Wu, S.-C.; Gschwend, P. M. Environ. Sci. Technol. **1986**, 20, 717-725.

24 Brusseau, M. L.; Jessup, R. E.; Rao, P. S. Environ. Sci. Technol. **1990**, 24, 727-735.

25 Witkowski, P. J.; Jaffe, P. R.; Ferrara, R. A. J. Contam. Hydro. **1988**, 2, 249-269.

26 Pignatello, J. J.; Ferrandino, F. J.; Huang, L. Q. Environ. Sci. Technol. **1993**, 27, 1563-1571.

27 Carroll, K. M.; Harkness, M. R.; Bracco, A. A.; Balcarcel, R. R. Environ. Sci. Technol. **1994**, 28, 253-258.

28 Cornelissen, G.; Rigterink, H.; Ferdinandy, M. M. A.; Noort, P. C. M. V. Environ. Sci. Technol. **1998**, 32, 966-970.

Chapter 9

Emissions of Polychlorinated Biphenyls from Combustion Sources

P. M. Lemieux, C. W. Lee, J. D. Kilgroe, and J. V. Ryan

Air Pollution Prevention and Control Division, U.S. Environment Protection Agency, Research Triangle Park, NC 27711

Polychlorinated biphenyls (PCBs) have been widely used in the past as industrial chemicals, particularly as additives in electrical transformer cooling oil. Growing evidence of PCBs' role as a persistent, bioaccumulative, human carcinogen has led to the banning of the production and use of PCBs as an industrial chemical in major industrialized countries including the United States. PCBs, however, are still being released into the environment as an unwanted by-product of combustion processes, particularly those associated with chlorinated materials. A subset of PCBs, the coplanar isomers, exhibit biological activity similar to that of polychlorinated dibenzo-p-dioxins and polychlorinated dibenzofurans (PCDDs/PCDFs), a widely recognized by-product of combustion processes. Significant progress has been made over the last 10 years investigating the fundamental PCDD/PCDF formation mechanisms, while emissions of PCBs from combustion devices have not been extensively investigated. This paper presents background information on some of the combustion sources that generate PCBs.

U.S. government work. Published 2001 American Chemical Society

INTRODUCTION

A number of persistent organic pollutants (POPs) are of national and international concern due to their persistence, their mobility, their ability to bioaccumulate, and their potential impacts on the health of humans, wildlife, and fish. Much of this concern involves 12 chemical classes that include polychlorinated biphenyls (PCBs), polychlorinated dibenzo-*p*-dioxins and -furans (PCDDs/PCDFs), and pesticides such as dichlorodiphenyltrichloroethane (DDT), chlordane, and heptachlor [1]. An international action plan for reduction and possible elimination of the so-called "dirty dozen" is the core of a global POP treaty currently under intensive negotiation at the United Nations [2]. The three chemical classes that are perhaps of greatest concern are PCDDs, PCDFs, and PCBs [3,4].

Formation and control of PCDDs/PCDFs from combustion sources have been a public environmental health concern because of reputed carcinogenic, teratogenic, endocrine-disrupting, and persistent and accumulative behavior of these compounds in biological systems. Of the 208 tetra-octa CDD/CDF isomers, 17 (those substituted in the 2,3,7,8 positions) are recognized for their toxicity.

From the 1940's to the 1970's, PCBs were widely used commercial chemicals. They were used primarily as heat transfer fluids and as pesticides. At one time, eight mixtures of PCBs were sold in the U.S. under the trade name Aroclor. Many other mixtures were manufactured elsewhere. In 1977 Congress passed the Toxic Substances Control Act (TSCA), providing EPA with broad "cradle-to-grave" regulatory authority for almost all existing and new chemicals manufactured, imported, or used in the U.S. [5]. In January 1978, EPA banned the manufacture, processing, distribution, or use of any PCBs in any manner other than entirely enclosed. However, the previous manufacture and use of PCBs produced a negative legacy. In 1991, EPA's Superfund Office estimated that PCBs were a major contaminate, accounting for about 34 million cubic yards of material at 20 % of the sites on the National Priorities List [5].

Certain polychlorinated biphenyl (PCB) isomers exhibit similar toxic effects as PCDDs/PCDFs. PCBs substituted with zero or one chlorine atom in the 2,2' or 6,6' (ortho) position on the phenyl ring and one or more meta (m) or para (p) chlorines on each ring (see Figure 1) can assume a planar configuration, leading to a molecule that is similar in structure and orientation to 2,3,7,8-tetrachlorodibenzo-*p*-dioxin (TCDD). These coplanar PCBs are also termed "dioxin-like" PCBs. Similar to PCDDs/PCDFs, each coplanar PCB isomer has been assigned a toxic equivalency factor (TEF) relative to 2,3,7,8-TCDD [6], which has been arbitrarily assigned a value of 1 as shown in Table 1 [7]. By summing the products formed by multiplying the concentrations of the various molecules for which a TEF has been assigned by its respective TEF, a toxic equivalency (TEQ) concentration is derived. The TEQ is used as an estimate of the total dioxin-like toxicity of a complex mixture. Human exposure to dioxin-like PCBs in the environment has been suggested to be very significant in a recent review [8], which showed that the human body burden of dioxin-like PCBs is between 50 and 70 % of the total TEQ, based on published European and North American studies. The review also suggested that dioxin-like PCBs dominate the total TEQ (up to 90% total TEQ) of several major food sources such as fish, oils, and fats consumed by humans.

Efforts to control the risks from PCBs have concentrated on exposure related to the manufacture and use of PCBs, with major emphasis on manufacturing process residues and wastewater discharges. It is generally believed that emission of PCBs from combustion sources results primarily from the incomplete destruction of wastes containing PCBs [4]. Few people, except combustion experts or individuals in the incineration industry, understand that PCBs can be synthesized during the combustion process via gas-phase or heterogeneous reactions.

There are very little available data reporting emissions of PCBs from combustion sources, since PCBs were extensively used as industrial chemicals, and their air emissions as trace combustion by-products were largely ignored [9]. Combustion sources in the U.S. do not typically require PCB emissions testing. Only in the UK was a national inventory of dioxin-like PCB sources developed [8]. Based on these limited data, PCBs appear to contribute significantly to the total TEQ emissions from combustion sources. For some sources, such as municipal waste combustors (MWCs), PCBs contributed a relatively small fraction (<5%) of the TEQ [8]. This observation on MWCs was duplicated by some Japanese researchers [10]. For other sources, however, such as cement kilns, PCBs contributed up to 60% of the TEQ [8].

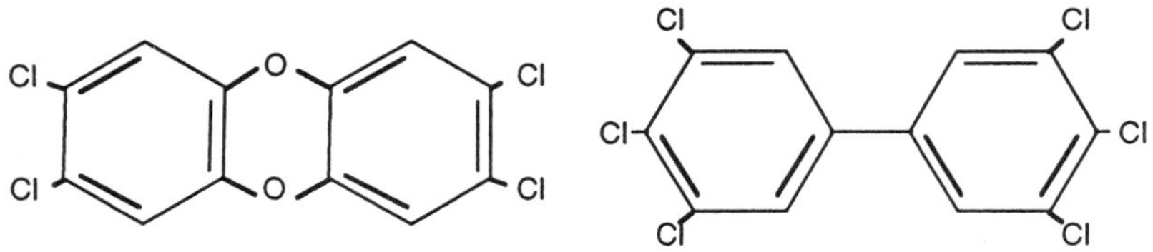

2,3,7,8-Tetrachlorodibenzo-*p*-dioxin   3,3',4,4',5,5'-Hexachlorobiphenyl

Figure 1. PCBs and Dioxin Structure

Table 1. Toxic Equivalency Factors for PCBs [6]

Type	Congener IUPAC No.	Structure	TEF
Non-ortho	77	3,3',4,4'-TCB	0.0001
	81	3,4,4',5 - TCB	0.0001
	126	3,3',4,4',5-PeCB	0.1
	169	3,3',4,4',5,5'-HxCB	0.01
Mono-ortho	105	2,3,3',4,4'-PeCB	0.0001
	114	2,3,4,4',5-PeCB	0.0005
	118	2,3'4,4',5-PeCB	0.0001
	123	2',3,4,4',5-PeCB	0.0001
	156	2,3,3',4,4',5-HxCB	0.0005
	157	2,3,3',4,4',5'-HxCB	0.0005
	167	2,3'4,4',5,5'-HxCB	0.00001
	189	2,3,3',4,4',5,5'-HpCB	0.0001

Metal reclamation and sintering plants have also been identified as an important source of coplanar PCBs [11]. PCB TEQs from combustor emissions appear to be dominated by the PCB-126 congener, one of the non-ortho PCBs, which also has the highest TEF (0.1) of all PCBs.

The largest body of PCB combustor emissions data from North America exists from some hazardous waste incinerator (HWI) and cement kiln trial burns performed under the Resource Conservation and Recovery Act (RCRA), from EPA's HWI database [12], and from some MWCs that were sampled during Environment Canada/U.S. EPA's National Incinerator Testing and Evaluation Program [13]. Recently developed RCRA guidance documents supporting current HWI regulations specify that certain sources must measure PCBs during emission tests to support risk assessments as part of the permitting process [14,15].

The limited available data suggest that PCBs may be formed by the same types of reactions that produce PCDDs/PCDFs [16]. PCDDs/PCDFs are generally believed to be formed via several proposed mechanisms: gas-phase formation [17]; heterogeneous formation from organic precursors [18]; and *de novo* synthesis from flyash-bound carbon [19]. Both of these latter mechanisms involve heterogeneous reactions with flyash-bound metals (such as copper) serving as catalytic sites for reactions involving various carbon-containing species in the gas and/or solid phases. These reactions occur downstream of the high temperature combustion zone at temperatures ranging from 250 to 700 °C. The formation reactions from organic precursors are believed to occur relatively quickly while the flue gases pass through the downstream regions of the combustion device, with reaction times on the order of seconds, while the *de novo* synthesis is believed to occur slowly, with reaction times on the order of minutes or hours, while the flyash is held up in the particulate control devices.

If both PCBs and PCDDs/PCDFs are generated through similar mechanisms, it would be expected that emissions of PCBs should correlate with emissions of PCDDs/PCDFs. This correlation would be expected to be very strong if there were a common rate-limiting step involved in the formation of both PCBs and PCDDs/PCDFs. This paper will examine that hypothesis.

128

APPROACH

PCB and PCDD/PCDF measurement data were gathered from various combustion sources in the literature and from risk assessment, trial burn, and risk burn reports from EPA's Regional Offices as well as from EPA's Hazardous Waste Combustor Database [12]. All data were placed into units of nanograms per dry standard cubic meter, corrected to 7 % oxygen (O_2), with the exception of the Environment Canada data, which were corrected to 12 % carbon dioxide (CO_2). All PCB data reflect total PCBs, and PCDD/PCDF data reflect the total of all tetra through octa chlorinated congeners of PCDD + PCDF. At this point, no TEQ calculations have been made since sufficient information was not available to make those calculations for all of the data sets.

The data are summarized in Table 2. Sources for the data in Table 2 [20-29] included test points from two different wet process cement kilns, data from several HWIs burning hazardous waste (HW), from a facility burning chemical demilitarization wastes (nerve gas), from a facility decontaminating soil at a Superfund site, and from several different types of MWCs burning municipal solid waste (MSW) or refuse-derived fuel (RDF). Additional measurements were included from some experiments evaluating open burning of household waste in barrels [29]. Samples taken at similar operating conditions for a single facility were averaged into a single entry in Table 2. However, if multiple operating conditions were sampled during a parametric test or trial burn, then each discrete operating condition was reported as an entry in Table 2. Also note that some of the HWIs included some quantity of PCB in the initial feed. The presence of PCB in the feed was not corrected for in any way.

RESULTS

PCBs were, for most cases, emitted at higher levels than PCDDs/PCDFs. The data from Table 2 are presented in Figure 2, grouped according to three general facility types: cement kilns, HWIs, MWCs, and open burning. There appears to be a very distinct trend with increasing PCB emissions as PCDD/PCDF emissions increase. This trend appears to be independent of the facility type, although the data

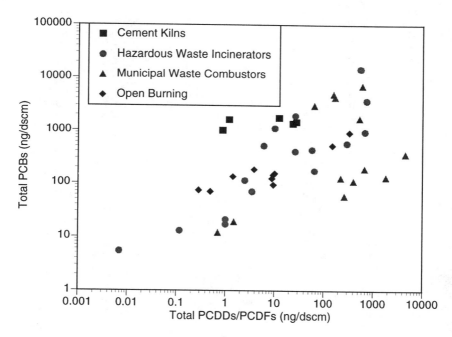

Figure 2. PCDDs/PCDFs vs PCBs

Table 2. PCB and PCDD/PCDF Data

Facility Type	Feed	PCB	Total	Total	Reference
Cement Kiln (Wet)	coal/ hazwaste	N	8.65E-01	9.98E+02	[20]
Cement Kiln (Wet)	coal/ hazwaste	N	1.17E+00	1.60E+03	[20]
HWI (Rotary Kiln)	soil	N	2.46E+00	1.12E+02	[21]
HWI (Rotary Kiln)	soil	N	3.45E+00	6.96E+01	[22]
HWI (Rotary Kiln)	hazwaste	Y	6.78E+02	9.39E+02	[12]
HWI (Rotary Kiln)	hazwaste	Y	7.28E+02	3.68E+03	[12]
HWI (Rotary Kiln)	hazwaste	Y	2.94E+02	5.70E+02	[12]
HWI (Liq. Injection)	hazwaste	Y	5.41E+02	1.47E+04	[12]
HWI (Liq. Injection)	hazwaste	Y	5.53E+02	1.46E+04	[12]
HWI (Rotary Kiln)	demil	N	1.00E+00	1.70E+01	[12]
HWI (Liq. Injection)	hazwaste	N	6.00E+00	5.13E+02	[12]
HWI (Rotary Kiln)	hazwaste	N	1.00E+00	2.10E+01	[12]
HWI (Rotary Kiln)	hazwaste	Y	1.00E+01	1.09E+03	[12]
HWI (Rotary Kiln)	hazwaste	Y	2.60E+01	1.92E+03	[12]
HWI (Rotary Kiln)	hazwaste	N	5.70E+01	4.32E+02	[12]
HWI (Rotary Kiln)	hazwaste	N	6.40E+01	1.73E+02	[12]
HWI (Rotary Kiln)	hazwaste	N	2.60E+01	4.03E+02	[12]
Cement Kiln (Wet)	coal/ hazwaste	N	1.22E+01	1.74E+03	[23]
Cement Kiln (Wet)	coal/ hazwaste	N	2.32E+01	1.35E+03	[23]
Cement Kiln (Wet)	coal/ hazwaste	N	2.72E+01	1.45E+03	[23]
HWI (Rotary Kiln)	demil	N	6.90E-03	5.40E+00	[24]
HWI (Moving	demil	N	1.16E-01	1.29E+01	[25]
MWC (Modular)	MSW	N	2.63E+02	5.80E+01	[26]
MWC (Modular)	MSW	N	1.57E+02	ND	[26]
MWC (Modular)	MSW	N	2.21E+02	1.26E+02	[26]
MWC (Stoker)	RDF	N	7.01E-01	1.19E+01	[27]
MWC (Stoker)	RDF	N	1.49E+00	1.91E+01	[27]
MWC (Mass Burn)	MSW	N	1.67E+02	4.30E+03	[28]
MWC (Mass Burn)	MSW	N	6.33E+01	3.00E+03	[28]
MWC (Mass Burn)	MSW	N	1.56E+02	4.90E+03	[28]
MWC (Mass Burn)	MSW	N	5.97E+02	7.00E+03	[28]
MWC (Mass Burn)	MSW	N	5.25E+02	1.70E+03	[28]
MWC (Stoker)	MSW	N	4.50E+03	3.60E+02	[10]
MWC (Stoker)	MSW	N	1.81E+03	1.30E+02	[10]
MWC (Stoker)	MSW	N	6.70E+02	1.90E+02	[10]
MWC (Stoker)	MSW	N	4.00E+02	1.10E+02	[10]
Open Burning	MSW	N	9.29E+00	1.43E+02	[29]
Open Burning	MSW	N	3.81E+00	1.83E+02	[29]
Open Burning	MSW	N	1.40E+00	1.32E+02	[29]
Open Burning	MSW	N	2.84E-01	7.34E+01	[29]
Open Burning	MSW	N	4.91E-01	6.92E+01	[29]
Open Burning	MSW	N	9.99E+00	1.54E+02	[29]
Open Burning	MSW	N	9.38E+00	9.28E+01	[29]
Open Burning	MSW	N	1.49E+02	5.19E+02	[29]
Open Burning	MSW	N	3.30E+02	9.05E+02	[29]
Open Burning	MSW	N	8.77E+00	1.22E+02	[29]

ND - none detected

from the cement kilns are found in the upper area of the cluster of data points. The MWC facility data and open burning data fit in quite well with the HWI data. Overall, PCB emissions exceeded PCDD/PCDF emissions by approximately a factor of 20, and this trend appeared to hold over five orders of magnitude in PCDD/PCDF emissions.

A study in Finland [30] compared levels of PCDD/PCDF TEQs and PCBs in human milk and found a similar trend, although the slope of the line was different. In the Finnish study, total PCBs were approximately three orders of magnitude higher than the PCDD/PCDF TEQs. Given that, depending on the source, total PCDDs/PCDFs are nominally a factor of 100 greater than the TEQ, this trend is strikingly similar. The high correlation coefficient ($R^2 = 0.93$) suggested the hypothesis that the sources of PCDDs/PCDFs and PCBs are the same in the Finnish study.

CONCLUSIONS

Data were gathered from various field samples where both PCDDs/PCDFs and PCBs were measured. PCBs were found in the stack in most cases, both in facilities that had PCBs initially in the feed and in those that did not have any PCBs in the feed. Stack concentrations of PCBs were generally higher than stack concentrations of PCDDs/PCDFs, and an apparent trend was observed. This trend was consistent across several orders of magnitude in PCDD/PCDF concentration as well as across several different facility types and feed stocks.

This trend suggests that the same fundamental mechanisms that contribute to the formation of PCDDs/PCDFs may result in formation of PCBs. It also similarly suggests that control strategies intended to reduce emissions of PCDDs/PCDFs may also serve to control PCBs. Although PCBs do not have as high a TEF as 2,3,7,8-TCDD, some of the PCBs have significant TEFs, and since they may be present at concentrations much higher than PCDDs/PCDFs, PCBs may contribute significantly to the overall TEQ for some sources.

ACKNOWLEDGMENTS

The authors would like to acknowledge the assistance of the following people from EPA's Office of Solid Waste and Regional Offices in locating the field data: Frank Behan, Stacy Braye, Gary Gross, Bob Holloway, George Peters, and Gary Victorine.

REFERENCES

1. Wania, F., and D. MacKay, "Tracking the Distribution of Persistent Organic Pollutants," *Environ. Sci. Technol.*, Vol. 30, No. 9, 1996.
2. Schmidt, C.W., "No POPs," *Environmental Health Perspectives*, 107, No. 1, 1999, pp. A25-A26.
3. U. S. EPA, "Health Assessment Document for 2,3,7,8-Tetrachlorodibenzo-p-dioxin (TCDD) and Related Compounds," Vol. 1, EPA/600/BP-92/001a (NTIS PB94-205465) (external review draft), Office of Health and Environmental Assessment, Washington, DC, June 1994.
4. U.S. EPA, "Deposition of Air Pollutants to the Great Waters: Second Report to Congress," EPA-453/R-97-011, Office of Air Quality Planning and Standards, Research Triangle Park , NC, June 1997.
5. Shifrin, N.S., and A. P. Toole, "Historical Perspective on PCBs," *Environmental Engineering Science*, Volume 15, Number 3, 1998.
6. Van den Berg, M., L. Birnbaum, A. Bosveld, B. Brunström, P. Cook, M. Feeley, J. Giesy, A. Hanberg, R. Hasegawa, S. Kennedy, T. Kubiak, J. Larsen, F. van Leeuwen, A. Liem, C. Nolt, R. Peterson, L. Poellinger, S. Safe, D. Schrenk, D. Tillitt, M. Tysklind, M. Younes, F. Waern, and T. Zacharewski, " Toxic Equivalency Factors (TEFs) for PCBs, PCDDs, PCDFs for Humans and Wildlife," *Environmental Health Perspectives*, Volume 106, No. 12, December 1998.
7. NATO, "International Toxicity Equivalent Factors (TEFs) Method of Risk Assessment for Complex Mixtures of Dioxins and Related Compounds," Report No. 176, Brussels: North Atlantic Treaty Organization, 1988.
8. Alcock, R.E., P.A. Behnisch, K.C. Jones, and H. Hagenmaier, "Dioxin-like PCBs in the Environment—Human Exposure and the Significance of Sources," *Chemosphere*, Vol. 37, No. 8, pp. 1457-1472, 1998.
9. Brown, J.F., Jr., G.M. Frame II, D.R. Olson, and J.L Webb, "The Sources of the Coplanar PCBs," *Organohalogen Compounds*, Vol. 26, pp. 427-430, 1995.
10. Kawakami, I., Y. Matsuzawa, M. Tanaka, S. Sakai, and M. Hiraoka, "Emission Level of Co-PCBs from MSW Incinerators," *Organohalogen Compounds*, Vol. 11, pp. 375-380, 1993.
11. Boers, J.P., E.W.B. de Leer, L. Gramberg, and J. de Koning, "Levels of coplanar PCB in flue gases of high temperature processes and their occurrence in environmental samples," *Fresenius J. Anal. Chem.*, 348: 163-166, 1994.
12. U.S. EPA, "Updated Hazardous Waste Combustor Database," Office of Solid Waste, Washington, DC, December 1996.
13. Finkelstein, A., and R.D. Klicius, "National Incinerator Testing and Evaluation Program: The Environmental Characterization of Refuse-derived Fuel (RDF) Combustion Technology, Mid- Connecticut Facility, Hartford, Connecticut," EPA-600/R-94-140 (NTIS PB96-153432), December 1994.

14. U.S. EPA (1998a), "Human Health Risk Assessment Protocols for Hazardous Waste Combustion Facilities," EPA-530-D-98-001A, Office of Solid Waste, Washington, DC, July 1998.
15. U.S. EPA (1998b), "Guidance on Collection of Emissions Data to Support Site-Specific Risk Assessments at Hazardous Waste Combustion Facilities," EPA-530-D-98-002, Office of Solid Waste, Washington, DC, August 1998.
16. Schoonenboom, M.H., P.C. Tromp, and K. Olie, "The Formation of Coplanar PCBs, PCDDs, and PCDFs in a Fly Ash Model System," *Chemosphere*, Vol. 30, No. 7, pp. 1341-1349, 1995.
17. Weber, R., and H. Hagenmaier, "Mechanism of Formation of Polychlorinated Dibenzo-p-dioxins and Dibenzofurans from Chlorophenols in Gas-Phase Reactions," *Chemosphere*, Vol. 38, No. 3, pp 529-549, 1999.
18. Gullett, B.K., P.M. Lemieux, and J.E. Dunn, "Role of Combustion and Sorbent Parameters in Prevention of Polychlorinated Dibenzo-p-Dioxin and Polychlorinated Dibenzofuran Formation During Waste Combustion," *Environ. Sci. and Tech.*, Vol. 28, No. 1, pp. 107-118, 1994.
19. Stieglitz, L., "Selected Topics on the De Novo Synthesis of PCDD/PCDF on Fly Ash," *Environmental Engineering Science*, Vol. 15, No. 1, 1998.
20. Weston, R.F., Keystone Cement Company Source Emissions Test Report, February 1997.
21. Midwest Research Institute, "Test Report for Risk Burn No. 1 on the Drake Chemical Superfund Site's Mobile On-site Hazardous Waste Incinerator, Volume 1--Technical Report (Draft)," MRI Project No. 3620-18, March 26, 1997a.
22. Midwest Research Institute, "Test Report for Risk Burn No. 2 on the Drake Chemical Superfund Site's Mobile On-site Hazardous Waste Incinerator, Volume 1--Technical Report (Draft)," MRI Project No. 3620-28, April 8, 1997b.
23. Radian International, "Lafarge Corporation, Paulding, Ohio, August 1998 Trial Burn Report," August 1998.
24. U.S. EPA (1998c), "GB Trial Burn Report for the Trial Burn of the MPF Incinerator, Johnston Atoll Chemical Agent Disposal System (JACADS) Johnston Island," (submitted to USEPA Region 9 by Program Manager for Chemical Demilitarization, Aberdeen Proving Ground, MD), July 1998.
25. U.S. EPA (1998d), "GB Trial Burn Report for the Trial Burn of the DFS Incinerator, Johnston Atoll Chemical Agent Disposal System (JACADS) Johnston Island," (submitted to USEPA Region 9 by Program Manager for Chemical Demilitarization, Aberdeen Proving Ground, MD), July 1998.
26. Environment Canada, "The National Incinerator Testing and Evaluation Program: Two-stage Combustion (Prince Edward Island)," Report EPS 3/UP/1, September 1985.
27. Environment Canada, "The National Incinerator Testing and Evaluation Program: The Environmental Characterization of Refuse-Derived Fuel (RDF) Combustion Technology," Report EPS 3/UP/1, December 1994.
28. Environment Canada, "The National Incinerator Testing and Evaluation Program: The Combustion Characterization of Mass Burning Incinerator Technology, Quebec City," Report IP-82, December 1987.
29. Gullett, B.K., P.M. Lemieux, C.C. Lutes, C.K. Winterrowd, D.L. Winters, "PCDD/F Emissions from Uncontrolled Domestic Waste Burning," 19th International Symposium on Halogenated Environmental Organic Pollutants and POPs, September 12 - 17, 1999, Venice, Italy.
30. Kiviranta, H., R. Purkunen, and T. Vartiainen, "Levels and Trends of PCDD/Fs and PCBs in Human Milk in Finland," *Chemosphere*, Vol. 38, No. 2, pp. 311-323, 1999.

Chapter 10

Butyltin Compounds in Freshwater Ecosystems

Bommanna G. Loganathan[1], Kurunthachalam Kannan[2],
David A. Owen[1], and Kenneth S. Sajwan[3]

[1]Department of Chemistry and Center for Reservoir Research,
Murray State University, Murray, KY 42071
[2]National Food Safety and Toxicology Center, Michigan State University,
East Lansing, MI 48824
[3]Department of Biology and Life Sciences, Savannah State University,
Savannah, GA 31404

Widespread use of butyltin (BT) derivatives (mono-, di-, and tributyltin) in industrial, domestic and consumer products has resulted in contamination of freshwater ecosystems. Butyltins are ubiquitous in wastewater effluents around the world. This chapter provides background information on environmental contamination by butyltin compounds with reference to sources, bioaccumulation features, and persistence in freshwater ecosystems. Concentrations of butyltins in freshwater organisms collected from several locations in the United States and other countries are compiled and compared with those recently measured in sediment and mussel tissues collected from the lowermost Tennessee River and Kentucky Lake. A major source of butyltin compounds in freshwater ecosystems is wastewater treatment plant effluents. Butyltin compounds are moderately persistent in water with half-lives in the range of several days to several months depending on ecosystem characteristics. Butyltins persist in sediments for several years and accumulate in a variety of aquatic organisms including fish-eating water birds and aquatic mammals. Monitoring studies are needed to evaluate the sources and effects of butyltin compounds in freshwater organisms.

Tin is a soft, white, silvery metal that is insoluble in water [1]. Tin can combine with carbon-containing materials to form organotin (OT) compounds. Depending on the number of organic moieties, OT compounds are classified as

© 2001 American Chemical Society

mono-, di- tri and tetra-organotins. The first OT compound was synthesized as early as 1852 by Löwig (see Snoeij et al [2]). However, uses of OT compounds were not known until the systematic investigations by Van der Kerk and co-workers during the 1950s [3,4]. The commercial uses of OT compounds have expanded rapidly during the last fifty years. Presently, tin is unsurpassed by any other metal in the number of its organic applications [5]. The annual world consumption of tin in all forms was about 200,000 tons in 1976, and of this total about 28,000 tons was in the form of OT compounds [6]. Due to the expansion of technical applications, the annual world production of OT compounds grew rapidly from 5000 tons in 1955 to 25,000 tons in 1975 [7], 35,000 tons in 1986 and 50,000 tons in 1992 [8]. The increasing annual usage of OT compounds, some of which are very toxic, attracted the attention of environmental health organizations in the 1970s. In particular, studies regarding environmental contamination and ecotoxicological effects of OT compounds have received serious attention after a tragic incident involving poisoning of 217 people and death of 100 people due to a drug that was sold in France for the treatment of staphylococcal skin infections. It was said to contain diethyltin diiodide, but the poisoning was most likely due to the contamination with triethyltin iodide [9,10].

The major OT compounds released into the environment are butyl-, phenyl- and octyltins. These OT compounds have been used as heat and light stabilizers in plastics and food packages, pesticides, paint and wood preservatives, marine antifouling agents, disinfectants and slime control in paper mills and as catalysts in the production of polyurethane foams and curing of silicone rubbers and epoxy resins [1]. Several aspects of the chemistry, applications and behavior of tin compounds have been reviewed by various researchers [6,11,12]. Of several OT compounds, tributyltin (TBT) has received considerable attention in the past decade due to its effect on aquatic organisms at relatively low concentrations (a few ng L^{-1} levels). TBT was first used in antifouling paints in Europe between 1959 and 1961 [13]. Alzieu et al. [14,15] and Alzieu and Heral [16] were among the first to investigate the toxicity of TBT in 1980, reporting deleterious effects (shell thickening and abnormalities) in Pacific oysters (Crossostrea gigas). Bryan and coworkers [17,18] established a link between TBT contamination and the sterilization of certain neogastropods, which resulted in the decline of populations. Deleterious effects of TBT have been observed in microalgae, bivalve mollusks, polychaetes, crustaceans, and fish at concentrations as low as 10-100 ng L^{-1} [17,18]. As a consequence, regulations on the use of TBT-based antifouling paints have been introduced in several countries since the late 1980s [12]. Despite a decrease in the TBT contamination in the aquatic environment after the regulations were introduced, concentrations persisted at levels considered chronically toxic to most susceptible organisms [19-21]. In addition, TBT is still used in vessels greater than 25 m long. Growing consumption of TBT lies in its use as a preservative for timber and wood, textiles, paper and leather. A small percentage of TBT also is used in dispersion paints and in a variety of other materials (e.g., PVC) as protection against microbial or fungal attack [22].

The uses of butyltin compounds are so diverse that they have been detected in a variety of aquatic matrices including sewage, sludge, sediments and water from both fresh water and marine systems. However, most studies have focused on the accumulation and toxic effects of butyltin compounds in coastal and marine environments [23-28]. BT contamination in freshwater ecosystems such as rivers and lakes has not been examined in detail. For instance, imposex in gastropods has been examined in marine waters but such studies have not been conducted in freshwaters. With increasing use of BT derivatives in industries located along the inland lakes and rivers, and in a variety of household products, it is important to monitor the concentrations of BT compounds in freshwater ecosystems.

In this chapter, we discuss persistent, bioaccumulative and toxic potentials of butyltins in freshwater ecosystems. The differences in the partitioning behavior between freshwater and marine water systems are also discussed. Concentrations of butyltins in freshwater organisms collected from several locations in the U.S. are compiled and compared with those measured in sediment and mussel tissues collected from the lowermost Tennessee River and Kentucky Lake.

Sources of Butyltins in Freshwater Ecosystems

Pioneering work of butyltin contaminations in freshwater systems was performed by Maguire and co-workers in Canada in the early 1980s (see Chau *et al.* [29] for a recent report). Their studies focused on butyltins in water and sediment of various lakes and rivers in Canada. While butyltin contamination arising from antifouling paints released due to pleasure boating activities was found to be a major source in several locations, these compounds also were found in rivers where there was minimal or no boating activities. In fact, fouling of boat hulls in freshwater is a less serious problem compared with that in seawater [30]. This provided evidence for the occurrence of other sources of butyltins in the environment. Monobutyltin (MBT) and dibutyltin (DBT) are ubiquitous pollutants in municipal wastewaters which originate mainly from the leaching of PVC pipes containing these compounds as stabilizers [31]. Disposal of municipal wastewater thus contributes to butyltin contamination in receiving surface waters such as lakes or rivers. Input of DBT and TBT into surface waters from applications other than antifouling paint on vessels (i.e., use of TBT as a slimicide in the cooling water of a thermoelectric power plant) was first reported in Italy in the late 1980s [32]. Several studies have reported the occurrence of butyltin compounds in wastewaters. For example, concentrations of butyltins in raw wastewater in Zurich, Switzerland, on six sampling days ranged from 136 to 564, 127 to 1026 and 64 to 217 ng/L, respectively for MBT, DBT and TBT [33]. Water treatment processes removed butyltin compounds to a certain extent but not completely. MBT, DBT and TBT were removed at a rate of 62, 80 and 66%, respectively by sedimentation [33]. The effluent from the treatment plant in Zurich contained butyltins from 7 to 47 ng L^{-1}. The sludge contained MBT, DBT and TBT concentrations in the range of 0.1-0.97, 0.41-1.24 and 0.28-1.51 μg g^{-1} dry wt, respectively. Similarly, studies in Canada and Germany have shown the presence of butyltin compounds in wastewater effluents [34,35]. MBT concentration of up to 22 μg L^{-1} was found in wastewater effluents in Canada. Presence of over hundred ng L^{-1} concentrations of butyltin compounds in Ganges river waters in India also has been shown [36]. Studies describing the occurrence of butyltins in wastewater effluents in the U.S. are not available. Nevertheless, the above evidence suggests that butyltins are widespread contaminants in wastewater, which eventually enters into lakes or rivers. Occurrence of butyltin compounds in non-navigable areas in certain rivers [37] as well as in aquatic organisms of certain non-navigable rivers suggest the release of these compounds from wastewater [38]. Other sources of butyltin compounds in freshwater systems are the use of unregulated antifouling paints, use in timber protection, and use in industrial applications as biocides or catalysts. The contamination of riverine environments by butyltin compounds strongly indicates that the occurrence of butyltins in surface waters is widespread.

Bioaccumulation of Butyltins in Freshwater Ecosystems

Speciation and partitioning behavior of OT compounds can vary between seawater and freshwater due to the differences in salinity and hydrogen ion concentration (pH). Knowledge of aqueous phase chemical speciation can provide important insights into the bioavailability, bioaccumulation and toxicity of butyltins. Organotins undergo pH dependent hydrolysis when introduced into water [39]. While several studies have examined the behavior and fate of OT compounds in seawater, a few studies have examined their behavior in freshwaters. Cations are formed in water at pH<pKa, and these monovalent organometallic cations behave as weak acids. TBT is present as cations (TBT-Sn$^+$) at low pH, while at higher pH, they are present as neutral hydroxide (TBT-SnOH). Besides pH, the dissociation reactions are influenced by water temperature, ionic composition and strength [39]. Relatively high Cl$^-$ concentrations in seawater favor the formation of TBTCl.

The pH-dependent speciation of butyltins has consequences for the bioaccumulation in aquatic organisms. Significant accumulation into biota occurs via uptake through membranes by hydrophobic mechanisms if dissolved butyltin species are uncharged. Neutral butyltins, including hydroxy complexes (TBTOH) and chloride complex (TBTCl) can readily accumulate relative to ionic forms. Studies have shown that the octanol-water partition coefficient (Kow) of TBTCl increased from 3.2 to 3.85 with an increase in pH from 5.8 to 7.8 [40]. The Kow of TBT also varies with salinity [41]. Fent and Looser [42] studied the effects of pH and humic acids on bioaccumulation and bioavailability of TBTCl in *Daphnia* and yolk sac larvae of fish *Thymallus thymallus*. Bioconcentration factors (BCF) in *Daphnia* and *Thymallus* were 198 and 2015, respectively. Uptake rates and bioaccumulation were significantly higher at pH 8.0, where TBT predominates as neutral TBTOH than at pH 6.0, where it predominates as cation. Bioaccumulation of TBT also decreased with increasing concentrations of dissolved organic carbon or humic acids. Tsuda *et al.* [43] examined BCF of TBT in guppies acclimated to either freshwater (pH 7.1-7.3) or artificial seawater (pH 8.0-8.2). At the pH used in their study, approximately 90% (freshwater) and 98-99% (seawater) of the compounds were in the neutral form as TBT hydroxide. Although several factors influence the bioaccumulation and bioavailability of trace organic contaminants in the environment, for BT compounds, it has been shown that the neutral species bioaccumulate relatively rapidly compared with the ionized species, in the absence of counterions [44]. The pH in the freshwater environment is somewhat lower compared with the pH in estuarine and marine waters. In general, for OT compounds with a pKa between 6 and 8.5, BCF and toxicity to aquatic organisms are predicted to become higher going from freshwater to seawater. Because of the differences, environmental quality guidelines established for TBT in seawater are lower (2 ng L^{-1}) than those for freshwater (20 ng L^{-1}).

A few studies have examined food chain transfer of butyltin compounds in freshwater ecosystems. Chau *et al.* [45] studied bioaccumulation of butyltin compounds by mussels in harbors. Guruge *et al.* [46] examined concentrations of butyltins in cormorants and their fish diet from Lake Biwa, Japan. Total butyltin concentrations of up to 1000 ng g^{-1} wet wt, were found in the liver of cormorants whereas those in fish ranged from 10-50 ng g^{-1}, wet wt. Biomagnification factors (BMF) of butyltins in cormorants, calculated based on the whole body concentrations, were in the range of 1.1-4.1. Similarly, Stäb *et al.* [47] measured BT concentrations in sediment, benthic organisms, fish and birds collected from a shallow freshwater lake in the Netherlands and showed the presence of high concentrations of butyltins in fish and birds. Sediment, benthic organisms, fish and dolphins collected from the Ganges River in India contained butyltins at noticeable concentrations. Total BT concentrations in sediment were 35 ng g^{-1} dry wt, whereas those in dolphin (*Platanista gangetica*) livers were up to 2000 ng g^{-1} wet wt [27]. River otters (*Lutra canadensis*) collected from various rivers and coastal bays in Washington and Oregon, USA, contained total butyltin concentrations of up to 2610 ng/g, wet wt, in the liver [38]. Butyltins also were found in the livers of cormorants and bald eagles from the Great Lakes [48]. The results provide strong evidence for bioaccumulation and biomagnification of butyltin compounds in freshwater ecosystems.

Persistence of Butyltins in Freshwater Ecosystems

The persistence of BTs or any chemical in an aquatic ecosystem is a function of its own physico-chemical properties and ecosystem-specific properties such as the concentration of suspended organic matter, the microbial communities, etc. [49]. In general, the persistence of BTs in aquatic systems is a function of physical (e.g., volatilization and adsorption to suspended solids and sediment), chemical (hydrolysis, photolysis) and biological (microbial degradation, uptake) removal mechanisms in addition to flow and other water characteristics. Volatilization of

TBT from natural freshwater-sediment mixtures is generally negligible over a period of 11 months, possibly because of adsorption of TBT to the sediment [50, 51]. Half-lives of TBT in water-sediment complex from a freshwater ecosystem in Canada were estimated to be greater than 11 months. Even in distilled water, there is generally negligible volatilization of TBT over a period of 2 months [50]. Photolysis of TBT was reported to occur in surface waters, but not in waters greater than 0.5 m deep [50]. The Sn-C bond is stable up to 200°C [13]. Therefore, abiotic degradation of TBT is thought to be of minor importance. A few studies also have reported microbial degradation of TBT [52,53]; However, at a concentration of several hundred parts-per-billion, butyltin compounds inhibit growth of heterotrophic bacteria [53-55]. Bacterial studies suggest that degradation in sediments, particularly in highly contaminated sediments is very slow. Because butyltins have been shown to exist at toxic levels in locations with intensive boating activities and in sewage sludge, the persistence of these compounds and their effects on abiotic and biotic degradation processes must be evaluated. Biodegradation of TBT in water and sediment appears to depend on temperature and microbial characteristics. Estimates of half-life of degradation of TBT in water ranges from 6-19 days in seawater in San Diego Bay [56] and 4 months in freshwater in Toronto Harbor [51]. In general, the half-life of TBT in waters varies from a few days to several months, depending on the ecosystem-specific properties. However, differences in the degradation rates of butyltins in freshwater and seawater cannot be generalized.

Several studies have used TBT as a model compound for studying environmental behavior including persistence. Very few studies have examined persistence of MBT or DBT. Because MBT and DBT are the primary pollutants in freshwaters that receive discharges from municipal wastewater treatment plants, research is needed to assess the persistence of MBT and DBT in freshwater systems. TBT in the environment is successively debutylated resulting in the formation of DBT and MBT and then to inorganic tin, which varies from location to location. Half-lives of TBT and DBT were 11 and 5 d, respectively, in river waters [57]. However, in water samples contaminated with great concentrations of butyltins, degradation was slow, resulting in longer half-lives [57].

Sediment appears to be the ultimate sink for butyltin compounds. Despite the ban on the use of TBT in antifouling paints, sediment concentrations have decreased very slowly [29]. TBT adsorbs to sediment very strongly [58]. The sorption process has been shown to reversible, indicating that TBT-contaminated sediments can act as sources for dissolved TBT [58]. In addition, it has been shown that the TBT sorption-desorption process is rapid, indicating that TBT, sediment and water may reach equilibrium in well-mixed environments [58]. The determined partition coefficients for butyltins in different sediments varied from 1700 to 29,000 for MBT, 2100 to 26,000 for DBT and 6200 to 55,000 for TBT [59]. Half-life of TBT in freshwater anaerobic sediment has been reported to be in the order of a few to ten years [60]. Assuming a first order kinetics, TBT half-lives were estimated to range from 0.9 to 5.2 years between different cores in the U.K. and New Zealand, and

140

those for DBT and MBT were assumed to be in the same range [61]. In general, degradation of butyltin compounds occur both aerobic and anaerobic conditions [33, 51]. However, under anoxic conditions degradation rate was slower [33]. The absence of oxygen and toxic effect on bacteria are probably responsible for the slow degradation. This suggests the reason for great concentrations of BTs in sediments despite partial regulations on the use TBT in antifouling paints. The estimated half-lives of 1 to 5 years in freshwater and marine sediments indicate the persistence of TBT. TBT and its derivatives have also been found in sediments deposited 15 to 20 years ago, which highlight the importance of the sedimentary sink for these compounds. Nevertheless, as mentioned above, sediments do not act solely as terminal sinks, but rather as a continuing source due to adsorption-desorption processes.

Butyltins in Freshwaters in the U.S.

Despite the widespread occurrence of BTs in wastewater and possibly in surface waters that receive effluents from water treatment plants, very few monitoring studies have been conducted in the U.S. Extensive butyltin monitoring studies have been conducted in inland freshwater ecosystems in Canada[29,62], the UK [63], Switzerland [64] and the Netherlands [65]. These studies have clearly suggested the widespread contamination of butyltins in freshwater systems. Results of a few available monitoring studies on BTs in the U.S. freshwaters have been compiled (Table I).

Although these studies indicate the occurrence of butyltins in freshwaters, most sampling locations are estuarine. Further studies are needed to evaluate butyltin contamination of freshwaters arising from the disposal of wastewaters. A small-scale butyltin contamination survey conducted in Tennessee River and Kentucky Lake, which are typical mid-western freshwater ecosystems, is discussed below to provide evidence for the occurrence of butyltins in inland lakes and rivers.

Butyltins in Lowermost Tennessee River and Kentucky Lake

Study Area

Kentucky Lake is one of the major man-made lakes in the U.S. It covers an area of 160,300 acres and constitutes the northernmost end of a major shipping route for large barges and small ships between the Gulf of Mexico and the Ohio River (Figure 1). Tennessee River downstream from Kentucky Lake (Kentucky dam tailwater) receives industrial wastewater from several industries located in the Calvert City Industrial Complex (CCIC). During the past two decades, mass mortality of mussels has been frequently encountered in these regional waters. In addition, the quality and quantity of shells harvested for the pearl industry also were

Table I. Reported Concentrations of Butyltins in Selected Freshwater or Estuarine Ecosystems in North America and Other Countries

Sample/Location	MBT	DBT	TBT	Ref
Water (ng L^{-1})				
Ganga River, India, 1995	2.1-70	1.7-101	2.9-19.8	[36]
Lake Michigan	14-762	NA	NA	[5]
Detroit River, 1982-1985	15	<3	1390	[62]
Buffalo River, 1982-1985	30	20	<3	[62]
Niagara River, 1982-1985	2690	<3	<3	[62]
Detroit and St.Clair Rivers	<3-148	<3-196	<3-171	[62]
Elizabeth River and Sarah Creek	<3-19	3.1-131	<3-158	[66]
Lake Lucerne, Switzerland, 1988-1990	2-73	5-102	100-752	[67]
Sediment (ng g^{-1} dry wt)				
Ganges River, India, 1995*	-	-	35	[30]
Lake Lucerne, Switzerland, 1988-1990	NA	NA	2043	[67]
Detroit and St.Clair Rivers	<3-44	<3-59	<3-170	[68]
Mussels (ng g^{-1} dry wt)				
Brielse meer, Netherlands, 1992	860	1740	11,500	[65]
Boterdiep, Netherlands, 1992	<6	<4	98	[65]
Lake Lucerne, Switzerland, 1988-1990	NA	NA	9350	[67]
Genoa, Italy, 1994-1995	80-260	1810-4940	250-1060	[69]
North River, Massachusetts Bay, 1990	<20	50	100	[70]
Connecticut River, Long Island Sound, 1990	<20	60	150	[70]
Houstonic River, Long Island Sound, 1990	<20	50	130	[70]
Shark River, New York Bight, 1990	30	140	320	[70]
Potomac River, Maryland,1990	10	50	180	[70]
Indian River, Florida, 1990	20	120	720	[70]
Savannah River, Tybee Island, 1990	<10	30	410	[70]
Pungo River, Pamlico Sound, 1990	<10	10	10	[70]
Lake Bodensee, Germany**	3-20	30-400	200-3000	[71]

*Total butyltins; ** Values in wet weight basis.

142

substantially reduced. These episodic events have raised considerable concern about the quality of water in the Lake and the Kentucky dam tailwater, thus, providing an ideal inland freshwater ecosystem for this study.

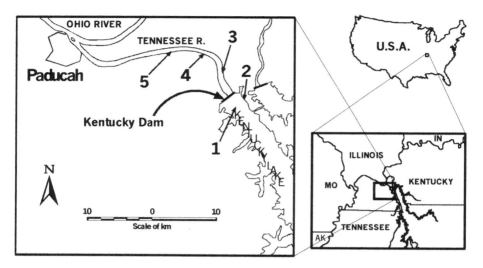

Figure 1. Map showing sediments and mussels sampling locations in the lowermost Tennessee River and Kentucky Lake, U.S.A.

Samples and Chemical Analysis

Sediment and mussel sampling was performed on May 21, 1997. Details of sampling locations are given in Figure 1 and Table II. Surface sediment (0-5 cm deep) samples were collected using PONAR grab sampler. Pre-cleaned I-Chem bottles were used to store the samples. Mussel samples were collected by SCUBA diving. The mussels were identified, measured for length, height and width, wet weight, and age were determined (Table III). Sites 1 and 2 samples contained washboard (*Megalonaias nervosa)* and ebonyshell (*Fusconaia ebena)* and Sites 3, 4 and 5 samples had mapleleaf (*Quadrula quadrula*) and threeridge (*Amblema plicata*) respectively. The mussel tissues were separated from shells. Individuals of same age and species were pooled and transferred to pre-cleaned glass jars and stored under -20°C until analysis. Sediment and mussel samples were freeze dried for over 60 h using Labconco FreeZone Freeze dry system Model 77535. BT derivatives were analyzed following the procedure described by Kannan *et al*. [27]. Briefly, acidified samples were extracted with 70 ml of 0.1% tropolone-acetone, and the solvent was transferred to 100 mL of 0.1% tropolone-benzene in a separatory funnel. Moisture in the organic extract was removed using 35 g of anhydrous sodium sulfate salt and then the concentrated extract was propylated with n-propylmagnesium bromide (ca 2 mol/L in TFH solution, Tokyo Kasei Kogyo Ltd., Japan) as a Grignard reagent. The derivatized extract was passed through a 6 g Florisil packed wet column for cleanup. The eluate from the Florisil column was rotary evaporated to 5 mL and injected into a gas chromatograph.

Table II. Details of sampling locations in the lowermost Tennessee River and Kentucky Lake.

Site No.	Location	Latitude	Longitude	Depth (m)	Substratum
1	TRM 23.2/L	36° 59' 56.08"N	88° 15' 50.75"W	18	Muddy
2	TRM 23.1/R	37° 0' 30.76"N	88°15' 11.86"W	4.5	Muddy-Gravel
3	TRM 20.9/R	37° 01' 41.37"N	88°17' 14.39"W	3.0	Sandy-Gravel
4	TRM 17.7/L	37° 03' 20.06"N	88°19' 50.75"W	3.0	Muddy
5	TRM 15.2/L	37° 03' 37.95"N	88°22' 23.82"W	3.6	Muddy-Sand Gravel

TRM: Tennessee River Mile; L: Left bank; R: Right bank

Chromatographic separation was performed using Hewlett-Packard 5890 series II gas chromatograph with flame photometric detector (FPD) and 30m x 0.25 mm i.d. DB-1 capillary column coated with 0.25µ film thickness. The flame photometer was operated using a hydrogen-air-nitrogen flame and was equipped with a 610 nm band-pass filter that is selective for tin-containing compounds. Butyltin trichloride (MBT), dibutyltin dichloride (DBT) and tributyltin chloride (TBT) of known amounts (100 ng) were spiked into tropolone-acetone and passed through whole analytical procedure, and were used as an external standard. The detection limits of MBT, DBT and TBT were 7, 2.4 and 0.5 ng g^{-1} wet wt. The average recovery rates for the analytes were between 90 and 110% for each compound.

Table III. Details of mussel samples collected from the lowermost Tennessee River and Kentucky Lake.

Site No.	Genus and Species	Age (yrs)	Length (mm)	Height (mm)	Width (mm)	Wet wt. (g)
1.	M. nervosa (3)	23-27	136-143	97-100	55-56.2	495-561
2.	F. ebena (1)	13	85.1	75.8	52	259
3.	Q. quadrula (4)	11-12	75.8-92	61.5-70.9	40.1-44.3	135-212
4.	A. plicata (5)	8-11	72.7-82.4	60.3-70.5	36.3-42.9	113-163
5.	A. plicata (6)	8-10	66.8-81.7	57-65.2	36.7-45.6	101-176

Values in parenthesis indicate number of specimens pooled for analysis.

Butyltin Concentrations in Sediments

BTs were detected in all of the sediments analyzed. Concentrations of BTs ranged from 6.8 to 356 ng g^{-1} dry wt. (Table IV), comparable to those reported for several marine sediments from the U.S. coastal areas (<5 to 282 ng Sn g^{-1}) away from intensive boating activities [72]. Sediment concentrations as high as 4000 ng g^{-1} dry wt. were reported from a marina in the southern Chesapeake bay [73]. Very limited information is available on the BT concentrations in rivers and lakes. Maguire *et al.* [37] and Maguire [74] reported sediment BT concentrations from Ontario harbors and rivers. Their reported concentration ranges in the top 2 cm sediments from were 100 to 400 ng g^{-1} dry wt. and were comparable to Kentucky Lake concentrations (Table IV). A wide range of concentrations of BTs in the lowermost Tennessee River and Kentucky Lake sediments suggests the presence of localized areas of contamination. Sediments from site #2 (TRM 23.1) contained the greatest concentrations and corresponded with barges parking sites in the area. MBT was the predominant compound in sites 1 and 2, and may have been derived from degradation of TBT in the sediments or by deposition from the overlying water.

Table IV. Concentrations (ng g^{-1} dry wt.) of butyltins in sediments collected from the lowermost Tennessee River and Kentucky Lake.

Site No.	Monobutyltin	Dibutyltin	Tributyltin	Total Butyltins
1.	63	4	24	91
2.	320	14	22	356
3.	<3	<1	6.8	6.8
4.	24	9.7	11	45
5.	5.1	<1	5.3	10

Butyltin Concentrations in Mussels Tissues

Total BT concentrations in mussel tissues varied between 32 and 107 ng g^{-1} dry wt. (Table V) and the butyltin composition was different at Kentucky Lake and lowermost Tennessee River sampling locations (Figure 2). The average BT concentrations in mussels from Kentucky Lake was less than those of mussels and oysters (*Mytilus edulis, Mytilus californianus, Crossostrea virginica*) from East (511 ng g^{-1} dry wt.) and West coasts (546 ng g^{-1} dry wt.) of the United States analyzed during 1987-1990 [70]. Stäb *et al.*[65] reported OT contamination of Dutch freshwaters using the Zebra mussel (*Dreissena polymorpha*) as bioindicator. The authors reported very high TBT (96 - 11,500 ng Sn g^{-1} dry wt.) body concentrations in

Table V. Concentrations of (ng g^{-1} dry wt.) of butyltins in mussels collected from the Lowermost Tennessee River and Kentucky Lake.

Site No.	Species	Monobutyltin	Dibutyltin	Tributyltin	Total Butyltins
1.	M. nervosa	17	14	61	92
2.	F. ebena	5.6	9.5	91	106
3.	Q. quadrula	14	6.4	12	32
4.	A. plicata	36	16	11	63
5.	A. plicata	50	18	39	107

the mussels from the Netherlands waters, despite the implementation of legislation restricting the use of TBT. As shown in the Fig. 2. MBT was the predominant compound at sites 3, 4 and 5 in the lowermost Tennessee River (where the river receives discharges from an industrial complex), while TBT accounted for the greater proportion of total BTs at sites 1 and 2 where large barges and ships traffic are higher. Presence of greater proportion of TBT indicates recent input, despite the ban on the use of TBT in the U.S. in 1989 [27]. Uhler *et al.* [70] also observed a higher proportion of TBT in mussel tissues from several estuaries and rivers from the U.S. (Table 1). Relatively greater concentrations of BTs in almost all of the mussel tissue samples than sediment BT levels indicate bioaccumulation of this chemical. BT compounds possess both lipophilic and ionic properties. The former promotes accumulation via solubilization in lipids, while the latter enable these compounds bind to macromolecules. These characteristics combined with poor metabolic capacity of bivalve mollusks to metabolize BTs result in the bioaccumulation of BTs by these organisms.

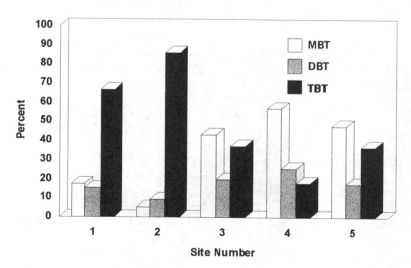

Figure 2. Percent composition of butyltins in mussel tissues from the lowermost Tennessee River and Kentucky Lake.

146

Leaching of tributyltin-containing anti-fouling paints in the ocean going ships and industries that produce or use TBT in this region may be sources of TBT, and discharge of municipal sewage and industrial waste waters in this watershed may account for the presence of the MBT and DBT compounds detected. Further studies with greater number of samples from the lowermost Tennessee River and Kentucky Lake are needed to elucidate distribution, source, bioaccumulation and toxic effects.

In general, it was expected that butyltins would degrade rapidly in the environment, however, evidence suggests that BT compounds are persistent and bioaccumulative. Studies with marine mammals also reveal that BT compounds biomagnify in the food chain. Thus, butyltins are persistent, bioaccumulative and toxic substances. Very few studies have been conducted in the freshwater systems and sewage disposal as a source will continue although the use of TBT is banned in antifouling paints. These results emphasize the need for future studies on butyltins in the freshwater ecosystems.

Acknowledgments

This research was supported by the Committee on Institutional Studies and Research (CISR) of Murray State University. The assistance provided by Dr. James Sickel, Mr. Darren P. Reed in sediment, mussel sampling and Mr. Carl Woods (MSU Science Resource Center) in preparation of this manuscript are gratefully acknowledged. Center for Reservoir Research (CRR) contribution number for this article is 058.

References

1. *Toxicological Profile for Tin*; U.S. Department of Health and Human Services: Public Health Service, Agency for Toxic Substances and Disease Registry, TP-91/27; 1992, 145 pp.
2. Snoeij, N.J.; Penninks, A.H.; Seinen, W. *Environ. Res.* **1987,** 44, 335-353.
3. Van der Kerk, G.J.M.; Luijten, J.G.A. *J. Appl. Chem.* **1954,** 4, 314-319.
4. Van der Kerk, G.J.M.; Luijten, J.G.A.; Noltes, J.G. *Angew. Chem.* **1958,** 70, 298-306.
5. Maguire, R.J. *Appl. Organomet. Chem.* **1987,** 1, 475-498.
6. Zuckerman, J.J.; Reisdorf, R.P.; Ellis III, H.V.; Wilkinson, R.R. In *Organometals and Organometalloids-Occurrence and Fate in the Environment;* Brinckman, F.E.; Bellama, J.M. (eds), American Chemical Society, ACS Symposium.ser.No.82, Washington, DC, 1978, pp 388-422.
7. Van der Kerk, G.J.M. *Chem. Tech.* **1978,** 8, 356-365.
8. Mercier, A.; Pelletier, E.; Hamel, J.F. *Aqua. Toxicol.* **1994,** 28, 259-273.

9.   Alajouanine, Th.; Dérobert, L.; Thiéffry, S. *Rev. Neurol.* **1958,** 98, 85-96.
10. Barnes, J.M.; Stoner, H.B. *Pharmacol. Rev.* **1959,** 11, 211-232.
11. Blunden, S.J.; Chapman, A.H. In *Organometallic Compounds in the Environment.* Craig, P.J., Ed.; ITRI (International Tin Research Institute) Publ.No. 665, Middlsex, UK, 1986; pp 111-159.
12. Fent, K. *CRC Crit. Rev. Toxicol.* **1996,** 26, 1-117.
13. Clark, E.A.; Sterritt, R.M.; Lester, J.N. *Environ. Sci. Technol.* **1988,** 22, 600-604.
14. Alzieu, C.; Heral, M.; Thibaud, J.; Dardignac, M.J.; Feuillet, M. *Rev. Trav. Inst. Peches   Marit.* **1982,** 45, 101-116.
15. Alzieu, C.; Thibaud, Y.; Heral, M.; Boutier, B. *Rev. Trav. Inst. Peches Marit.* **1980,** 44, 301-348.
16. Alzieu, C.; Heral, H. In *Ecotoxicological Testing for the Marine Environment. Vol. 2;* Personne, G.; Jaspers, E.; Claus, C., Eds.; Institute for Marine Sciences Research, Bredene, Belgium, 1984, pp 187-196.
17. Bryan, G.W.; Gibbs, P.E.; Huggett, R.J.; Curtis, L.A.; Bailey, D.S.; Dauer, D.M. *Mar. Pollut.   Bull.* **1989,** 20, 458-462.
18. Bryan, G.W.; Gibbs, P.E. In *Metal Ecotoxicology, Concepts & Applications;* Newman, M.C.; McIntosh, A.W., Eds.; Lewis Publishers: Chelsea, MI, 1991, pp 323-361.
19. Alzieu, C.; Michel, P.; Tolosa, I.; Bacci, E.; Mee, L.D.; Readman, J.W. *Marine Environ. Res.* **1991,** 32, 261-270.
20. Bryan, G.W.; Burt, G.R.; Gibbs, P.E.; Pascoe, P.L. *J. Mar. Biol. Assoc. U.K.* **1993,** 73, 913-929.
21. Kannan, K.; Guruge, K.S.; Thomas, N.J.; Tanabe, S.; Giesy, J.P. *Environ. Sci. Technol.* **1998,** 32, 1169-1175.
22. Evans, C.J.; Karpel, S. . *J. Organomet. Chem. Libr.* **1985,** 16, 178-188.
23. Langston, W.J.; Bryan, G.W.; Burt, G.R.; Gibbs, P.E. *Functional Ecol.* **1990,** 4, 433-443.
24. Kannan, K.; Tanabe, S.; Iwata, H.; Tatsukawa, R. *Environ. Pollut.* **1995,** 90, 279-290.
25. Kannan, K.; Yasunaga, Y.; Iwata, H.; Ichihashi, H.; Tanabe, S.; Tatsukawa, R. *Arch. Environ. Contam. Toxicol.* **1995,** 28, 40-47.
26. Kannan, K.; Corsolini, S.; Focardi, S.; Tanabe, S.; Tatsukawa, R. *Arch. Environ. Contam. Toxicol.* **1996,** 31, 19-23.
27. Kannan, K.; Senthilkumar, K.; Loganathan, B.G.; Takahashi, S.; Odell, D.K.; Tanabe, S. *Environ. Sci. Technol.* **1997,** 31, 296-301.
28. Kannan, K.; Falandysz, J. *Mar. Pollut. Bull.* **1997,** 34, 203-207.
29. Chau, Y.K.; Maguire, R.J.; Brown, M.; Yang, F.; Batchelor, S.P. *Water Qual. Res. J. Canada* **1997,** 32, 453-521.
30. Kannan, K.; Senthilkumar, K.; Sinha, R.K. *Appl. Organomet. Chem.* **1997,** 11, 223-230.

148

31. Snyder, S.A.; Snyder, E.; Villeneuve, D.; Kannan, K.; Villalobos, A.; Blankenship, A.; Giesy, J.P. Instrumental and bioanalytical measures of endocrine disruptors in water. ACS annual meeting, Boston. 1999.

32. Bacci, E.; Gaggi, C. *Mar. Pollut. Bull.* **1989**, 20, 290-292.

33. Fent, K.; Muller, M.D. *Environ. Sci. Technol.* **1991**, 25, 489-493.

34. Chau, Y.K.; Zhang, S.; Maguire, R.J. *Sci. Total Environ.* **1992**, 121, 271-281.

35. Donard, O.F.X.; Quevauviller, P.H.; Bruchet, A. *Wat. Res.* **1993**, 27, 1085-1089.

36. Ansari, A.A.; Singh, I.B.; Tobschall, H.J. *The Sci. Total Environ.* **1998**, 223, 157-166.

37. Maguire, R.J.; Chau, Y.K.; Bengert, G.A.; Hale, E.J. *Environ. Sci. Technol.* **1982**, 16, 698-702.

38. Kannan, K.; Grove, R.A.; Senthilkumar, K.; Henny, C.J.; Giesy, J.P. *Arch. Environ. Contam. Toxicol.* **1999**, 36, 462-468.

39. Tobias, R.S. In *Organometals and Organometalloids: Occurrence and Fate in the Environment,* Brinkman, F.E.; Bellama, J.M., Eds., ACS Symposium Series; 1978; pp.130-148.

40. Tsuda, T.; Aoki, S.; Kojima, M.; Harada, H. *Comp. Biochem. Physiol.* **1990**, 95, 151-153.

41. Laughlin, Jr. R.B.; Guard, H.E.; Coleman, III, W.M. *Environ. Sci. Technol.* **1986**, 20, 201-204.

42. Fent, K.; Looser, P.W. *Wat. Res.* **1995**, 29, 1631-1637.

43. Tsuda, T.; Aoki, S.; Kojima, M.; Harada, H. *Wat. Res.*, **1990**, 24, 1373-1376.

44. van Wezel, A.P. *Environ. Rev.*, **1998**, 6, 123-137.

45. Chau, Y.K.; Young, P.T.S.;Bengert, G.A.;Yaromich, J. *Chemical Speciation and Bioavailabilty.* **1989**, 1, 151-156.

46. Guruge, K.S.; Tanabe, S.; Iwata, H.; Tatsukawa, R.; Yamagishi, S. *Arch. Environ. Contam. Toxicol.* **1996,** 31, 210-217.

47. Stäb, J.A.; Traas, T.P.; Stroomberg, G.; van Kesteren, J.; Leonards, P.; van Hattum, B.; Brinkman, U.A.Th.; Cofino, W.P. *Arch. Environ. Contam. Toxicol.* **1996**, 31, 319-328.

48. Kannan, K.; Senthilkumar, K.; Elliott, J.E.; Feyk, L.A.; Giesy, J.P. *Arch. Environ. Contam. Toxicol.* **1998,** 35, 64-69.

49. Baughman, G.L.; Lassiter, R.R. In   Estimating the Hazard of Chemical Substances to Aquatic Life.   Cairns, Jr, J.; Dickson, D.L.; Maki, A.W. Eds. ASTM (American Society for Testing and Materials) STP 657, 1978, pp. 35-54.

50. Maguire, R.J.; Carey, J.H.;   Hale, E.J. *J. Agr. Food Chem.* **1983**, 31, 1060-1065.

51. Maguire, R.J.; Tkacz, R.J. *J. Agr. Food Chem.* **1985,** 33, 947-953.

52. Lee, R.F.; Valkirs, A.O.;   Seligman, P.F. *Environ. Sci. Technol.* **1989**, 23, 1515-1518.

53. Errécalde, O.; Astruc, M.; Maury, G.; Pinel, R. *Appl. Organomet. Chem.* **1995**, 9, 23-28.

54. Cooney, J.J. *J. Ind. Microbiol,* **1988**, 3, 195-204.
55. Miller, M.E.; Cooney, J.J. *Arch. Environ. Contam. Toxicol.* **1994**, 27, 501-506.
56. Seligman, P.F.; Valkirs, A.O.; Lee, R.F. *Environ. Sci. Technol.,* **1986**, 20, 1229-1235.
57. Hattori, Y.; Kobayashi, A.; Nonaka, K.; Sugimae, A.; Nakamoto, M. *Wat. Sci. Tech.,* **1988**, 20, 71-76.
58. Unger, M.A.; MacIntyre, W.G.; Huggett, R.J. *Environ. Toxicol. Chem.* **1988**, 7, 907-915.
59. Stang, P.M.; Seligman, P.F. In: *Proc. Inter.Organotin Symp.Oceans'87,* Halifax, Nova Scotia, Canada, Sept 28-Oct 1, vol.4, 1987, pp. 1386-1391.
60. Dowson, P.H.; Bubb, J.M.; Williams, T.P.; Lester, J.N. *Wat. Sci. Tech.,* **1993a**, 28, 133-137.
61. Dowson, P.H.; Bubb, J.M.; Lester, J.N. Environ. Monit. Assess. **1993b**, 28, 145-160.
62. Maguire, R.J.; Tkacz, R.J.; Chau, Y.K.; Bengert, G.A.; Wong, P.T.S. *Chemosphere* **1986,** 15, 253-274.
63. Dowson, P.H.; Pershke, D.; Bubb, J.M.; Lester, J.N. *Environ. Pollut.* **1992**, 76, 259-266.
64. Fent, K.; Hunn, J. *Environ. Toxicol. Chem.* **1995**, 14, 1123-1132.
65. Stäb, J.A.; Frenay, M.; Freriks, I.L.; Brinkman, U.A.Th.; Cofino, W.P. *Environ. Toxicol. Chem.* **1995,** 14, 2023-2032.
66. Unger, M.A.; MacIntyre, W.G.; Greaves, J.; Huggett, R.J. *Chemosphere* **1986**, 15, 461-470.
67. Fent, K.; Hunn, J. *Environ. Sci. Technol.* **1991**, 25, 956-963.
68. Maguire, R.J.; Tkacz, R.J.; Sartor, D.L. *J. Great Lakes Res.* **1985**, 11, 320-327.
69. Rivaro, P.; Frache, R. Leardi, R. *Chemosphere* **1997**, 34, 99-106.
70. Uhler, A.D.; Durell, G.S.; Steinhauer, W.G.; Spellacy, A.M. *Environ. Toxicol. Chem.* **1993,** 12, 139-153.
71. Kalbfus, W.; Zellner, A.; Frey, S.; Stanner, E. In Gewässergefährdung durch organozinnhaltige antifouling- anstriche. 1991. TEXTE 44/91. Umweltbundesamt, Berlin, Germany.
72. Wade, T.L.; Garcia-Romero, B.; Brooks, J.M. *Chemosphere* **1990,** 647-662.
73. Espourteille, F.A.; Greaves, J.; Huggett, R.J. *Environ. Toxicol. Chem.* **1993,** 305-314.
74. Maguire, R.J.; *Environ. Sci. Technol.* **1984,** 18, 291-294.

Chapter 11

Methylmercury Bioaccumulation Dependence on Northern Pike Age and Size in 20 Minnesota Lakes

Gary E. Glass[1,2], John A. Sorensen[2], and George R. Rapp, Jr.[2]

[1]National Health and Environmental Effects Research Lab., Mid-Continent Ecology Division, U.S. Environmental Protection Agency, Duluth, MN 55804 (218–726–8909; gglass@d.umn.edu)
[2]Archaeometry Lab., University of Minnesota at Duluth, Duluth, MN 55812

Mercury accumulation in northern pike muscle tissue (fillets) was found to be directly related to fish age and size. Measurements were made on 173 individual northern pike specimens from twenty lakes across Minnesota. Best fit regressions of mercury fillet concentration (wet wt.) vs. length, weight and age were found using power (log-log transforms) as compared with exponential or linear forms. The resulting equations (after averaging the individual equations for the 20 cases) are: [Hg], ppb fillet = $0.50 \cdot (\text{length, cm})^{2.17} = 7.3 \cdot (\text{weight, g})^{0.68} = 110 \cdot (\text{age, yr})^{1.00}$, with $r^2 = 0.76$, 0.72, and 0.76, respectively. Lengths, weights, and ages ranged over 22-110 cm, 60-7,500 g, and 0.3-11 yr, respectively, and [Hg] ranged from 50 to 2,000 ppb. Inter-lake comparisons were made using computed mercury levels for standard length, weight and fish age for each of the 20 lakes studied. Age based comparisons were found to correlate significantly with methylmercury concentrations in lake water, and with total mercury concentrations in lake plankton. Methylmercury bioaccumulation factors (BAFs, total Hg in fillets divided by total methylmercury in lake water) ranged from $0.3 - 14 \times 10^6$ for individual specimens. Methylmercury BAFs showed a similar dependency on fish size and age as [Hg] in fillet, i.e., methylmercury BAFs = $3,030 \cdot (\text{length, cm})^{2.17} = 47,100 \cdot (\text{weight, g})^{0.68} = 752,000 \cdot (\text{age, yr})^{1.00}$, where values represent the means of twenty different lake equations. Fish mercury levels indicate that each lake differs in its own characteristics controlling bioaccumulation processes and transport through the aquatic food chain. Assuming no size or age dependency of fish mercury concentrations and BAF's may result in impact overestimates for younger fish and underestimates for older fish from mercury contamination in aquatic ecosystems.

© 2001 American Chemical Society

Growing levels of mercury in the atmosphere from fossil fuel combustion, and more recently, from solid waste incineration, are the leading causes of anthropogenic sources and increases in environmental exposure. A recent comprehensive study by the USEPA (1) resulted in a large volume of information documenting the well known and less well known features of the mercury environmental and public health problem. One of the main exposure routes of mercury to man and wildlife is through consumption of freshwater fish. Mercury concentrations in fish are the leading cause for restricting fish consumption due to exceedence of toxic levels by human and wildlife consumers (2-4).

It is generally recognized that mercury levels in fish are related to fish size (5-7), and are usually expressed as concentration at a given size of specimen. Comparisons between different waterbodies are usually made by computing the mercury level expected to be present in fish of the same size (8). Plots of mercury muscle concentrations versus length are generally curvilinear, and log-log transformations are necessary to observe linear functionality with a normal distribution of data. An average sized 3 yr old fish is commonly chosen for making comparisons across sites and waterbodies. For northern pike and walleye, these lengths are approximately 55 and 39 cm, respectively (8), and 1 kg weight for northern pike (9).

Length is generally assumed to represent fish age in making comparisons, and standard sizes of fish have been used in comparing fish across different water bodies to "control" for differences in age and, therefore, exposure. However, this assumption is seldom tested as fish age data usually lag the reporting and interpreting of fish mercury levels for a given survey. The focus of this report is to: 1) review the assumption that northern pike size is a reasonable indicator of age, 2) test if fish size or age is best related to mercury exposure, and 3) define the relationships between mercury levels and BAFs in muscle tissue versus fish size and age.

Lake Selection and Sampling Methods

The 20 Minnesota lakes (Figure 1) selected for this analysis are from various regions of the state and were part of a larger set of 80 lakes studied to determine trends in mercury deposition and lake quality (10). These lakes are valued for their fishery resource and have been surveyed previously for mercury contamination levels (as early as 1975). The 20 lakes selected for this analysis were sampled for additional measurements of lake water chemistry (total methylmercury concentrations) to assess fish mercury bioaccumulation.

Lake Selection

Study lakes were selected based on the following considerations:
1) presence of northern pike fish populations, fish population surveys, and fish mercury residue data (primarily available for NE Minnesota);
2) existence of past water quality data; and
3) to be of significant resource value and representative (e.g. no point sources or extraordinary anthropogenic inputs) of other lakes in the region.

Figure 1. Minnesota map of twenty lakes selected for methylmercury bioaccumulation study.

Sampling and Sample Processing

Fish Sampling
A total of 173 northern pike were sampled from 20 lakes across Minnesota by staff of the Department of Natural Resources (MDNR) and Voyageurs National Park (VNP) during May - Sept. of 1995 and 1996. Fish were collected using gill and/or trap nets (11), tagged with an identification number, placed in a plastic bag, and kept in a cooler until transferred to the MDNR or VNP local area headquarters where they were frozen. Fish were then sent by courier or 1-day parcel delivery to the University of Minnesota - Duluth (UMD) Archaeometry Laboratory where they were received frozen, logged in, and stored frozen until processing.

The goal of the fish sampling was to collect at least 10 individual fish of lengths that were evenly distributed across the available pool of fish lengths sampled.

Fish Processing
Fish weights and lengths and sex were determined in the field and in the laboratory during processing. This provided a cross-check of fish identification. Laboratory processing included: scaling the fish (saving scale samples and cliethra for aging); removing a fillet, slicing it in strips, and refreezing; grinding while frozen, and separately storing the bulk ground sample and a small aliquot (5-10 g) for analyses in zip-loc bags.

Water Sampling
Water sampling for the study lakes was accomplished in the fall (when lakes were mixed) of 1996 by UMD personnel. Some lakes were sampled twice (spring and fall of 1996) in order to investigate variabilities associated with dates of sampling. Field measurements and sampling were conducted in a deep portion (not necessarily the deepest) of the lake, according to methods described elsewhere (8, 10).

Plankton Sampling
Plankton was sampled using a Wisconsin style net (Wildco Wildlife Supply Co.) and a Minnesota plankton bucket with check valve (J. Shapiro, U. of Minnesota, Mpls, MN) according to methods described elsewhere (8, 10).

Methods of Analyses

Mercury Analyses

Total mercury was analyzed in fish fillets, plankton, and unfiltered water samples using cold vapor atomic absorption spectrometry (EPA Method 245 (12, 13), Perkin Elmer model 403 spectrometer) using methods previously reported (8, 14). Although the analyses were conducted for total mercury, it is generally accepted that the form of mercury in freshwater fish muscle (fillet) tissue is 95+% that of mono-methylmercury(II) (15) and for plankton, 10-30% methylmercury (16). All fish were analyzed as individuals rather than composites. A subset of water samples was analyzed for methylmercury using phase ethylation (17) followed by fluorescence detection (EPA Method 1631[18]). Fish and plankton mercury levels are reported on a wet weight and dry weight basis, respectively.

Accuracy of all mercury measurements was checked using spikes of known concentrations and NIST (National Institutes of Science and Technology) certified tissue samples. Laboratory precision was checked by analyzing at least 10% of all samples in duplicate.

Fish Aging

Fish ages were determined by the various MDNR area offices and VNP staff personnel that collected the fish. Annular lines on scales was the primary method used, however, VNP personnel used both scales and cliethra. Numerous cross checks with experienced personnel were employed to assure quality of this data set.

Regression and Statistical Analyses

Relationships between variables were quantified using linear regressions and Pearson correlation coefficients. Variable transformations (log, exponential, or untransformed) are indicated for each result presented. Statistical analyses were made with Macintosh computers using Excel and Systat software.

Results and Discussion

N. Pike Growth - Age Relationships

The growth relationships of northern pike length, weight and age data are presented in Figure 2. The relationship between fish length and weight is very tight with a log-log regression r^2 of 0.98 for the combined 20 lakes. The same value (0.98) is obtained when the r^2 results for the individual lake regressions are averaged. Similar length and weight vs. age regression results show significantly weaker relationships with combined lake r^2 values of 0.69 and 0.65, respectively. The regressions (log-log) for individual lakes show higher r^2, with average values of 0.83 and 0.82, respectively. For selected ages between 1 and 7 yr (Fig. 2, center), the range of fish lengths between different lakes can vary by a factor of about two. This indicates that fish length alone is not necessarily a good predictor of fish age.

The relationship between fish weight and age is more variable than fish length. For ages between 1 and 7 yr the range of observed weights is greater than a factor of two and may be as high as five for lakes of extreme productivity. Estimates of fish age based on weight are clearly less reliable than fish length (Fig. 2, bottom). Of the mathematical models considered, the best at fitting the fish length vs. weight relationship was found to be the power regression (linear regression of log transformed variables). The r^2 statistic indicating the "best fit" averaged 0.98, 0.97, and 0.93, respectively, for power, exponential, and linear regressions.

Fillet Mercury Concentration vs. N. Pike Growth, Age

The regression equation form that best describes the mathematical relationship between mercury fillet concentrations vs. fish length, weight, and age was found to be the power regression. This is indicated by the r^2 statistic, the "best fits" for each of the individual lakes, using three mathematical models tested, summarized in Table I. Of the three models tested, the power format gives a slightly better overall performance, and is better in a majority of the individual cases, than both the exponent and linear formats. The equations for each of the mathematical models showing the relationships between mercury concentrations and northern pike size and age are given in Table I and plotted for the individual lakes in Figure 3. These results show the different fish mercury concentration vs. fish growth responses that result from mercury contamination of the food chain.

The rates of mercury uptake or bioaccumulation by northern pike are different across the 20 lakes studied. The differences in slope for mercury concentration vs. age (log-log) shown in Table I vary by a factor of 5 from the lowest to the highest. The causal mechanisms explaining the differences in response and bioaccumulation are due to differences in watershed mercury loading rates, terrestrial and aquatic mercury methylation rates and factors controlling them, and food chain/lake productivity which may tend to dilute the exposure for a given production of methylmercury. A study of controlling mechanisms was a major objective of the eighty lakes study (10) and an in-depth analysis is currently underway (19).

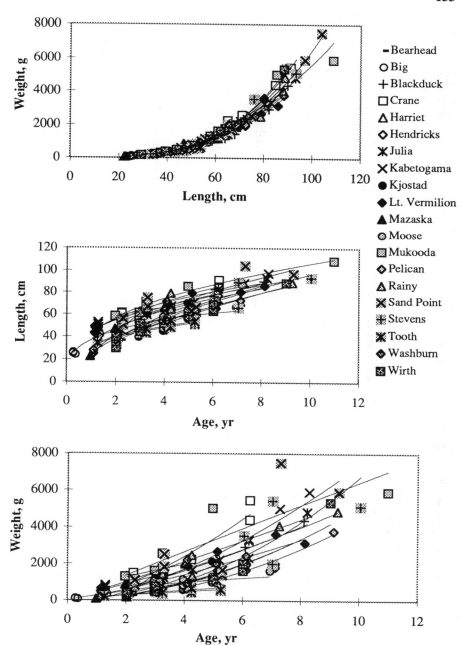

Figure 2. Northern pike growth measures: length, weight, and age relationships for 20 Minnesota lakes. Individual lake trend lines (log-log) are shown.

Table I. Fillet Mercury Concentration Dependence on N. Pike Growth Measures: Results are Averages Across Regressions for Individual Lakes Using Equations with Power, Exponential, and Linear Forms[a]

Power:

	n	\multicolumn{3}{Log[Hg] vs. Log(length)}			\multicolumn{3}{Log[Hg] vs. Log(weight)}			\multicolumn{3}{Log[Hg] vs. Log(age)}	
		slope	intcpt	10^{intcpt}	r^2	slope	intcpt	10^{intcpt}	r^2	slope	intcpt	10^{intcpt}	r^2
mean	9	2.17		0.50	0.76	0.68		7.34	0.72	1.00		109.8	0.76
median	10	2.15		0.08	0.81	0.69		4.65	0.78	1.06		86.4	0.80
stdev	2	0.76				0.26				0.31			
min	4	0.94		0.00	0.20	0.29		0.04	0.13	0.30		26.3	0.20
max	10	3.88		4.18	0.95	1.24		31.86	0.94	1.48		230.7	0.98

Exponential:

		\multicolumn{3}{Log[Hg] vs length}			\multicolumn{3}{Log[Hg] vs weight}			\multicolumn{3}{Log[Hg] vs age}	
		slope	intcpt	10^{intcpt}	r^2	slope	intcpt	10^{intcpt}	r^2	slope	intcpt	10^{intcpt}	r^2
mean		0.038		47.47	0.76	0.00068		192.1	0.66	0.29		137.1	0.74
median		0.034		35.67	0.78	0.00038		157.0	0.66	0.27		132.0	0.79
stdev		0.016				0.00065				0.14			
min		0.022		1.82	0.19	0.00021		26.9	0.11	0.10		27.3	0.18
max		0.083		129.90	0.97	0.00270		410.4	0.91	0.73		290.0	0.99

Linear:

		\multicolumn{3}{[Hg] vs length}			\multicolumn{3}{[Hg] vs weight}			\multicolumn{3}{[Hg] vs age}	
		slope	intcpt	10^{intcpt}	r^2	slope	intcpt	10^{intcpt}	r^2	slope	intcpt	10^{intcpt}	r^2
mean		16.1	-509		0.74	0.28	136		0.69	115	-13		0.75
median		10.6	-381		0.78	0.17	126		0.73	88	-27		0.79
stdev		13.7	476			0.44	124			79	109		
min		3.2	-1770		0.14	0.05	-51		0.06	27	-187		0.15
max		61.7	8		0.92	2.11	362		0.99	292	282		0.99

[a]Abbreviations and units: n = number of fish; intcpt = intercept; [Hg], ng Hg/g; length, cm; weight, g; age, yr.

157

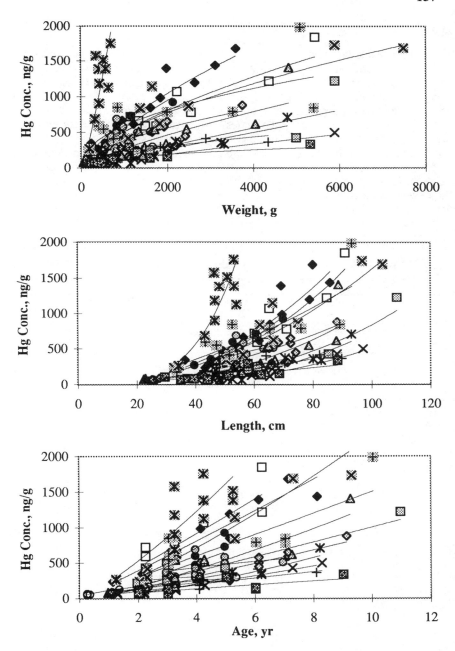

Figure 3. Northern pike fillet mercury contamination concentrations (ppb, ww) vs. fish growth measures for 20 Minnesota lakes. Individual lake trend lines (log-log) are shown.

Definition of Standard Fish Sizes for Contamination Comparisons

Once the best model for describing the relationship between mercury concentrations in fillets as a function of fish size and age measurements has been selected (log-log), then the question of how best to describe or represent the status of mercury contamination in a given lake and across lakes may be asked. Should mercury contamination concentrations be compared for the same length, same weight, or same age fish across lakes or other water bodies? An example of each case for two standard size selections are presented in Table II for the 20 lakes studied. The ages 3 and 6 yr were selected to be most representative of the data set, and computed the length and weight of fish (power regression) for each lake having these ages, and averaged their values. These average length and weight values were then used with the individual lake's regression equation to compute the mercury content for a fish of that "standard" size. The results of the calculations are given in Table II and show significant variations between each of the three growth characteristics.

Table II. Calculated Mercury Concentrations (ng Hg/g, ww, Fillet) and Methylmercury BAFs for N. Pike Using Power Regression Equations

regression description	standard size or age	[Hg] statistics (20 regressions for 20 lakes) for standard size or age (ng/g)					coef. of var. (%)
		average	median	stdev	min	max	
[Hg] vs length	53.9 cm	359	233	353	93	1710	98
[Hg] vs length	73.6 cm	796	456	1140	209	5490	143
BAF vs length	53.9 cm	2340	2120	1960	616	9990	84
BAF vs length	73.6 cm	5136	3783	6511	1109	32095	127
[Hg] vs weight	986 g	440	252	609	100	2930	138
[Hg] vs weight	2688 g	1070	527	2160	212	10100	202
BAF vs weight	986 g	2840	2220	3480	660	17100	123
BAF vs weight	2688 g	6840	3870	12500	1130	59300	183
[Hg] vs age	3 yr	321	257	202	107	862	63
[Hg] vs age	6 yr	666	505	471	191	2020	71
BAF vs age	3 yr	2140	1990	1030	569	5040	48
BAF vs age	6 yr	4380	3570	2480	1020	11800	57

A popular choice in the past has been to select a standard length (8, 20) to represent the basis of comparison between different sites or different times. However, the results of this study indicate that age may be the better choice. A correlation analysis among calculated mercury concentrations in standard size fish based on length, weight, and age shows that all three indices are related and show the same general trends with correlation coefficients ranging from 0.78 to 0.92. A better test, however, would be to have an independent measure of a parameter related to mercury exposure against which to test the calculated mercury concentrations in standard size or age fish.

Measurements of other exposure levels of mercury contamination are concentrations in water and in plankton. Such measurements are quite independent of fish measurements, and were made on plankton as part of the eighty lake study (10) and on lake water for methylmercury as part of this analysis. The results (using untransformed data) show that standard fish mercury concentrations based on fish age (3, 6 yr) correlate stronger with both plankton mercury concentrations (r = 0.7, 0.6) and methylmercury water concentrations (r = 0.5, 0.5) than do calculated fish mercury concentrations based on a standard fish length or weight. It is also interesting to note that the higher methylmercury correlations for plankton compared to water are consistent with ingestion as the primary exposure route for fish in these lakes.

Mercury Bioaccumulation Dependency

Early studies of waterborne toxic, bioaccumulative, persistent chemical pollutants focused on using the simplest model for representing a fish in aquatic ecosystems. This "bag of oil" fish model gave rise to numerous but similar approaches of considering all pollutant uptake as an "equilibrium partitioning" event, calculating "bioaccumulation factors" (BAF [dimensionless, and equivalent to ml/g or L/kg], the ratio of pollutant concentration in fish [ng/g] to that in water [ng/g] resulting from environmental exposure) and interpreting BAFs as constants related to partition coefficients between water and oil (hexane and octanol, most commonly). Little or no consideration has been given to pharmacological or metabolic mechanisms that actually control the uptake of methylmercury in fish and incorporate substantial concentrations in the muscle tissue of fish (1). The BAF approach is typically used for organic pollutants, and has been recently extended to organic pollutants in Great Lakes sediments (21). A more recent modeling concept involves the kinetic approach where trophic transfer of the contaminant is considered and bioaccumulation can be more accurately described (22). However, much work is needed to evaluate model coefficients.

To evaluate the functional dependency of methylmercury bioaccumulation factors on fish size and age, methylmercury concentrations were measured in the lake water from which the fish were sampled. Since the most meaningful BAF for mercury would involve using the same chemical species present in water as is present in fish muscle tissue (15), total methylmercury concentrations are the measurement of choice to be used to evaluate the nature of the BAF dependency with fish size and age. Methylmercury BAFs were calculated by dividing each fish mercury concentration by the lake water total methylmercury concentration. Sample results are presented in Table III for the standard sized fish. Comparisons between the two standard sizes immediately show the trend toward larger BAFs for older (larger) fish for each of the 20 study lakes. This observation is consistent with slow elimination rates of methylmercury yielding longer equilibration times for a kinetic model (22). In addition, it is also consistent with the idea that larger fish are exposed to larger doses of mercury because they consume larger and more contaminated prey.

The BAFs for each lake were regressed (using log-log transformed data) against fish length, weight, and age. The resulting equations are presented in Table III. Since each of the fish concentration values within a given lake are divided by the

Table III. Methylmercury BAFs Dependence on N. Pike Growth: Length, Weight, and Age Relationships[a]

Lake Name	Log(BAF) vs Log(length) slope	intcpt	10^{intcpt}	r^2	Log(BAF) vs Log(weight) slope	intcpt	10^{intcpt}	r^2	Log(BAF) vs Log(Age) slope	intcpt	10^{intcpt}	r^2
Bearhead	2.10	2.87	735	0.72	0.74	4.37	23180	0.66	0.67	6.00	991000	0.61
Big	1.69	3.53	3390	0.81	0.56	4.82	66800	0.83	0.55	6.17	1480000	0.93
Blackduck	1.80	3.06	1150	0.70	0.55	4.57	37300	0.63	1.25	5.46	288000	0.78
Crane	2.33	2.27	187	0.86	0.67	4.33	21400	0.87	1.06	5.95	883000	0.81
Harriet	1.72	3.27	1880	0.71	0.37	5.18	150000	0.41	0.30	6.22	1660000	0.59
Hendricks	3.88	-0.74	0.2	0.75	1.20	2.47	294	0.80	1.42	5.60	397000	0.98
Julia	2.89	1.07	12	0.82	0.96	3.21	1610	0.78	1.48	5.39	248000	0.91
Kabetogama	2.72	1.08	12	0.91	0.86	3.24	1750	0.94	0.95	5.53	341000	0.91
Kjostad	2.33	2.32	210	0.90	0.77	4.08	12100	0.82	1.15	5.65	445000	0.36
Lt.Vermilion	2.19	2.63	424	0.86	0.70	4.36	22800	0.86	1.06	5.93	858000	0.95
Mazaska	0.94	4.37	23300	0.87	0.29	5.11	130000	0.88	0.82	5.66	452000	0.87
Moose	1.26	4.22	16600	0.20	0.38	5.29	194000	0.13	0.45	6.15	1410000	0.20
Mukooda	2.63	1.66	46	0.95	0.91	3.46	2900	0.78	1.06	5.94	866000	0.97
Pelican	1.81	3.32	2090	0.75	0.56	4.82	66300	0.68	1.14	5.74	544000	0.59
Rainy	2.48	2.06	115	0.57	0.78	4.08	12100	0.60	1.28	5.83	678000	0.76
Sand Point	2.39	2.31	206	0.83	0.75	4.23	17000	0.79	1.19	5.99	968000	0.91
Stevens	1.66	3.48	3040	0.78	0.48	4.94	87400	0.73	1.08	5.72	523000	0.75
Tooth	3.74	0.53	3.4	0.80	1.24	3.53	3380	0.69	1.23	6.12	1310000	0.79
Washburn	1.55	3.53	3370	0.88	0.51	4.71	51200	0.86	1.02	5.67	471000	0.91
Wirth	1.33	3.57	3720	0.62	0.42	4.61	40700	0.63	0.84	5.35	226000	0.67
average	2.17		3026	0.76	0.68		47100	0.72	1.00		751800	0.76
median	2.15		580	0.81	0.69		22960	0.78	1.06		611400	0.80
stdev	0.76				0.26				0.31			
min	0.94		0.2	0.20	0.29		294	0.13	0.30		226000	0.20
max	3.88		23300	0.95	1.24		194000	0.94	1.48		1660000	0.98

[a]Abbreviation: intcpt = intercept.

methylmercury concentration for that lake, the slopes and r^2 are the same as for regressions using the fillet mercury concentrations vs. fish growth measures. The individual plots of methylmercury BAFs vs. northern pike growth are presented in Figure 4 where it is observed that the methylmercury BAFs increase with increasing fish growth measure and are more linear with fish age than with fish length or weight.

It is interesting to compare coefficients of variations (CV = std. dev./mean), given in Table II, for mercury concentrations and BAFs that have been normalized to standard sized fish. The first observation is that the lowest CVs (e.g. 63% for Hg conc. and 48% for BAF at age 3) occur for the aged based normalizations. This further supports the evidence that age is a better choice than length or weight for modeling mercury accumulation in fish. The second observation is that the coefficient of variation for methylmercury BAFs is lower than that for mercury concentration. This clearly indicates that methylmercury concentrations in water have some explanatory power regarding the variability of mercury in fish tissue.

The importance of fish mercury concentration and mercury BAF size and age relationships are illustrated by the following comparisons. The average mercury concentration was calculated using all sampled fish (n = 173) for the 20 study lakes by two methods: 1) a simple arithmetic average of the mercury concentrations [equivalent to assuming no size dependency], and 2) a mass weighted mean [equivalent to a composite sample of all fish; the proper method for the case of size dependency]. The results indicated that the simple arithmetic mean mercury concentration was 469 ng Hg/g, compared with the mass weighted mean concentration, 681 ng Hg/g, a 45% higher concentration and body burden in the overall fish population. Methylmercury BAFs for 10 yr old fish are, on average, 10 times those for 1 yr old fish. Thus, the earlier concept of assuming no size or age dependency of fish mercury residues and BAF's (3, 4) may have resulted in impact overestimates for younger fish and underestimates for older fish from mercury contamination in aquatic ecosystems.

Acknowledgments

The regional staff of the Minnesota Department of Natural Resources played a key role in the success of this project by their cooperation and application of their extensive experience in obtaining the necessary specimens required to build the foundation for this bench-mark study. We thank L. Kallemeyn of the VNP for help in fish sampling and aging, K. Schmidt and T. Pierce for fish analyses; for water sampling and/or fish processing S. Cummings, R. Hedin, L. Heinis, C. Lippert, and R. Rude; J. Reed, D. Busick, and D. Sorensen for assistance in data base and manuscript preparation; and the USEPA for salary (GEG), space, and laboratory equipment support for this project. Financial support was appropriated by the Minnesota Legislature, (ML 95 Chp. 220, Sect. 19, Subd. 5(g) and ML 96 Chp. 407, Sec. 8, Subd. 9), from the Future Resources Fund and Great Lakes Protection Account, as recommended by the Legislative Commission on Minnesota Resources. The identification of commercial products is for descriptive and documentation purposes only and does not constitute endorsement. The contents of this manuscript do not necessarily reflect the official views of the USEPA, and no official endorsement should be inferred.

162

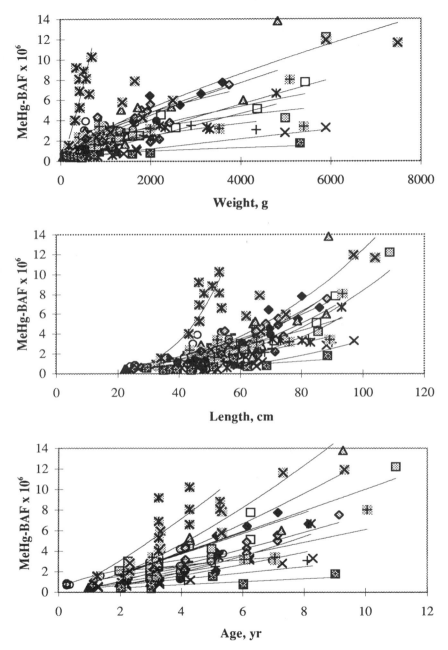

Figure 4. Northern pike methylmercury bioaccumulation factors vs. fish growth measures for 20 Minnesota lakes. Individual lake trend lines (log-log) are shown.

Literature Cited

1. *Mercury Study Report to Congress*. Volume VI: An ecological assessment for anthropogenic mercury emissions in the United States. Office of Air Quality Planning and Standards/Office of Research and Development, USEPA 1997, http://www.epa.gov/oar/mercury.html
2. *Minnesota Fish Consumption Advisory*. Minnesota Department of Health, Minneapolis, MN. 1999, 96 pp.
3. *National Study of Chemical Residues in Fish*. EPA 823-R-92-008a, b, Vols. I, II. USEPA, Office Science and Technol. Washington, D.C. 20460 Sept. 1992
4. *Final Water Quality Guidance for the Great Lakes System*. USEPA Final Rule, 40 CFR Pts. 9, 122, 123, 131, 132. [FLR-5173-7] RIN 2040-AC08 Fed. Reg. Vol. 60, No. 56 pgs. 15366-15425 Thurs. Mar. 23, 1995.
5. Fagerstrom, T., Asell, B., Jernelov, A. *Oikos* **1974**, 25,14-20.
6. Olsson, M. *Ambio* **1976**, 5, 73-76.
7. Barber, R. T.; Whaling, P. J. *Environ. Sci. & Technol.* **1984**, 18, 552-555.
8. Sorensen, J. A., Glass, G. E., Schmidt, K. W., Huber, J. K., and Rapp, G. R., Jr. *Environ. Sci. Technol.* **1990**, 24, 1716-1727.
9. Hakanson, L. *Environ. Pollut.* Ser. B 1, **1980**, 285-304.
10. Glass, G. E., Sorensen, J. A., Rapp, Jr., G. R. Mercury Deposition and Lake Quality Trends, Project I-11/I-15, Legislative Commission on Minn. Resources. St. Paul, MN 1998, pp 150.
11. *Manual of Instructions for Lake Survey* Special Publication No. 147. Minnesota Department of Natural Resources, St. Paul, MN 1993.
12. *Determination of Metals in Environmental Samples* EPA 600/4-91-010, USEPA, Cincinnati, OH, 1991.
13. *Methods for the Determination of Metals in Environmental Samples*. EPA/600/R-94/111. Supplement I to EPA 600/4-91-010, USEPA. Cincinnati, OH, 1994.
14. Glass, G. E.; Sorensen, J. A.; Schmidt, K. W.; Rapp, Jr., G. R. *Environ. Sci. & Technol.* **1990**, 24, 1059-1069.
15. Bloom, N. S. *Can. J. Fish. Aquat. Sci.* **1992**, 49, 1010-1017.
16. Watras, C.J.; Bloom, N. S. *Limnol. Oceanogr.* **1992**, 37, 1313-1318.
17. Liang, L.; Horvat, M.; Bloom, N. S. *Talanta*, **1994**, 41, 371-379.
18. Method 1631: Mercury in Water by Oxidation, Purge and Trap, and Cold Vapor Atomic Fluorescence Spectrometry. EPA 821-R-95-027. Office of Water, Engineering. and Analysis Div., USEPA, Washington, D. C. 20460, 1995.
19. Glass, G. E.; Sorensen, J. A.; *Can. J. Fish. and Aquat. Sci.* **1999**, in prep.
20. Parks, J.W.; Craig, P.C.; Ozburn, G.W. *Can. J. Fish. and Aquat. Sci.* **1994**, 51, 2090-2104.
21. Cook, P. M.; Burkhard, L. P. Development of bioaccumulation factors for protection of fish and wildlife in the Great Lakes EPA 823-R-98-002 Office of Water, USEPA, Washington, DC. 1998, 3, 19-27.
22. Reinfelder, J.; Fisher, N.; Luoma, S.; Nichols, J.; Wang, W. *Science of the Total Environ.* **1998**, 219, 117-135.

Chapter 12

Selective Persistence and Bioaccumulation of Toxaphene in a Coastal Wetland

Keith A. Maruya[1], Walter Vetter[2], Stuart G. Wakeham[1],
Richard F. Lee[1], and Leo Francendese[3]

[1]Skidaway Institute of Oceanography,
10 Ocean Science Circle, Savannah, GA 31411
[2]Friedrich–Schiller Universitat Jena, Dornburger Str. 25, Jena, Germany
[3]U.S. Environmental Protection Agency, Region 4, Atlanta Federal Center,
61 Forsyth Street, Atlanta, GA 30303

Surface sediment, crustacean (*Palaemonetes* sp.) and fish (*Fundulus* sp.) from an estuarine marsh that received discharge from a toxaphene production facility contained several chlorobornanes (CHBs) that are predicted to be recalcitrant based on their chlorine substitution pattern. The two predominant CHBs in all samples were 2-*exo*,3-*endo*, 6-*exo*,8,9,10-hexachlorobornane (B6-923 or Hx-Sed) and 2-*endo*,3-*exo*,5-*endo*,6-*exo*,8,9,10-heptachlorobornane (B7-1001 or Hp-Sed), considered to be reductive dechlorination products of prominent components of technical toxaphene (CTTs). Also prominent were several Cl_7-Cl_9 bornanes, including B8-1414/1945, B8-2229, B8-1413, and B9-1679 and at least 2 unidentified heptachlorobornanes. The CHB profiles and estimated concentrations in *Palaemonetes* and *Fundulus* corresponded to that of matched sediment, indicating that residues of toxaphene associated with contaminated sediment are taken up and accumulated by these abundant, estuarine food web organisms.

Toxaphene is the trade name coined by Hercules Inc. for their chlorinated hydrocarbon biocide, a complex mixture of polychlorinated monoterpenes that was produced at their Brunswick, Georgia, USA plant from 1948-1980. Toxaphene was produced by reacting chlorine gas with technical camphene (a derivative of α-pinene extracted from pine resin) in the presence of ultraviolet radiation and catalysts. The resulting formulation was composed mostly of chlorobornanes (CHBs), compounds with 5 to 11 chlorine atoms on a saturated C10 bicyclic hydrocarbon skeleton (*1*). The average chlorine content of toxaphene was 68-70%. The primary commercial application for toxaphene in the U.S. was as an agricultural pesticide, particularly for cotton production in the southern states during the 1960s and 70s. Due to environmental concerns, the use of toxaphene was severely restricted in the U.S. in 1982 and was banned for all registered uses in 1990 (*2*). Several "toxaphene like"

© 2001 American Chemical Society

mixtures (e.g. Strobane, Melipax) have also been produced both in the U.S. and abroad. The world production of toxaphene and its relatives was recently estimated at well over 1 million metric tons (*3*).

Due to their high degree of chlorination, original components of technical toxaphene (CTTs) are poorly soluble in water and thus can be expected to accumulate in soils, sediments and biota, particularly in those that are rich in organic matter and/or lipids. However, it is increasingly apparent from samples worldwide that only certain residues of toxaphene persist, or at the very least are highly enriched, in the environment (*4*). For example, three CHBs (B8-1413 or Parlar 26, B9-1679 or Parlar 50 and B9-1025 or Parlar 62) were reported to comprise much greater than 50% of total CTTs in a composite marine food fish sample from northwestern Europe (*5*). Similarly, accumulation of these same CHBs were noted in animals high on the Arctic food web (*6*).

In contrast, profiles of CHBs were found to be much more complex in media from toxaphene-treated lakes (*7*) or those in close proximity to known point sources, such as the Terry/Dupree tidal creek system in Brunswick (8, *this study*) and Lake Nicaragua, the site of another Hercules toxaphene plant (*9*). In this study, we characterized the profiles and computed estimated concentrations of toxaphene residues in sediments and biota from the Brunswick site, focusing on the identification of prominent CHBs in these samples and a comparison of source (sediment) and receptor (marsh biota) profiles/levels.

Materials and Methods

Study Site — Terry/Dupree Creek Tidal Marsh, Brunswick, Georgia, U.S.A.

Dupree Creek is a short (~2 km) tidal creek that drains a small (~5 km^2) coastal wetland characterized by emergent, salt marsh macrophytes (*Spartina and Juncus* spp.) and a large tidal range (up to 3 m). This marsh is typical of the southeastern U.S. coastline and is home to a variety of invertebrates, fish, marine and fish-eating terrestrial mammals, and birds. Dupree Creek feeds into Terry Creek approximately 100 m from the Hercules plant cooling water discharge canal, the original source of toxaphene contamination in the marsh. Terry Creek is connected to St. Simons Sound, an coastal embayment of the Atlantic Ocean. Maintenance dredging in Terry and Dupree Creeks has occurred periodically over the years, with much of the dredged sediment deposited onto islands within the marsh complex.

Sample Collection and Processing

Sediment, grass shrimp (*Palaemonetes* sp.) and composites of whole fish (*Fundulus* sp.) were collected during November and December 1997. Sediments were collected to a depth of ~15 cm with a clean stainless steel spoon at low tide at 6 stations from the head of Dupree Creek to the confluence of Terry Creek with the

Back River/St. Simons Sound (Figure 1). The distance between the discharge canal (Station 2) and the Back River site (Station 6) is ~2 km. Sediment grab samples were placed in a clean glass bowl, mixed thoroughly, placed into precleaned glass jars and immediately stored in ice-filled coolers. *Fundulus* were collected using baited minnow traps located within 0.25 m of where sediment was collected at 3 of the 6 stations. *Palaemonetes* were collected at all 6 stations using a dip net. After returning to the boat dock, all fish were transferred to plastic buckets containing aerated, clean seawater and held overnight (~12 hrs). Composites of whole *Palaemonetes* and *Fundulus* were wrapped in clean aluminum foil or placed in clean glass jars and stored on ice. Upon return to the lab, iced samples were placed in a freezer.

Ten gram aliquots of freeze dried sediment were ground with kiln-fired Na_2SO_4 and extracted in a Soxhlet apparatus with 400ml CH_2Cl_2 for $\geq$ 16 hrs. The resulting CH_2Cl_2 extract was reduced in volume over a heated water bath using a Kuderna-Danish type concentrator and exchanged to hexane. This extract was then treated for elemental sulfur using acid-activated copper powder until no characteristic black precipitate (CuS) was observed. The Cu-treated hexane extract was applied to a glass column slurry-packed with 2.0g of silica gel and three fractions of increasing polarity were collected. After pre-separation of nonpolar analytes including PCBs with 50-70 ml hexane (fraction 1 or F1), toxaphene components were eluted in the second (F2) fraction with 150 ml 20% (v:v) CH_2Cl_2:hexane. Each fraction was reduced and exchanged to hexane using a TurboVap II (Zymark Inc., Hopkington, MA). A 40 mg aliquot of freeze-dried sediment was analyzed for TOC after decarbonization with 2.0 M HCl using a Fisons NA1500 Elemental Analyzer.

Thawed composites of *Palaemonetes* (20-30 individuals) and *Fundulus* (3-5 individuals) from each station were wet homogenized with kiln-fired Na_2SO_4 and Soxhlet extracted as previously described for sediments. Individual *Fundulus* specimens ranged in size from "small" (4.0 cm; 1.0 g) to "large" (8.2 cm; 8.8 g). No attempt was made to measure or size-segregate individual grass shrimp. After concentration, tissue extracts were evaporated to dryness in a fume hood. The resulting residue was weighed on a microbalance to the nearest 0.1 mg as an estimate of extractable lipid. This residue was re-dissolved in hexane and applied to a glass column packed with 18.0 g of Florisil, activated/deactivated as described previously (*10*). After elution of PCBs with hexane (F1), CHBs were captured in the F2 using 150 ml 20% (v:v) CH_2Cl_2:hexane.

Instrumental Analysis

Sediment and tissue extracts were analyzed on three separate instruments: (1) a Varian 3400CX gas chromatograph with dual electron capture detectors (GC-ECD); (2) a Hewlett-Packard (HP) 5890 Series II gas chromatograph coupled to a Finnigan INCOS 50 single quadrupole mass spectrometer; and (3) a HP 5890 Series II GC coupled to a HP 5989B MS Engine. The operating conditions for the GC-ECNIMS analyses are described in detail elsewhere (*11, 12*). Individual congener and total toxaphene (ΣTOX) concentrations were determined using GC-ECD as follows.

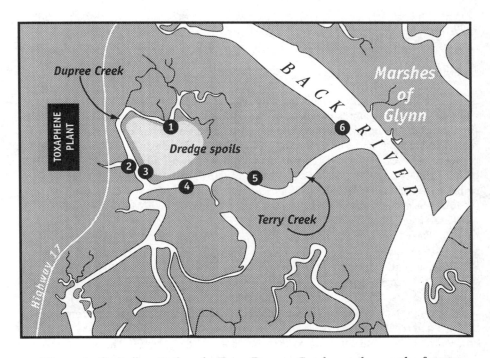

Figure 1. Sampling stations in Terry/Dupree Creek marsh near the former toxaphene plant in Brunswick, Georgia, USA.

Congener-specific concentrations were based on 5 point calibration curves (2-100 pg) of the 22 component "TM2" and 4 component "BgVV4" standard solutions (Dr. Ehrenstorfer, Augsburg, Germany and Promochem, LLC, Laramie, WY, USA, respectively). TM2 contains Cl_6-Cl_{10} homologs including all components in BgVV4 - - B8-1413 (P26), B7-515 (P32), B9-1679 (P50) and B9-1025 (P62). ΣTOX was estimated by integrating a 30 min window that corresponded to the complex series of peaks observed for injections of technical toxaphene (TTX) product standard (supplied by J. Hoffman of Hercules Inc., Wilmington, DE, USA) diluted in hexane, applying the total response factor based on known total mass of TTX injected, and subtracting peak areas that co-eluted with other persistent organochlorine compounds such as DDTs and chlordanes.

Quality Control

Procedural blanks, spiked matrices, and replicate samples were analyzed using the same procedures outlined above. Procedural blanks consisted of ~50g kiln-fired Na_2SO_4. Intertidal surface sediment and muscle of red drum collected from reference locations and fortified with TTX and BgVV4 were analyzed as matrix spike recovery samples. The GC-ECD was monitored for sensitivity and retention time drift by injecting a mid-level BgVV4 standard every 10-15 samples. GC-MS systems were calibrated daily using difluorotriphenylphosphine (DFTPP) and perfluorokerosene (PFK). Mass sensitivity and resolution criteria in accordance with EPA Method 525 were used to monitor instrument performance.

Based on GC-ECD analysis, the recoveries of B8-1413, B7-515, B9-1679 and B9-1025 spiked into red drum muscle were 90, 93, 98, and 100%, respectively. Similarly, for spiked sediment, recoveries were between 90-100%. The recovery of spiked TTX ranged between 70-80% for tissue and sediment. B8-1413 typically eluted in the first fraction (F1) apparently due to its relative nonpolarity whereas the other 3 components in BgVV4 eluted in the second (F2) fraction. Mean recoveries of DBOFB and α-HCH added as F1 and F2 surrogates, respectively, were 78% and 88%. Procedural blanks indicated no toxaphene like contamination. The mean percent difference for estimated ΣTOX, B6-923 and B7-1001 concentrations in duplicate samples of site sediment, *Palaemonetes* and *Fundulus* composite samples were 8.9, 24 and 17%, respectively.

Results and Discussion

Water content for sediments was fairly uniform at $71 \pm 1.8\%$ (range: 69-74%). Total organic carbon (TOC) ranged between 4.4-6.5% ($x \pm \sigma$: $5.0 \pm 0.71\%$) on a dry weight basis (Table I). These parameters are typical of fine-grained salt marsh muds. The sediments from the discharge canal (station 2) may be slightly enriched in TOC, as might be expected since the effluent is generally warmer and of lower salinity than the ambient creek water. Lipid content on a wet weight basis was $1.9 \pm 0.39\%$

169

Table I. Estimated toxaphene concentrations[a] (ng/g) in Terry/Dupree Creek sediments and biota.

Station	Matrix	%TOC or LIPID	ΣTOX	Hx-Sed	Hp-Sed	P31/32	P40/41	P44	P51	P62
1	Sediment	4.4	3900	147	190	15.5	16.8	41.8	40.4	21.4
	Shrimp	2.6	1200	61.7	75.4	6.37	10.5	39.1	14.6	11.1
	fish (n=3)	2.6	4500±1200	202±26	179±22	12.1±2.3	26.6±4.2	70.6±13.3	38.8+3.7	≤11.2
2	Sediment	6.5	8100	188	279	48.2	66.6	139	125	64.6
	shrimp (n=2)	2.0	1400±90	38.9±7.3	38.9±4.3	8.25±1.61	11.4±2.2	39.3±7.0	20.4±3.0	13.5±1.2
	Fish	3.5	21000	235	268	76.2	135	450	275	127
3	Sediment	5.0	4000	108	158	22.9	29.6	63.4	54.9	31.8
	Shrimp	2.0	730	20.2	17.8	2.60	4.41	21.5	9.29	12.9
4	Sediment	5.3	2000	103	128	6.28	8.94	15.7	17.4	8.42
	Shrimp	1.9	520	20.4	18.0	2.41	3.41	13.3	4.55	4.26
5	Sediment	4.6	1700	81.4	86.8	6.23	8.82	19.9	16.6	13.9
	Shrimp	1.4	150	7.40	6.99	<0.32	0.45	2.38	0.81	0.60
6	Sediment	4.5	840	62.9	56.6	1.94	3.52	6.04	5.90	2.19
	Shrimp	1.5	≤100	2.84	2.97	≤19.3	0.13	1.16	2.66	≤22.9
	Fish	2.4	≤570	52.1	48.3	≤1.06	3.38	9.77	4.50	≤1.31

[a] sediment – dry wt.; tissue – wet wt.; < not detected; ≤ detected but less than method detection limit

(range: 1.4-2.6%) and 2.5 ± 0.44% (range: 2.0-3.5%) for *Palaemonetes* and *Fundulus*, respectively (Table I). Lipid content appeared to decrease from the head of Dupree Creek (station 1) out to the confluence of Terry Creek and the Back River (station 6).

Chlorobornane Identification and Estimated Concentrations

In all sediment, *Palaemonetes* and *Fundulus* extracts, a single Cl_6, several Cl_7 and Cl_8, and up to three Cl_9 CHBs were prominent in GC-ECNIMS chromatograms. This typical CHB pattern is illustrated for sediment from station 1 in Figure 2. Several unidentified major and minor chlorobornanes, particularly in the Cl_6-Cl_8 homologous series, were also detected. Two of the more abundant compounds were B6-923 (Hx-Sed), and B7-1001 (Hp-Sed), both thought to be reductive dechlorination metabolites of major CTTs (*13, 14*). The most prominent Cl_8 peak has been identified as the co-eluting pair B8-1414/B8-1945 (Parlar 40/41). The prominent Cl_9 compound was identified as B9-1679 (Parlar 50). Also confirmed by independent GC-ECNIMS analysis of these extracts were B7-515 (Parlar 32), B8-1413 (Parlar 26), B8-2229 (Parlar 44), B8-806/809 (Parlar 42a/b), B8-786 (Parlar 51), B8-1412 and B9-1025 (Parlar 62). Structural formulas and other names for these congeners are given in Table II.

Individual and total toxaphene concentrations estimated from GC-ECD analyses are summarized in Table I. As expected, total and individual toxaphene concentrations in sediment and biota were highest in the discharge canal (station 2) and decreased with distance away from the source. The levels in samples from the upper reach of Dupree Creek (station 1) were also relatively high; this is likely due to toxaphene in leachate/runoff that drains the upland area containing dredged materials (Figure 1). Individual and ΣTOX concentrations in biota also mirrored those in sediment, indicating that local sediments were the predominant source of available toxaphene residues in these organisms. The general hierarchy for individual CHB concentrations was B7-1001 ~ B6-923> B8-2229 ~ B8-786 > B7-515 ~ B9-1025. The predominance of B7-1001 and B6-923 in our samples was consistent with the pattern seen in sediments and fish from toxaphene-treated Canadian lakes (*7, 14*). B8-1413 and B9-1679 were found to be the 2 major residue components in marine mammals (*17*) and fish (*18*). Unfortunately, co-eluting interferences (other CHBs) prevented accurate estimation of their concentrations in our samples. Maximum concentrations of B8-1413 and B9-1679 were however, less than either Hx- or Hp-Sed. Other prominent CHBs in our samples, including B8-1414, B8-1945, B8-2229 and B9-1025, have also been reported in biological tissue, e.g. Antarctic seal blubber (*11*).

Selective Recalcitrance and Bioaccumulation

Explanations for the apparent recalcitrance of chlorobornanes indicated herein have invoked the "bridge and exo" rule and the stability of a given

Mummichogs

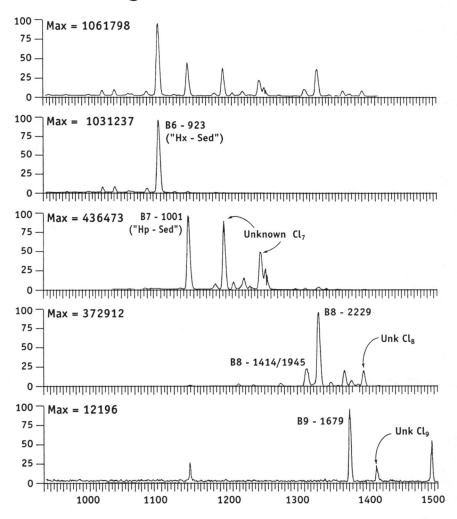

Figure 2. GC-ECNIMS profile of Cl$_6$-Cl$_9$ bornanes in mummichogs (Fundulus sp.). Note the prominence of B6-923, B7-1001 and several unidentified chlorobornanes.

Table II. Names and structural formulas for prominent chlorobornanes.

IUPAC	Andrews & Vetter (15)	Parlar (16)	Others (2,14)
Hexachlorobornanes			
2-exo,3-endo,6-exo,8,9,10-	B6-923		Hx-Sed
Heptachlorobornanes			
2-endo,3-exo,5-endo,6-exo-	B7-1001		Hp-Sed
2,2,5-endo,6-exo,8,9,10-	B7-515	32	tox B
Octachlorobornanes			
2-endo,3-exo,5-endo,6-exo,8,8,10,10-	B8-1413	26	T2, TOX8
2-endo,3-exo,5-endo,6-exo,8,9,10,10-	B8-1414[a]	40	
2-exo,3-endo,5-exo,8,9,9,10,10-	B8-1945	41	
2,2,5-endo,6-exo,8,8,9,10-	B8-806	42a	tox A$_1$
2,2,5-endo,6-exo,8,9,9,10-	B8-809	42b	
2-exo,5,5,8,9,9,10,10-	B8-2229	44	
2,2,5,5,8,9,10,10-	B8-786	51	
Nonachlorobornanes			
2-endo,3-exo,5-endo,6-exo,8,8,9,10,10-	B9-1679	50	tox A$_c$, T12, TOX9
2,2,5,5,8,9,9,10,10-	B9-1025	62	

[a] co-elutes with B8-1945

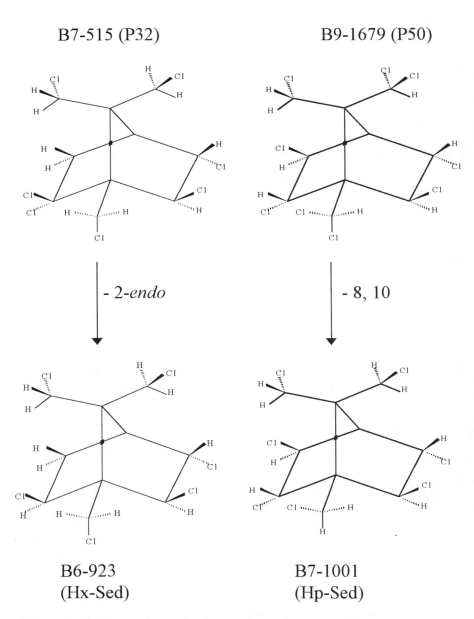

B7-515 (P32)

B9-1679 (P50)

- 2-*endo*

- 8, 10

**B6-923
(Hx-Sed)**

**B7-1001
(Hp-Sed)**

Figure 3. Common structural features of prominent, recalcitrant chlorobornanes (B6-923 and B7-1001) and possible reductive dechlorination pathways of major components of technical toxaphene (B7-515 and B9-1679).

conformation based on chlorine substitution patterns and ring strain (*4*). For example, the structure of B7-1001 reveals (i) alternating exo-endo chlorines on the 6-membered carbon skeleton and (ii) no geminal dichlorosubstituents (Figure 3). These features are apparent in other congeners prominent in our samples, including B6-923. Also, removal of geminal dichloro substituents from carbon positions 2, 8, 9 or 10 from several "precursor" CTTs in our samples (e.g. B7-515 or B9-1679) would explain the formation of B7-1001 and/or B6-923 (Figure 3). The apparent accumulation of B6-923 and B7-1001 is consistent with additional evidence of in situ reductive dechlorination reported elsewhere (*12*) and indicates their stability under the highly reducing conditions associated with anaerobic salt marsh sediments.

Accumulation of toxaphene residues in Terry/Dupree Creek poses a potential ecological and human health risk that is presently unknown. The toxic potential of recalcitrant CHBs such as B6-923 and B7-1001 has not been well characterized due in part to the lack of purified standards. Our future work will focus on the identification of unknown congeners, further characterization of the sedimentary source of toxaphene in this marsh, and the primary uptake and elimination mechanisms for prominent CHB residues in target estuarine species.

Literature Cited

1. Saleh, M. A. *Toxaphene: Chemistry, Biochemistry, Toxicity and Environmental Fate*; Ware, G. W., Ed.; Springer-Verlag: New York, p 1-85.
2. Agency for Toxic Substances and Disease Registry (ATSDR). Toxicological Profiles for Toxaphene (update); U.S. Dept. Human Health Services: Atlanta, GA, 1996.
3. Voldner, E. C.; Li, Y. F. *Chemosphere* **1993**, *27*, 2073-2078.
4. Vetter, W; Scherer, G. *Chemosphere* **1998**, *37*, 2521-2539.
5. Alder, L.; Veith, B. *Fresenius J. Anal. Chem.* **1996**, *354*, 81-92.
6. Bidleman, T. F.; Walla, M. D.; Muir, D. C. G.; Stern, G. A. *Environ. Toxicol. Chem.* **1993**, *12*, 701-709.
7. Miskimmin, B. M.; Muir, D. C. G.; Schindler, D. W.; Stern, G. A.; Grift, N. P. *Environ. Sci. Technol.* **1995**, *29*, 2490-2495.
8. Maruya, K.A.; Francendese, L. Analysis of toxaphene residues in sediment and mummichogs (Fundulus sp.) from Terry/Dupree Creek, Final Report – Phase II. U.S. Environmental Protection Agency, Atlanta, GA, USA. October 5, 1999.
9. Calero, S.; Fomsgaard, I.; Lacayo, M. L.; Martinez, V.; Rugama, R. *Int. J. Environ. Anal. Chem.* **1993**, *53*, 297-305.
10. Maruya; K. A.; Lee, R. F. *Environ. Toxicol. Chem.* **1998**, *12*, 2463-2469.
11. Vetter, W.; Krock, B.; Luckas, B. *Chromatographia* **1997**, *44*, 65-73.
12. Maruya, K. A.; Wakeham, S. G.; Francendese, L. *Analysis of Toxaphene Residues in Sediment and Fundulus from Terry/Dupree Creek*, U.S. Environmental Protection Agency, Atlanta, GA, 1998.
13. Fingerling, G; Hertkorn, N; Parlar, H. *Environ. Sci. Technol.* **1996**, *30*, 2984-2992.

14. Stern, G. A.; Loewen M. D.; Miskimmin, B. M.; Muir, D. C. G.; Westmore, J. B. *Environ. Sci.Technol.* **1996**, *30*, 2251-2258.
15. Andrews, P.; Vetter, W. *Chemosphere* **1995**, *31*, 3879-3886.
16. Burhenne, J.; Hainzl, D.; Xu, L.; Vieth, B.; Alder, L.; Parlar, H. *Fresenius J. Anal. Chem.* **1993**, *346*, 779-785.
17. Vetter, W.; Scherer, G.; Schlabach, M.; Luckas, B.; Oehme, M. *Fresenius J. Anal. Chem.* **1994**, *349*, 552-558.
18. Alder, L.; Beck, H.; Khandker, S.; Karl, H.; Lehmann, I. *Chemosphere* **1997**, *34*, 1389-1400.

Chapter 13

Microbial Degradation of Polychlorinated Dibenzo-*p*-dioxins and Polychlorinated Dibenzofurans

John R. Parsons and Merel Toussaint

Department of Environmental and Toxicological Chemistry, ARISE, University of Amsterdam, Nieuwe Achtergracht 166, 1018 WV Amsterdam, The Netherlands

Polychlorinated dibenzo-*p*-dioxins (PCDDs) and polychlorinated dibenzofurans (PCDFs) are two of the most notorious classes of persistent, bioaccumulative and toxic chemicals [1]. They are produced inadvertently as trace contaminants during waste incineration, chemical syntheses and other industrial processes. The extremely high toxicity of the 2,3,7,8-chlorinated PCDDs and PCDFs has driven a lot of research on the sources, environmental distribution and effects of these compounds. However, comparatively little attention has been paid to the microbial degradation, or biodegradation, of these compounds, despite the fact that this could potentially have an important impact on their environmental fate. In fact, it is often assumed that PCDDs and PCDFs are not degraded at all by microorganisms.

A few researchers have, however, devoted efforts to investigating the potential for microbial degradation of PCDDs and PCDFs. This is of importance to understand the long-term fate of these compounds and may also may help to develop methods to remediate soils and sediment contaminated with these compounds. We here review the current state of knowledge concerning the microbial degradation of PCDDs and PCDFs, describing both studies of degradation in pure cultures of microorganisms aimed at elucidating the biochemistry of degradation, as well as studies of degradation under 'real world' conditions. Because of the fundamentally different reaction mechanisms involved, biodegradation by aerobic bacteria is discussed

© 2001 American Chemical Society

separately from that under anaerobic conditions. For a more extensive overview the reader is referred to a recently published book devoted to this subject [2].

Biodegradation of PCDDs and PCDFs under aerobic conditions

PCDDs and PCDFs are generally considered as xenobiotic compounds, i.e. they are not part of the natural environment. It may therefore be expected that their structure is foreign to the biosphere and that microorganism may therefore not have developed enzymes capable of attacking these compounds. However, similar ring systems do occur in a few natural products and dibenzofuran itself has been present as a constituent of oil tars for millions of years [3]. Furthermore, there is evidence that traces may have always been formed naturally, for example in forest fires [4]. Therefore it is not unlikely that enzymes able to degrade such structures have evolved.

Degradation of PCDDs and PCDFs in soils

In the 1970s, as a result of increasing concern about the effects of the most toxic PCDD congener, 2,3,7,8-TCDD, the fate of this compound was studied in soil. Kearney *et al.* [5] were the first to report the degradation of 2,3,7,8-TCDD in soil in 1972. About half the dioxin added could be recovered after one year and there was evidence for the formation of a polar metabolite as well as a small amount of CO_2 from the dioxin. Similarly slow degradation was also reported for 2,3,7,8-TCDD in freshwater sediment [6] and 1,3,6,8-TCDD in both soil and sediment [7]. In the latter case, traces of unidentified polar metabolites were formed in both soil and sediment.

The potential to mineralise dibenzo-*p*-dioxin (DD) is present in different agricultural soils contaminated with PCDDs and PCDFs [8]. Degradation of DD is stimulated by addition of amendment of the soil with compost or wheat straw and also by inoculation of the soil with lignolytic white-rot fungi. However, no data are given on the possible effects on the PCDDs and PCDFs present in the soil. Addition of the dibenzofuran (DF)- and DD-degrading bacterium *Sphingomonas* sp. RW1 to soil also stimulated the degradation of soil-spiked DD and DF [9]. Degradation was accompanied by a short-lived exponential increase in the cell density of this strain. The rates of degradation of DD, 2-CDD and DF in soil microcosms depends on the cell density in which strain RW1 was inoculated and on the organic matter content of the soil [10]. The pseudo-first-order rate constant for 2-CDD correlated inversely with soil organic matter and total organic carbon content.

Degradation in pure cultures of bacteria

The evidence that microbial degradation of 2,3,7,8-TCDD was possible in soil led to the first studies using pure cultures of bacteria. Several strains isolated from 2,3,7,8-TCDD-contaminated soil showed evidence for degradation [11]. A polar metabolite identified as a hydroxylated derivative was identified using GC-MS. An unidentified hydroxylated metabolite, as well as *cis*-1,2-dihydroxy-1,2-dihydrodioxin, were produced from dibenzo-*p*-dioxin (DD) by a naphthalene-degrading *Pseudomonas* strain [12]. A biphenyl-degrading *Beijerinckia* strain produced similar metabolites form DD and monochlorinated dioxins [13]. These strains are not able to use these dioxins as growth substrates.

The *Alcaligenes* sp. strain JB1 (later reclassified as a *Burkholderia* strain) also grows on biphenyl and transforms DD and dibenzofuran (DF) and some of their mono-, di- and trichlorinated derivatives [14,15]. Degradation rates were much higher in continuous cultures than in batch cultures and decreased with increasing chlorination. As was the case for the biphenyl-degrading strains mentioned above, strain JB1 also produced hydroxylated metabolites of the chlorinated dioxins, although the subsequent identification of a chlorotrihydroxydiphenyl ether and 4-chlorocatechol showed that the dioxin ring system was cleaved by this strain [16]. Similarly, degradation of 2-chlorodibenzofuran produced a chlorotrihydroxybiphenyl and 5-chlorosalicylic acid.

Equivalent metabolites are produced by strains which can grow on dibenzofuran, such as *Brevibacterium* sp. DPO 1361 [17], *Terrabacter* sp. DPO360 [18], *Sphingomonas* sp. HH69 [19] and *Sphingomonas* sp. RW1 [20]. The latter strain can grow on both DF and DD. The pathways these strains use to degrade DD and DF have been studied extensively and are shown in Figures 1 and 2. In both cases, degradation starts with a so-called angular dioxygenation producing an unstable hemiacetal, followed by cleavage of the ether bond to give a trihydroxybiphenyl or a trihydroxydiphenyl ether.

Mono- and dichlorinated dioxins and dibenzofurans are degraded by these strains, but not completely and with accumulation of chlorinated catechols and salicylates as they lack the specialised enzymes which are required to degrade these compounds efficiently [21]. More highly chlorinated congeners are not attacked. Mineralisation of 4-chlorodibenzofuran was demonstrated in a mixed culture consisting of *Sphingomonas* sp. strain RW1 and *Burkholderia* sp. strain JWS, which is able to grow on chlorosalicylates [22]. An extensive study in which several strains were used to study the influence of chlorine substitution pattern on the biodegradation of all 210 PCDD and PCDF congeners showed that those with five or more chlorines were not degraded [23]. In the case of the mono- to tetrachlorinated congeners, chlorination in the 1,4, 6 or 9 positions generally retarded degradation.

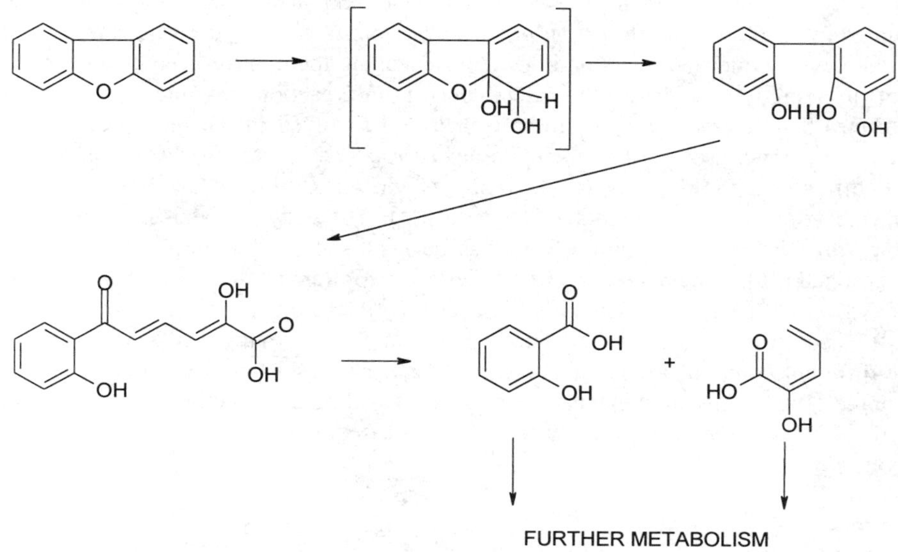

Figure 1. Microbial degradation pathway of dibenzofuran.

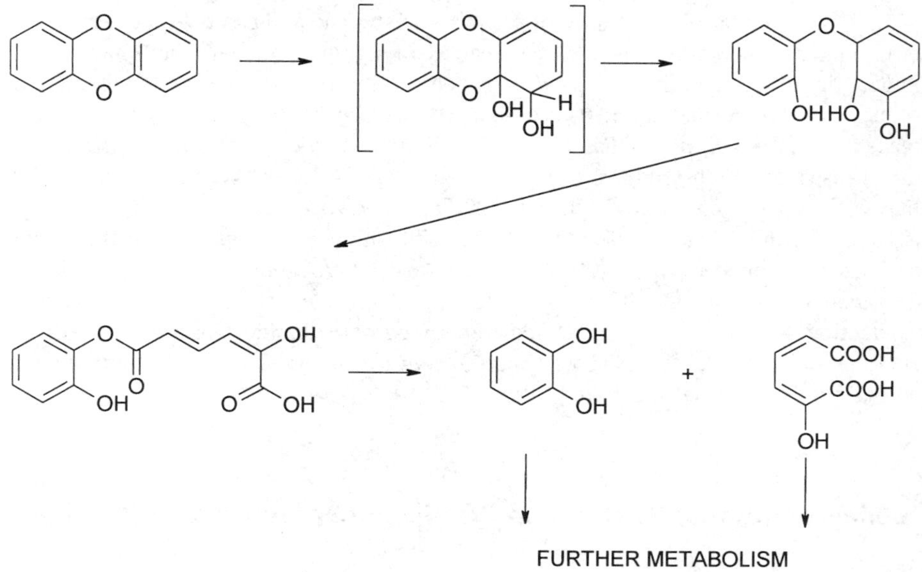

Figure 2. Microbial degradation pathway of dibenzo-p-dioxin.

The biochemistry of the biodegradation of dioxins and dibenzofurans has been studied in detail in some strains, in particular in *Sphingomonas* sp. strain RW1. Some of the key enzymes involved, and the genes encoding them, have been characterised. Not surprisingly considering the resemblance in the reactions involved, some of the enzymes bear a great deal of similarity to those involved in the biodegradation of biphenyls. When growing on dibenzofuran, *Terrabacter* sp. strain DPO360 expresses two different 2,3-dihydroxy-1,2-dioxygenase enzymes which resemble equivalent enzymes from biphenyl-degrading bacteria [24]. The enzyme catalysing the initial attack on DF and DD in strain RW1 has been purified and some of its properties determined [25]. The enzyme, termed either dibenzofuran 4,4a-dioxygenase or dioxin dioxygenase [26], is a multicomponent system similar to other aromatic dioxygenases and is also active towards biphenyl and other aromatic substrates. The genes encoding the different components of this enzyme are not clustered but are dispersed on the genome, suggesting that the catabolic diversity of this and other *Sphingomonas* strains arises from their poorly evolved and regulated, but dynamic, genetic organisation [26].

Biodegradation by white-rot fungi

White-rot fungi are able to degrade lignin, which is resistent to attack by most microorganisms, using an extra-cellular free radical-based biodegradation system [27]. The peroxidases in this system are fairly unspecific and can degrade a variety of environmental contaminants. Mineralisation of lignin or contaminants takes place under lignolytic conditions when nitrogen is limiting. Some contaminants are reductively dechlorinated under nitrogen-sufficient conditions, but this does not take place for PCDDs and PCDFs. The white-rot fungus *Phanerochaete chrysosporium* mineralises 2,7-dichlorodibenzo-*p*-dioxin through 1,2,4-trihydroxybenzene under lignolytic conditions [28]. Cultures of *Phanerochaete sordida* YK-624 in low-nitrogen medium degraded a mixture of 2,3,7,8-substituted tetra- to octachlorinated dioxins and dibenzofurans [29]. Between 40 and 76% of these congeners was degraded within 14 days. Metabolites identified included chlorocatechols, but CO_2 production was not measured. Production of carbon dioxide was reported for the degradation of 2,3,7,8-TCDD by *Phanerochaete chrysosporium*, but this amounted to only 2.5% of the initial substrate [30].

Biodegradation of PCDDs and PCDFs under anaerobic conditions

Biodegradation in sediments

Sediments are an important sink for dioxins and dibenzofurans in the aquatic environment because of the strong sorption to particulate matter of these hydrophobic compounds. While there are many data on concentrations of PCDDs and PCDFs in sediments, the possibility of reductive dechlorination reactions similar to those demonstrated for other chlorinated compounds has not been recognised until recently. The first evidence for the reductive dechlorination of PCDDs and PCDFs in sediments came from a study in which concentrations of chlorinated aromatic contaminants in sediment cores from Lake Ketelmeer, a sedimentation area of the River Rhine, were compared with those in samples from the same sediment layer archived in 1972 [31]. Significant decreases in levels of highly chlorinated benzenes and biphenyls were found in samples taken in 1988 and 1990 and concomitant increases in levels of lower chlorinated congeners. In the case of PCDDs and PCDFs, changes were less clear, but there were significant decreases in 1,2,3,4,7,8- and 1,2,3,6,7,8-hexachlorodioxins and 1,2,3,7,8- and 2,3,4,7,8-pentachlorodibenzo-furans. At the same time, there were significant increases in concentrations of 2,3,7,8-TCDD and 1,2,3,7,8,9-HxCDD. Examples of the trends in PCDDs and PCDFs are shown in Figures 3 and 4, respectively. Half lives of 12 to 13 years were estimated for the four congeners showing significantly reduced concentrations.

Time trends of PCDDs and PCDFs were also studied in sediment samples from the Baltic Proper [32]. Sediment layers deposited between 1882 and 1962 were analysed. Half lives between 16 and 1845 years were estimated for a large number of congeners. However, as no account was taken of changing emissions during this period, these half lives were considered as minimum values by the authors. Much lower half lives of 4.4 and 6.2 years were found for 1,3,6,8-TCDD, 1,3,7,9-TCDD, 1,2,3,4,6,7,8-HpCDD and OCDD added to sediments in an experimental lake enclosure in Western Ontario [33]. However, as there was no change in congener profile and lower chlorinated congeners were not analysed, it is not clear whether concentrations were reduced as a result of biodegradation or partitioning and diffusion to the dissolved phase.

Reductive dechlorination by bacteria present in sediments

Indications for the reductive dechlorination of chlorinated aromatic compounds such as PCBs and chlorobenzenes in anaerobic sediments has been confirmed in laboratory experiments in which these compounds were incubated either in sediment suspended in an anaerobic medium or in a suspension of bacteria eluted from sediment. Adriaens et al. added a mixture of penta-, hexa- and heptachlorinated PCDDs and PCDFs to a suspension of PCB-contaminated sediment [34,35]. Reductions in the concentrations of these congeners and the formation of less chlorinated ones was observed over a period of two years with half lives of 1 to 4 years. Dechlorination was preferentially in the non-lateral positions. Dechlorination also took place in sterilised sediment, but at lower rates.

Rapid dechlorination was observed for 1,2,3,4-TCDD added to a methanogenic culture enriched from Lake Ketelmeer sediments [36]. This culture, which also dechlorinated chlorobenzenes and PCBs, transformed 1,2,3,4-TCDD with a half life of 15.5 days to products which included 1,2,4- and 1,2,3-trichlorodioxins, 1,3- and

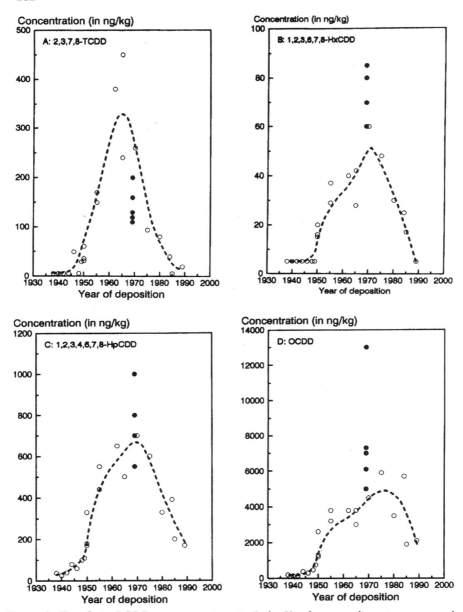

Figure 3. Trends in PCDD concentrations in Lake Ketelmeer sediment core samples (open symbols) and top layer samples from 1972 (filled symbols) vs. the estimated year of deposition. (Reproduced from reference 31. Copyright 1993 Society of Environmental Toxicology and Chemistry.)

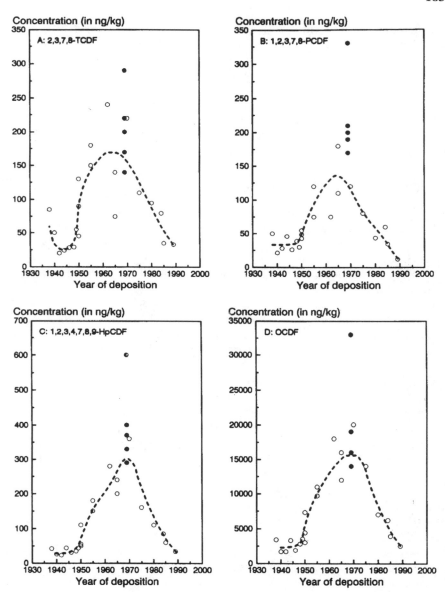

Figure 4. Trends in PCDF concentrations in Lake Ketelmeer sediment core samples (open symbols) and top layer samples from 1972 (filled symbols) vs. the estimated year of deposition. (Reproduced from reference 31. Copyright 1993 Society of Environmental Toxicology and Chemistry.)

2,3-dichlorodioxins and 2-chlorodioxin. These products are formed by the removal of both lateral and non-lateral chlorines. Subsequent work including incubations with the trichlorodioxins led to the identification of more dechlorinated products [37], showing that dechlorination took place by the pathway shown in Figure 5. There is no evidence as yet that the last chlorine is removed to produce DD in this culture.

Anaerobic bacteria enriched from sediment of the River Saale (Germany) dechlorinated 1,2,3,4-TCDD to form 1,2,3- and 1,2,4-tri- and 1,3-dichlorinated congeners [38], but cultures from other sources such as PCDD/F contaminated soil and sewage sludge did not show this activity. More highly chlorinated PCDD and PCDF congeners were dechlorinated in spiked suspensions of PCB-contaminated Hudson River sediment and creosote-contaminated aquifer sediment [39,40]. The half lives of the spiked penta, hexa- and heptachlorinated congeners was estimated to be between 1 and 4.1 years. Traces of less chlorinated congeners were identified. The observation of (slower) dechlorination in autoclaved suspensions suggests that abiotic mechanisms may also be important. Reduced rates in long term incubations were ascribed to the effects of resistent sorption (see below). Freshly spiked OCDD was dechlorinated to monochlorinated dioxins as end product in an anaerobic microbial consortium eluted from Passaic River sediments [41]. Two distinct dechlorination pathways could be distinquished: *peri* dechlorination of 2,3,7,8-substituted congeners to produce 2,3,7,8-TCDD and mixed *peri*-lateral dechlorination of non-2,3,7,8-substituted congeners. As eluted from the sediment, the microorganisms contained 2,3,7,8-TCDD which was converted stoichiometrically to tri- and monochlorinated congeners during the incubations.

The observation mentioned above that reductive dechlorination takes place under sterile conditions is a common one, not only for dioxins but also for other chlorinated compounds. It is not always possible to make a clear distinction between biochemical and chemical mechanisms of reductive dechlorination. It is possible that dechlorination in sterile systems is catalysed by microbially produced redox active cofactors such as vitamin B_{12}, which have been proposed as the biochemical catalyst. Alternatively, humic acid constituents may also play a role in abiotic reductive dechlorination. Regardless of the mechanisms involved in extracellular abiotic reductions, microorganisms must be involved indirectly since anaerobic conditions in sediments arise as a result of microbial activity. Abiotic reductive dechlorination of OCDD, OCDF and 1,2,3,7,8-PnCDD can be catalysed by vitamin B_{12} with Ti(I) citrate as reductant and by the humic acid models resorcinol, catechol and 3,4-dichlorobenzoate as well as by zerovalent zinc [42]. However, the reactions were in most cases very slow (for example a 20% reduction in OCDD concentration after 3 months in the vitamin B_{12} catalysed system. Since vitamin B_{12} has been proposed as a possible catalysed in the microbial reductive dechlorination reactions, the abiotic vitamin B_{12} system could be regarded as a so-called biomimetic model for the biochemical reductive dechlorination reaction. Recent results from a study of the reductive dechlorination of 1,2,3,4-TCDD in a system consisting of vitamin B_{12} and

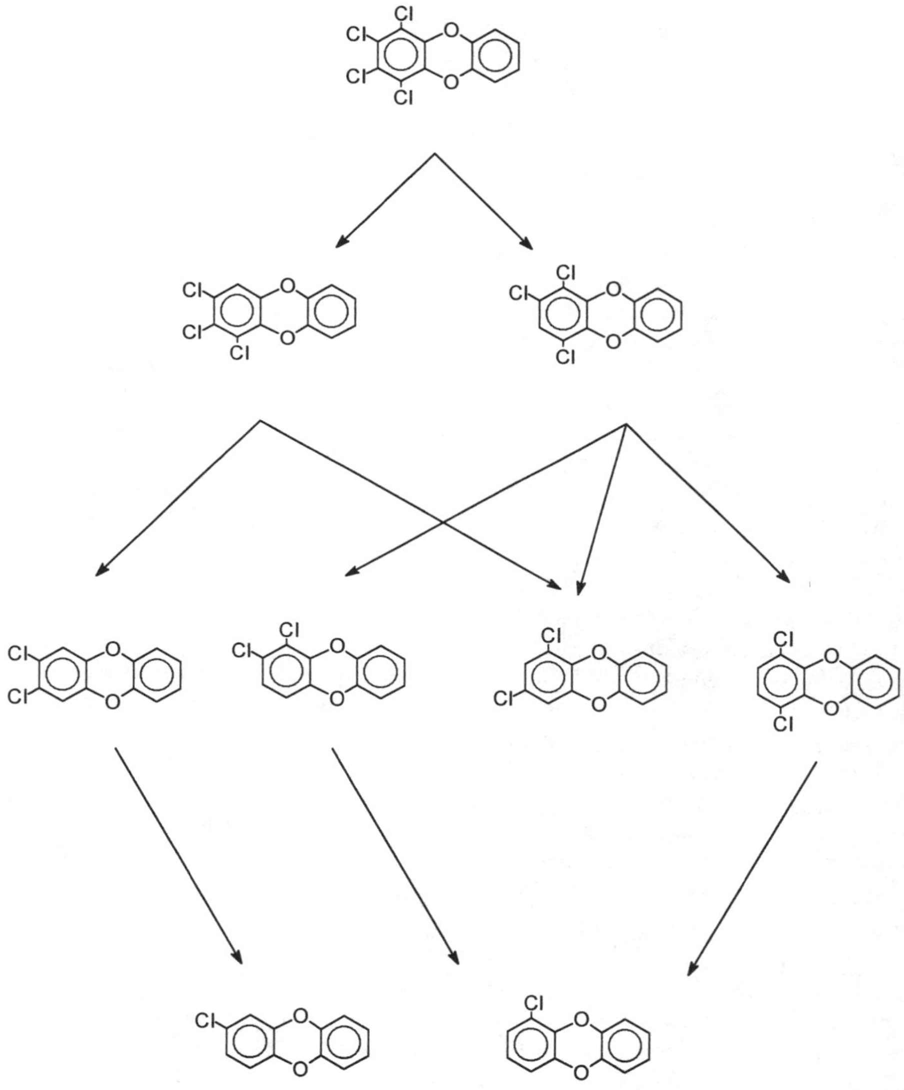

Figure 5. Pathway for the dechlorination of 1,2,3,4-TCDD in a methanogenic culture enriched from Lake Ketelmeer sediment.

Ti(I) citrate show that the major reduced products formed are similar to those observed in anaerobic microbial cultures [43], although there appear to be differences in their relative yields. Such biomimetic model systems could be used to explore the chemistry of reductive dechlorination of dioxins, dibenzofurans and other organochlorines.

Bioavailability of PCDDs and PCDFs for microorganisms

PCDDs and PCDFs are highly hydrophobic, with log K_{ow} values between 3.68 and 8.75 and correspondingly high sediment-water partition coefficients [44]. Since it is generally assumed that uptake of substrates by microorganisms takes place from the dissolved phase, sorption to sediment and soil is likely to have an important impact on the biodegradation of PCDDs and PCDFs [45]. The extent to which the strongly sorbed fractions are available to be taken up and degraded by microorganisms will depend not only on the partition coefficient but also on the rate at which desorption takes place to replenish the dissolved fraction. This must be sufficiently high to enable the rate of uptake and degradation of substrate to satisfy the minimum requirements to support microbial activity [46]. However, in the case of dioxins this may not be such a rigid requirement as these are often degraded cometabolically and thus another substrate may be utilised to satisfy this requirement.

There has been relatively little work done on the bioavailability of sediment-sorbed dioxins and dibenzofurans for biodegradation. The rates of cometabolism of lightly chlorinated dioxins in three-day-old spiked sediments were comparable to those in the dissolved phase [47]. However, rates were lower in sediment suspensions which had been spiked 24 days before inoculation with bacteria. The total amounts degraded were in all cases greater than the initially dissolved fractions, indicating that the sorbed phases were available for degradation. In a model system in which 3-CDF was sorbed to teflon granules, the consumption of 3-CDF by cells attached to the granules drove desorption of 3-CDF [48]. It was proposed that the rate at which a sorbed substrate becomes available for organisms is partially determined by microbial factors such as the ability of the degrading organisms to reduce dissolved substrate concentrations and adherence of the organisms to the sorbent.

Time-dependent decreases of extraction efficiencies of highly chlorinated PCDDs and PCDFs in spiked autoclaved sediments were ascribed to diffusion of the compounds into the sediment particles [49]. The decrease in extractable congeners was slightly faster in biologically active sediment which was attributed to microbial transformation. This was confirmed by the formation of lesser chlorinated congeners.

Sediment pore water is rich in dissolved organic matter (DOM) and this may have an important effect on the bioavailability of dioxins and dibenzofurans in sediments. Although DOM is often considered a sorbent which reduces the bioavailability of sorbing chemicals in the aqueous phase, it may also have a solubilising effect on compounds present in the solid phase. The preliminary results of a study of the reductive dechlorination of 1,2,3,4-TCDD in the presence of increasing concentrations of DOM showed no reduction of the rate of dechlorination at DOM concentrations between 10 and 400 mg/l [49]. In fact, the rate appeared to be highest at a DOM concentration of 100 mg/l, suggesting that DOM may have had a solubilising effect on the very high concentration of the dioxin used in this study.

Conclusions

As we have outlined above, there is ample evidence that even highly chlorinated dioxins and dibenzofurans are susceptible to microbial degradation if the conditions are suitable. This means in short that for the highly chlorinated congeners conditions should be reducing, such as in anaerobic sediments, whereas for mineralisation of the lightly chlorinated congeners conditions should be aerobic. However, it is clear that in the environment PCDDs and PCDFs are highly persistent. The most important reason for this discrepancy is probably related to the very limited bioavailability of these compounds in the environment.

References

1. Schecter; A. *Dioxins and Health*, Plenum Press, New York, 1994.
2. *Biodegradation of Dioxins and Furans;* Wittich, R.-M., Ed.; Springer, Berlin, 1998.
3. Wittich, R.-M. In *Biodegradation of Dioxins and Furans;* Wittich, R.-M., Ed.; Springer, Berlin, 1998, pp 1-28.
4. Alcock, R.E.; Jones, K.C. *Environ. Sci. Technol.* **1996**, 30, 3133-3143.
5. Kearney, P.C.; Woolson, E.A.; Ellington, C.P. Jr. *Environ. Sci. Technol.* **1972**, 6, 1017-1019.
6. Ward, C.T.; Matsumura, F. *Arch. Environ. Contam. Toxicol.* **1978**, 7, 349-357.
7. Muir, D.C.G.; Yarechewski, A.L.; Corbet, R.L.; Webster, G.R.B.; Smith, A.E. *J. Agric. Food Chem.* **1985**, 33, 518-523.
8. Rosenbrock, P.; Martens, R.; Buscot, F.; Zadrazil, F.; Munch, J. C. *Appl. Microbiol. Biotechnol.* **1997**, 48, 665-670.
9. Megharaj, M.; Wittich, R.-M.; Blasco, R.; Pieper, D. H. *Appl. Microbiol. Biotechnol.* **1997**, 48, 109-114.
10. Halden, R. U.; Halden, B. G.; Dwyer,D. F. *Appl. Environ. Microbiol.* **1999**, 65, 2246-2249.
11. Philippi, M.; Schmid, J.; Wipf, H. K.; Hütter, R. A. *Experientia* **1982**, 38, 659-661.
12. Klecka, G. M.; Gibson, D. T. *Biochem. J.* **1979**, 180, 639-645.
13. Klecka, G. M.; Gibson, D. T. *Appl. Environ. Microbiol.* **1980**, 39, 288-296.
14. Parsons, J. R.; Storms, M. C. M. Chemosphere **1989**, 19, 1297-1308.

15. Parsons, J. R.; Ratsak, C.; Siekerman, C. *Organohalogen Compounds* , **1990**, 1, 377-380.

16. Parsons, J. R.; de Bruijne, J. A.; Weiland, A. R. *Chemosphere* **1998**, 37, 1915-1922.

17. Strubel, V.; Engesser, K.-H.; Fischer, P.; Knackmuss, H.-J. *J. Bacteriol.* **1991**, 173, 1932-1937.

18. Schmid, A.; Rothe, B.; Altenbuchner, J.; Ludwig, W.; Engesser, K.-H. *J. Bacteriol.* **1997**, 179, 53-62.

19. Fortnagel, P.; Harms, H.; Wittich, R.-M.; Krohn, S.; Meyer, H.; Sinnwell, V.; Wilkes, H.; Francke, W. *Appl. Environ. Microbiol.* **1990**, 56, 1148-1156.

20. Wittich, R.-M.; Wilkes, H.; Sinnwell, V.; Francke, W.; Fortnagel, P. *Appl. Environ. Microbiol.* **1992**, 58, 1005-1010.

21. Wilkes, H.; Wittich, R.-M.; Timmis, K. N.; Fortnagel, P.; Francke, W. *Appl. Environ. Microbiol.* **1996**, 62, 367-371.

22. Arfmann, H.-A.; Timmis, K. N.; Wittich, R.-M. *Appl. Environ. Microbiol.* **1997**, 63, 3458-3462.

23. Schreiner, G.; Wiedmann, T.; Schimmel, H.; Ballschmiter, K. *Chemosphere* **1997**, 34, 1315-1331.

24. Schmid, A.; Rothe, B.; Altenbuchner, J.; Ludwig, W.; Engesser, K.-H. *J. Bacteriol.* **1997**, 179, 53-62.

25. Bünz, P. V.; Cook, A. M. *J. Bacteriol.* **1993**, 175, 6467-6475.

26. Armengaud, J.; Happe, B.; Timmis, K. N. *J. Bacteriol.* **1998**, 180, 3954-3966.

27. Aust, S. D.; Stahl, J. D. In *Biodegradation of Dioxins and Furans;* Wittich, R.-M., Ed.; Springer, Berlin, 1998, pp 61-73.

28. Valli, K.; Wariishi, H.; Gold, M. H. *J. Bacteriol.* **1992**, 174, 2131-2137.

29. Takada, S.; Nakamura, M.; Matsueda, T.; Kondo, R.; Sakai, K. *Appl. Environ. Microbiol.* **1996**, 62, 4323-4328.

30. Bumpus, J. A.; Tien, M.; Wright, D.; Aust, S. D. *Science* **1985**, 228, 1434-1436.

31. Beurskens, J. E. M.; Mol, G. A. J.; Barreveld, H. L.; van Munster, B.; Winkels, H. J. *Environ. Toxicol. Chem.* **1993**, 12, 1549-1566.

32. Kjeller, L.-O.; Rappe, C. *Environ. Sci. Technol.* **1995**, 29, 346-355.

33. Segstro, M. D.; Muir, D. C. G.; Servos, M. R.; Webster, G. R. B. *Environ. Toxicol. Chem.* **1995**, 14, 1799-1807.

34. Adriaens, P.; Grbic-Galic, D. *Chemosphere* **1994**, 29, 2253-2259.

35. Adriaens, P.; Fu, Q.; Grbic-Galic, D. *Environ. Sci. Technol.* **1995**, 29, 2252-2260.

36. Beurskens, J. E. M.; Toussaint, M.; de Wolf, J.; van der Steen, J. M. D.; Slot, P. C.; Commandeur, L. C. M.; Parsons, J. R. *Environ. Toxicol. Chem.* **1995**, 14, 939-943.

37. Toussaint, M.; Beurskens, J. E. M.; de Wolf, J.; Grooteman, M. N.; van der Steen, J. M. D.; Slot, P. C.; Parsons, J. R. in preparation.

38. Ballerstedt, H.; Kraus, A.; Lechner, U. *Environ. Sci. Technol.* **1997**, 31: 1749-1753.

39. Adriaens, P.; Grbic-Galic, D. *Chemosphere,* **1994**, 29:9-11.

40. Adriaens, P.; Fu, Q.; Grbic-Galic, D. *Environ. Sci. Technol.,* **1995**, 29: 2252-2260.
41. Barkovskii, A. L.; Adriaens, P.; *Appl. Environ. Microbiol.* **1996**, 62: 4556-4562.
42. Adriaens, P.; Chang, P. R.; Barkovskii, A. L. *Chemosphere* **1996**, 32: 433-441.
43. Parsons, J. R.; Cannegieter, C.; Hekster, F.; Tambach, T.; te Kloeze, A.-M. *Amer. Chem. Soc. Div. Environ. Chem. Preprints of Extended Abstracts* **1999**, 39: 84-86.
44. Govers, H.A.J.; Krop, H.B. *Chemosphere* **1998**, 37: 2139-2152.
45. Mihelcic, J.R.; Lueking, D.R.; Mitzell, R.J.; Stapleton, J.M. *Biodegradation* **1993**, 4: 141-153.
46. Harms, H. In *Biodegradation of Dioxins and Furans;* Wittich, R.-M., Ed.; Springer, Berlin, 1998, pp 135-163.
47. Parsons, J. R. *Chemosphere* **1992**, 25: 1973-1980.
48. Harms, H.; Zehnder, A. J. B. *Appl. Environ. Microbiol.* **1994**, 60: 2736-2745.
49. Toussaint, M.; Krop, H. B.; Grooteman, M. N.; van Breugel, M.; de Vries, P.; Parsons, J. R. *Novel Approaches forBioremediation of Organic Pollution*; Fass, R.; Flashner, Y.; Reuveny, S., Eds; Kluwer/Plenum, New York, 1999, pp 189-195.

Food Chain Transfer
and Exposure

Chapter 14

The Intake and Clearance of PCBs in Humans: A Generic Model of Lifetime Exposure

Ruth E. Alcock, Andy J. Sweetman, and Kevin C. Jones

Department of Environmental Science, Institute of Environmental and Natural Sciences, Lancaster University, Lancaster LA1 4YQ, United Kingdom

ABSTRACT

We have developed a model which successfully reconstructs the lifetime PCB-101 burden of the UK population for individuals born between 1920 and 1980. It not only follows burdens and clearance of persistent organic contaminants throughout a human lifetime - taking changes in age and body composition into account - but also, importantly, incorporates changing environmental concentrations of the compound of interest. Predicted results agree well with available measured lipid concentrations in human tissues. Its unique construction takes into account both changing environmental levels of PCBs in principal food groups and changing dietary habits during the time period. Because environmental burdens of persistent organic contaminants have changed over the last 60 years, residues in food will also have mirrored this change. Critically in this respect, the year in which an individual was born determines the shape and magnitude of their exposure profile for a given compound. Observed trends with age represent an historical legacy of exposure and are not simply a function of equal yearly cumulative inputs. We can demonstrate that the release profile of PCB-101 controls levels in the food supply and ultimately the burden of individuals throughout their life. This effect is expected to be similar for other PCB congeners and persistent organic compounds such as PCDD/Fs. Models of this type have important applications as predictive tools to estimate the likely impact of source-reduction strategies on human tissue concentrations.

INTRODUCTION

Foods, particularly those which are animal-based, represent the most important source of human exposure for many persistent organic compounds (1; 2; 3). However, intake is just one of many factors which controls the body burden of these compounds. The effect of *cumulative* exposure, and factors such as absorption efficiency, potential formation/biotransformation in the gastro-intestinal tract and rates of metabolism and depuration influence tissue concentrations. Our current

192   © 2001 American Chemical Society

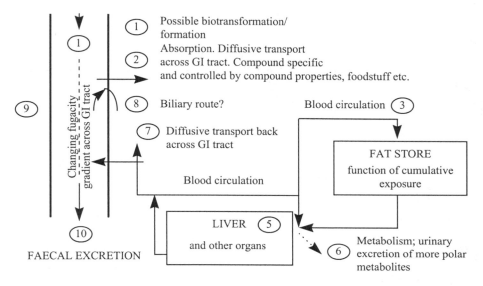

Figure 1. Schematic representation of key processes of compound transfer within the body.

understanding of these transport and fate processes within the body and of the rates/efficiencies of absorption and metabolism are relatively limited. These and other potentially important processes are illustrated in Figure 1.

Physiologically based pharmacokinetic (PB-PK) models have been used to characterize the ability of an organism to convert an external dose of compound to tissue concentrations. These are usually equilibrium models that describe partitioning within body tissues and organs over relatively short periods of time, without the need to incorporate changes in body weight and composition throughout life (e.g. ref 4). Recently, more dynamic models have been developed which describe burdens of TCDD during an entire human lifespan (5) and - more specifically - uptake and elimination of TCDD in newborns and infants (6). Using these studies as a basis, we have developed a model which not only follows burdens and clearance of persistent organic contaminants throughout a human lifetime - taking changes in age and body composition into account - but also, importantly, incorporates changing environmental concentrations of the compound of interest.

Initially, we have selected PCBs, specifically PCB-101, as the target compound for this work. Valuable data is available in the UK on both historical inputs of this particular congener into the environment (see Ref 7) and measured dietary intake information. Recently, there has also been a renewed interest in the fate, effects and toxicity of PCBs. The UK Committee on Toxicity of Chemicals in Food, Consumer Products and the Environment (COT) has recently recommended the use of Toxicity Equivalent Factors (TEFs) to assess the potential toxicity of complex mixtures of dioxin *and* selected polychlorinated biphenyls (PCBs) congeners present in food (8).

Their significance to ΣTEQ in a range of foods and human tissues has also recently been highlighted (9). The model approach is illustrated by simulating body and tissue burdens of PCB-101 in humans for individuals/populations born from 1920 to 1980 in the UK. Importantly, the framework is considered to be generic and applicable to a wide range of persistent organic compounds for which a historical input profile can be generated.

MODEL CONSTRUCTION

General introductory remarks In general terms, modeling lipophilic compounds throughout a lifetime relies on a straightforward mass balance of lipids within the body. As the model uses a time-scale of years rather than days, rapid processes such as absorption can be summarized using simple relationships and short-term fluctuations, for example changes in diet (i.e. in the contaminant quality of foods) are 'smoothed' out. As a result, the amount of compound present anywhere in the body is determined by the lipid content of the material concerned and an age dependent function allowing for growth/increasing body fat composition and subsequent compound dilution with age. Since many parameters required for physiologically based pharmacokinetic (PB-PK) models are irrelevant over the timescale of a year (e.g. blood flow rates, organ volumes, partition coefficients), the minimum unit of time the model incorporates, they can be effectively ignored for a model of this type.

Slob (10) demonstrated the importance of modeling the exposure of whole populations, rather than individuals to chemicals in food. We incorporated these principles here by using average body weight and body fat statistics and indications of the expected range in values for a given age taken from ICRP (11). Based on the assumption that lipophilic compounds will partition into the body lipids of organs and tissues equally, the whole organism can be treated as a single compartment containing a given amount of lipid i.e. concentrations based on a lipid weight will be approximately equal in all body compartments at steady state (12).

Body weight and lipid composition
The human body is simply described by weight and by total body fat composition. Body weight and body fat composition are age-dependent and since these differ between men and women are modeled separately. Within a given age cohort of a population there are males and females with a range of bodyweight and fat compositions i.e. contaminants can be accumulated in different lipid volumes.

Food intake
Importantly, we were able to incorporate several measured data series into the model dietary intake function. One of the most comprehensive contemporary dietary exposure studies in terms of both food composition and recent temporal exposure information is the UK Total Diet Study (TDS) for foods collected in 1982 and 1992. These data represent composite samples representative of UK national dietary habits and provide valuable information on both the concentration of PCBs and changes in dietary composition over the last 17 years.

We were able to incorporate changing dietary composition data for major food groups in the UK over longer timescales using measured annual average UK food consumption statistics from 1940 to 1990 from the UK Household Food Consumption and Expenditure Survey (13). This includes food groups which are known to dominate human exposure to persistent organic pollutants such as meat, dairy produce and fats and fish.

Food intake also varies in terms of not only the mass of food ingested with age but also the type of foods ingested with age. These factors were built into the model using measured intake information taken from The British Adult Study and the school children and toddlers surveys to allow intakes for different age groups to be calculated (see Ref 2 for details). Age dependent dietary intakes were incorporated into the model framework based on the measured age-dependent information for the UK population within the following age classes; toddlers age 1.5 to 2.5, toddlers age 2.5 to 3.5, toddlers age 3.5 to 4.5. school children age 10-15 and adults age 16+ (see Ref 2 for details). Dietary intakes by school children (age 10-15) were estimated using consumption data from a 7-day weighed diary study of 10-11 and 14-15 year olds conducted in 1983. Intakes for toddlers were estimated using data from a dietary and nutritional study of children aged 1.5 to 4.5. Table 1 presents measured dietary intake of PCB-101 and PCB-TEQ in the UK for different age groups in 1982 and 1992 on a per kg bodyweight basis. Dietary intake of PCB-TEQ consumed by adults fell from 2.7 pg TEQ/kg bw/day in 1982 to 0.9 pg TEQ/kg bw/day in 1992. In the case of school children, estimated dietary intakes have fallen from 3.1 in 1982 to 1.0 in 1992.

Table 1: Estimated upper bound UK dietary intake of PCB-TEQ and PCB-101 for average consumer adults, schoolchildren and toddlers estimated from the TDS surveys in 1982 and 1992 (WHO 1994 TEF values used).

Age Group	1982		1992	
	PCB-TEQ	PCB-101	PCB-TEQ	PCB-101
	pg TEQ/kg bw/day	ng/kg bw/day	pg TEQ/kg bw/day	ng/kg bw/day
Toddlers (1.5-2.5)	6.7	1.0	2.4	0.36
Toddlers (2.5-3.5)	6.2	0.93	2.2	0.33
Toddlers (3.5-4.5) (boys)	6.1	0.90	2.1	0.32
Toddlers (3.5-4.5) (girls)	5.7	0.84	2.0	0.3
Children (age 10-15)	3.1	0.4	1.0	0.17
Adults (age 16+)	2.7	0.4	0.9	0.14

Toddlers and young children ingest a greater amount of PCBs per kg body weight compared to adults. For example, in 1982 a young toddler (age 1.5- 2.5 years) was estimated to consume 1.0 ng/kg bw/day of PCB -101, while an average adult consumer ingested 0.4 ng/kg bw/day. These findings are similar to those of a study in which dietary intake of PCDD/Fs was ~3.6 fold higher in young children (aged 2-5 years) than adults (14). In combination, these 'food' factors allow us to have the necessary information to build into the model changes in quantity and type of major food groups consumed by UK adults, young children and infants since 1940.

COMMENTS ON THE PROCESSES OF ABSORPTION, METABOLISM AND CLEARANCE

The literature contains very little information on the rates/efficiencies of absorption, metabolism and clearance in relation to the fate of individual PCB congeners from food. It should be stressed that in combination, they are complex processes which are at present poorly understood and therefore difficult to begin to model (see Figure 1). It is important to highlight that data reported as simple 'loss' of compounds over time from the human body does not allow the adequate separation of absorption, metabolism and clearance. It is a useful first step to define these processes as they are used here. Absorption relates to the amount of ingested contaminant which passes from the gastrointestinal tract across the gastrointestinal wall and into the body. Metabolism essentially involves enzyme-mediated loss of contaminant in the body and clearance is defined here as non enzyme mediated elimination from the body. The following section briefly reviews our current state of knowledge of each process.

Absorption Reported mass balance studies rely on the ratio of fecal excretion and dietary intake to calculate net absorption ratios for persistent organic pollutants (eg 14; 15). However, the process of absorption across the gut is thought to be diffusive and hence driven by concentration gradients (or more preferably fugacity gradients). Studies have shown that in younger people, as a result of their low contaminant burden, there is net diffusion from the gut contents into the gut tissue. However, in older people with higher contaminant burdens the diffusion gradient is reduced and may even be reversed resulting in net negative absorption. As a result, absorption is one part of a complex continuum of processes and controlling factors which we envisage will be incorporated as a critical model refinement in the future. A recent study at Lancaster University investigating human net absorption of PCBs from food found that between 70-90% of PCB-101 ingested via the diet was absorbed in male volunteers (16). For this model, based on this measured data we have assumed gastrointestinal adsorption of PCB-101 to be 80% while that of infants in the first year of life is assumed to be 95%.

Metabolism Only a very limited amount of data is available in the literature on PCB metabolism from the human body. Reported half-lives range from between 5 to 10 years, although these figures have been measured based on accidental/occupational high exposure studies which may not be representative of background exposure. PCB congener clearance rates developed by Brown (17) for occupationally exposed populations predict a value of 10 years for PCB-101. Our selected value of 10 years appears to be a reasonable approximation in the absence of measured human data for PCB-101.

An important finding demonstrated by Kreutzer et al., (6) was that the half life of TCDD metabolism decreases during a human lifetime. This is probably applicable to other persistent organic compounds with long half-lives and is used to describe PCB-101 metabolism throughout life in this model. The method is briefly described as follows. Enzyme mediated metabolism is assumed to occur exclusively in the liver, the mass/activity of enzyme per unit liver is modeled to be proportional to liver volume. Metabolism is assumed to follow first order kinetics. Equilibrium among all tissues is assumed, and the amount of compound metabolized per unit time is proportional to its concentration in body lipids. The following approach is taken from Kreutzer et al (6) and allows calculation of metabolic half life taking into account the age dependent volumes of body lipids and liver.

The half life of metabolic elimination ($tm_{1/2}$) is given by:

$$tm_{1/2} = tm_{1/2\,40} * V/V_{40} * (V_{L40}/V_L)^{2/3} \hspace{2cm} \text{Equation 1}$$

where, $tm_{1/2\,40}$ is the reference metabolic half life of elimination in a 40-year-old human subject (e.g. 10 years), V is the total volume of body lipids, V_{40} the total volume of body lipids of a reference 40 -year-old subject, V_L is the total liver volume (calculated from liver weights given in ICRP for males and females), V_{L40} the liver volume of the 40-year-old reference subject. Liver surface is proportional to 2/3 power of liver volume (V_L). Because metabolism rates are strongly influenced by the accumulation of total body lipids, half lives were modeled for PCB-101 to be ~2 years in the first year of life increasing to ~10 years in adults age 40.

Clearance Non enzyme mediated elimination from the body includes external skin and internal (gut wall) skin cell shedding and reverse diffusion. The loss of skin from the body surface amounts to only between 0.5 and 1 g lipid day[-1] while the loss of body lipids associated with fecal excretion (as gut epithelium) amounts to 5-7 g lipid day[-1]. Fecal clearance of non-metabolised PCDD/Fs contributed between 37% (for 2,3,7,8-TCDD) and 90% (for OCDD) of the total elimination of PCDD/Fs in a recent mass balance study of occupationally exposed men in Germany (18). In another mass balance study, a net excretion of PCDD/Fs from the body was also found in the oldest group of volunteers (15). This implies that as contaminant levels in the food supply decrease, concentration gradients between the gut contents and gut tissues may be leading to net excretion. It suggests that the gastrointestinal pathway plays an important role in the clearance of PCDD/F congeners. Similar processes could be applied to other organic pollutants such as PCBs, although at present information and understanding of this topic is very limited. We have therefore not attempted to incorporate these processes here.

THE PCB CASE-STUDY - RE-CONSTRUCTING THE INPUT CURVE

Van der Molen et al. (5) highlighted the potential importance of changing levels of TCDD in foods and used a time trend correction function to allow for a changing intake history. Valuable time trend information can be obtained from sediment core profiles and environmental monitoring data which can provide an estimate of the timing and magnitude of peak contaminant environmental levels. Ideally, we need information to estimate contaminant levels in major food groups over time and to combine this with measured changes in dietary composition so that a predicted input profile can be generated. Recent trends in dietary PCB exposure can be inferred using the UK Total Diet Study (TDS) samples collected in 1982 and 1992. These also correct dietary intake for young children and infants (see earlier). In summary, average dietary intake of PCBs has declined from 1.0 µg/person/day in 1982 to 0.34 µg/person/day in 1992 (2). No comparable comprehensive data-set of PCB residues in foods is available prior to the early 1980s, but an indication of broad changes in dietary exposure can be inferred using human tissue PCB data from the late 1960s/early 1970s (19). Unfortunately, this is expressed only as 'total PCB' and is of little use in validating congener specific model scenarios. We therefore have to 're-construct' a historical input profile for PCBs.

We have developed a method of predicting PCB concentrations in foods from modeled and measured time trend data for environmental media in the UK and our knowledge of foodchain transfer of individual PCB congeners. Sediment and archive soil data suggest peak PCB inputs in the late 1960/early-1970s followed by a decrease to the present day (20), a pattern supported by recent trends in a broad range of environmental media (e.g. 21; 22; 23) including human tissues such as breast milk (e.g. 24; 25).

A recent study using a Mackay fugacity fate model has predicted historical PCB air concentrations in the UK environment based on a reconstructed PCB release profile (26). This was generated using PCB production/use data for individual Aroclor formulations. Profile predictions were confirmed by UK measured soil (27) and sediment profile data (28). Concentrations of PCBs in air supply those found on vegetation and in turn grazing animals receive PCBs primarily through ingestion of grass and silage (29; 30; 31). As a consequence, atmospheric concentrations will ultimately exert a strong influence on human tissue concentrations. Beginning with a modeled historical air concentration profile we have the necessary starting information to be able to relate air concentrations to those in human foods.

First, the derivation of PCB-101 concentration curves for principal food groups was conducted as follows:
(i) The predicted historical air concentration profile for PCB-101 from Sweetman and Jones (26) was used to generate a cows milk concentration profile using a recently reported relationship between air PCB concentrations and milk PCB concentrations. These were checked and calibrated against measured data by Sweetman and Jones (26). Thomas et al. (29) found that under normal UK bovine husbandry conditions

the PCB concentration in pg. PCB-101. per g of milk fat is the equivalent of that in 11 m^3 of air (i.e. the air/milk transfer factor for PCB-101 is 11 m^3/g). Further studies by Thomas et al. (30) have shown that PCB transfer from the air to grass is rapid, providing important evidence that concentrations in this major food group closely follow trends in air concentrations.

(ii) Predicted air concentrations were also used to predict feed intake (i.e. pasture grass) and subsequently meat concentrations following the well documented relationships used in the EUSES (31) model based on a bioaccumulation factor (BAF) approach. In the absence of measured transfer factor information this method was considered to be the best approximation of beef concentrations.

$$BAF_{meat} = 10^{-7.6 + \log Kow} \qquad \text{Equation 2}$$

$$C_{meat} = BAF_{meat} \text{ *feed intake} \qquad \text{Equation 3}$$

(iii) Similarly, fish tissue concentrations were predicted by multiplying a congener-specific bioconcentration factor (BCF) for fish by a predicted concentration in surface water (data estimated using Ref 26).

$$\log BCF_{fish} = 0.85 \text{ *} \log Kow - 3.7 \qquad \text{Equation 4}$$

$$C_{fish} = BCF_{fish} \text{ * concentration in surface water} \qquad \text{Equation 5}$$

Summation of each of these concentration curves for individual food groups gives the annual average yearly intake for a given PCB congener. Despite the extensive use of predicted data, the relationships used have been applied successfully in contemporary situations and this model-based approach is considered to be a best estimate of UK dietary intake in the past, in the absence of measured data. Predicted PCB concentrations for each food group and the resulting estimated adult PCB-101 intake between 1940 and 2000 are shown in Figure 2.

MODELING LIFETIME TISSUE CONCENTRATIONS FOR INDIVIDUALS BORN AT DIFFERENT TIMES.

The model can demonstrate changing PCB-101 body burdens and lipid concentrations throughout a lifetime, incorporating growth, dietary intake and metabolism. Importantly, the year of birth can be selected to examine the influence of peak exposure on body lipids. Figure 3 shows the changing lifetime body lipid concentrations of PCB-101 for males born in 1922, 1930, 1950, 1960, and 1970. It demonstrates that older individuals have accumulated a greater body burden i.e. their cumulative exposure has been greatest through life because they have experienced the rise and peak in PCB concentrations in their diet. Individuals born in the 1970s, exhibit highest concentrations in tissues early in their life. Conversely, individuals born in the 1920s and 1930s experience low tissue concentrations early in their life, peak tissue concentrations in their forties and fifties and a declining body burden into their sixties and seventies when PCB levels were declining in many environmental media. However, Figure 3 clearly shows the importance of the birth year and not age *per se* on the contaminant status of body lipids. Age-related increases of lipophilic contaminants in human tissues are usually modeled using linear functions. However, scatter plots of concentration data tend to exhibit a strong linear trend in the first 20-30 years of life, followed by a much broader scatter of points from age 40 onwards. Concentrations are not increasing as a simple function of age, but rather represent the sum of an input profile over time, which itself has varied due to changes in compound inputs. Importantly, observed trends with age represent a historical legacy of exposure and are not simply a function of equal yearly cumulative inputs.

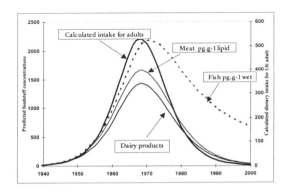

Figure 2: Predicted concentrations of PCB-101 in meat, fish and dairy products and calculated dietary intake of PCB-101 for adults in the UK from 1940 to 2000 (pg kg^{-1} bw d^{-1})

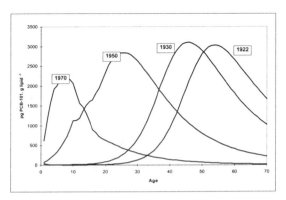

Figure 3: Predicted tissue concentrations of PCB-101 for different age classes in the UK born from 1920 to 1990 (pg PCB-101 g lipid^{-1})

This model framework can be used to predict the expected range in lipid concentrations (pg/g fat weight basis) for a given population at a point in time as shown in Figure 4. If a representative sample of the population was taken in 1990 in the UK, made up of individuals aged between 20 and 70, the model predicts that lipid concentrations would range between about 0.5 and 2 ng g⁻¹ and follow a broadly asymptotic function. We are able to compare these predictions with measured adipose tissue data collected from 55 males in 1990. These data have been averaged over 5 year age classes and plotted on Figure 4. Measured PCB-101 concentrations in these samples ranged from 0.8 to 5.3 ng g⁻¹ lipid with an average of 2.0 ng g⁻¹ (12). The agreement between measured and predicted model data is encouraging, although for individuals under 35 the model appears to under predict tissue levels.

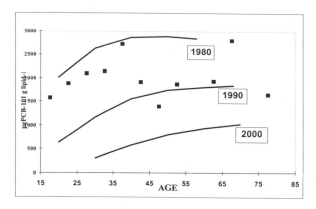

Figure 4: Expected range in lipid concentrations for a given sampling year (pg g lipid⁻¹ weight basis). Data points represent 1990 UK survey data of male adipose tissue (Ref 11)

If sampling occurred 10 years earlier in 1980, tissue concentrations for all age classes would be higher and the range in expected concentrations would be slightly narrower as a higher proportion of younger individuals which had 'seen' higher PCB concentrations would be present in this population. It is also possible to predict the range in concentrations for a sampling event in the future. The range in concentrations for a sampling event in the year 2000 has also been added to Figure 4. The predicted range of values becomes smaller for this population aged between 30 and 70 as an increased proportion of the population have been exposed to lower dietary PCB intakes as environmental levels continue to decline.

CONCLUDING REMARKS

We have described a model which successfully reconstructs the lifetime PCB-101 burden of the UK population for individuals born between 1920 and 1980. Predicted results agree well with available measured lipid concentrations in human tissues. Its unique construction takes into account both changing environmental levels of PCBs in principal food groups and changing dietary habits during the time period. We can demonstrate that the release profile of PCB-101 controls levels in the food supply and ultimately the burden of individuals throughout their life. This effect is expected to be similar for other PCB congeners and persistent organic compounds such as PCDD/Fs. Models of this type have important applications as predictive tools to estimate the likely impact of source-reduction strategies on human tissue concentrations.

Acknowledgments

Ruth Alcock would like to thank the UK Environment Agency and NERC for supporting a Post-Doctoral Fellowship Award to investigate sources and fate of organic contaminants in the environment. We would also like to thank the UK Ministry of Agriculture, Fisheries and Food (MAFF) and the UK Department of the Environment, Transport and the Regions (DETR) for supporting PCB research at Lancaster University.

LITERATURE CITED

1. Beck, H. et al., 1994. PCDD and PCDF exposure and levels in Germany. *Environ. Health Perspectives.* 102, 173-185.
2. Ministry of Agriculture, Fisheries and Food. (MAFF) 1997. Dioxins and polychlorinated biphenyls in food and human milk. Food Surveillance Information Sheet No 105. MAFF Food Contaminants Division, London
3. Hallikainen, A. and T. Vartiainen 1997. Food control surveys of PCDDs and PCDFs and intake estimates. *Food Additives and Contaminants.* 14. 355-366.
4. Lawrence, G. and A.P. Gobas. 1997. A pharmacokinetic analysis of interspecies extrapolation in dioxin risk assessment. *Chemosphere*, 35, 427-452.
5. Van der Molen, G.W., S. Kooijman and W. Slob. 1996. A generic toxicokinetic model for persistent lipophilic compounds in humans: An application to TCDD. *Fundamental and Applied Toxicol.,* 31, 83-94.
6. Kreutzer, P.E., G.A. Csanady, C. Baur, W. Kessler, O. Papke, H. Greim and J.G. Filser. 1997. 2,3,7,8-TCDD and congeners in infants. A toxicokinetic model of human lifetime body burden by TCDD with special emphasis on its uptake by nutrition. *Arch Toxicol.,* 71, 383-400.
7. Sweetman, A.J. and K.C. Jones. 1999. Modelling historical emissions and environmental fate of PCBs in the UK. ACS Publication Persistent Bioaccumulative Toxic Chemicals: Fate and Exposure. In press.
8. COT (Committee on Toxicity of Chemicals in Food, Consumer Products and the Environment). 1997. Statement on the Health Hazards of Polychlorinated Biphenyls. UK Department of Health. London.
9. Alcock, R.E., P. Behnisch, H. Hagenmaier and K.C. Jones. 1998. Dioxin-like PCBs in the environment *Chemosphere*, 33, 1252-1264.
10. Slob, W. 1993. Modeling long-term exposure of the whole population to chemicals in food. *Risk Analysis*, 13, 525-530.
11. ICRP (International Commission on Radiological Protection) 1992. *Report on the Task Group on Reference Man.* Pergamon Press.

12. Duarte-Davidson. R. et al., 1994. PCBs and other organochlorines in human tissue samples from the Welsh population: 1 - adipose. *Environ. Poll.* 84, 69-77.
13. Ministry of Agriculture, Fisheries and Food. (MAFF) 1991. Household Food Consumption and Expenditure 1990 (National Food Survey) HMSO, London.
14. Schrey, P. Wittsiepe, J. and F. Selenka.1996. Dietary intake of PCDD/Fs by small children measured by the duplicate method. *Organohalogen Compds.* 30,166-171.
15. Schlummer, M., G. A. Moser and M.S. McLachlan. 1998. Digestive tract absorption of PCDD/Fs, PCBs and HCB in humans: mass balance and mechanistic considerations. *Toxicol and Applied Pharmacol.* 152, 128-137.
16. Juan, Ching-Yi, G.O. Thomas and K.C. Jones. 1999. Unpublished data.
17. Brown, J. 1994. Determination of PCB metabolic, excretion and accumulation rates for use as indicators of biological response and relative risk. *Environ. Sci Technol.,* 28, 2295-2305.
18. Rohde, S., G.A. Moser, O. Papke and M.S. McLachlan. 1999. Clearance of PCDDD/Fs via the gastrointestinal tract in occupationally exposed persons. *Chemosphere,* 38, 3397-3410.
19. Abbot, D.C., G.B. Collins, R. Goulding and R.A. Hoodless. 1981. Organochlorine pesticide residues in human fat in the United Kingdom 1976-7. *British Medical Journal* 283, 1425-1428.
20. Alcock, R.E. and K.C. Jones. 1995. The influence of multiple sewage sludge amendments on the PCB content of an agricultural soil over time. *Environ. Toxicol. Chem. 14*, 553-560.
21. Coleman, P.J. et al., 1997. Observations on PAH, PCB and PCDD/F trends in UK urban air, 1991-1995. *Environ. Sci Technol.,* 31, 2120-2124.
22. Kjeller, L-O. and C. Rappe. 1995. *Environ. Sci Technol.,* 29, 346-355.
23. Becher, G. et al., 1995. PCDDs, PCDFs and PCBs in human milk from different parts of Norway. *J. of Toxicol Environ. Heath,* 46, 133-148.
24. Lunden, A. and K. Noren 1998. PCNs and other OC contaminants in Swedish human milk 1972-1992. *Arch Environ. Contam Toxicol.* 34, 414-423.
25. Liem, A.K.D. et al., 1995. PCBs, PCDD/Fs and OCs in human milk in the Netherlands - levels and patterns. *Organohalogen Compounds* 26, 69-74.
26. Sweetman, A.J. and K.C. Jones. 1999. Modelling historical emissions and environmental fate of PCBs in the UK . *Environ. Sci Technol* submitted.
27. Lead, W.A., Steinnes, E, Bacon, J.R. and Jones, K.C. 1997. PCBs in UK and Norwegian soils: spatial and temporal trends. *Sci. Tot. Environ.* 193, 229-236.
28. Gevao, B., J. Hamilton Taylor and K.C. Jones. 1998. Depositional time trends and remobilisation of PCBs in lake sediments. Environ. Sci. Technol., 31, 3274-3280.
29. Thomas, G.O., A.J. Sweetman, R. Lohmann and K.C. Jones. 1998a. Derivation and field testing of air-milk and feed-milk transfer factors for PCBs. *Environ. Sci Technol* 32, 3522-3528.
30. Thomas, G.O., A.J. Sweetman, W. Ockenden, D. Mackay and K.C. Jones. 1998b. Air-Pasture transfer of PCBs. *Environ. Sci Technol* 32,936-942.
31. European Union System for the Evaluation of Substances (EUSES) 1997. Joint Research Center, European Commission.

Chapter 15

Persistence of DDT Residues and Dieldrin off a Pesticide Processing Plant in San Francisco Bay, California

David Young[1], Robert Ozretich[1], Henry Lee II[1], Scott Echols[2], and John Frazier[2]

[1]Western Ecology Division, U.S. Environmental Protection Agency, Newport, OR 97330
[2]CH2M Hill, Corvallis, OR 97330

This paper reports concentrations of DDT residues and dieldrin in surficial sediment, unfiltered near-surface water, intertidal mussels, and epibenthic and pelagic organisms collected from Richmond Harbor more than 25 years after removal of the predominant local source of these contaminants. High concentrations were measured in most of the samples collected. Despite the large concentration gradients away from the source zone, relatively little variability in the   tissue/water bioconcentration factors for the mussel, a benthic Gobiid fish, and the pelagic shiner surfperch was observed.   This indicates that these organisms are potentially useful bioindicators of organochlorine contamination in estuarine ecosystems.

Disclaimer: This information has been funded wholly (or in part) by the U. S. Environmental Protection Agency. It has been subjected to the Agency's peer and administrative review, and it had been approved for publication as an EPA document. Mention of trade names or commercial products does not constitute endorsement or recommendation for use.

Between 1947 and 1966, United Heckathorn and other companies operated a pesticide formulation plant in Richmond, California. The operations resulted in discharges of DDT and dieldrin to the shoreline and the poorly flushed waters of Richmond Harbor of San Francisco Bay (Figure 1). In 1990 U.S. EPA listed the site (since remediated by removal of contaminated sediment) on the National Priorities List, and a study of contamination levels there was conducted by EPA during October 1991 and February

© 2001 American Chemical Society

1992 *(1)*. Here we report results of that study related to the persistence of residues of these organochlorine pesticides in the sediment and water of Richmond Harbor, and their bioconcentration by estuarine organisms.

Methods

Sample Collection and Preparation

Sediment: During the October survey, surficial sediment (0-10 cm) was collected in a 0.1m² van Veen grab sampler at 21 stations within the three inner channels and the outer channel (Figure 1). The sediment samples were collected with pre-cleaned glass corers from the grab and transferred to glass jars with Teflon-lined caps. The jars were placed immediately on " gel" ice in ice chests, returned to the laboratory in a cool state and were processed within 24-36 hours of collection. Aliquots were taken for determination of percent solids (by drying overnight at 105° C), total organic carbon - TOC (by acidifying and high-temperature combustion), and DDT and dieldrin residues. The latter sample (~2 g wet wt.) was spiked with surrogate internal standards (4,4' DDE-d8 and ^{13}C-heptachlor epoxide), desiccated with sodium sulfate, sonicated in acetonitrile and passed through a C-18 solid phase cartridge *(2)*. To maintain the analytical degradation of DDT to DDD and DDE to less that 10%, the biogenic material remaining in the extracts was further reduced by passage through a column of silica gel *(1)*. The resulting extracts were refrigerated until GC/MS analysis.

Water: Near-surface water was collected within ~ 1 meter of the intertidal mussel sampling sites in three inner channels of Richmond Harbor - Lauritzen Channel, Santa Fe Channel, and Richmond Channel (Figure 1). During the October survey, at each site a single sample was collected on three separate days during lower, intermediate, and higher tidal stages. During the February survey, triplicate samples were collected simultaneously at each site on an intermediate tide. Pre-cleaned capped 0.5-liter glass bottles with Teflon-lined caps were immersed from the bow of a small boat, triple-rinsed in the sample without contacting the air-water interface, filled completely while underwater, and capped tightly before withdrawal. The bottles were kept cool in an ice chest, and that evening were air-shipped to the EPA laboratory in Newport, OR. The following morning, water was decanted down to the bottle's shoulder and surrogate internal standards (4,4' DDE-d8 and ^{13}C-heptachlor epoxide) were added to each unfiltered sample, which was shaken and then topped with the extraction solvent (12 ml of 10 % isooctane in hexane). These samples were extracted for 12 to 18 hours by gentle swirling on a shaker table in the dark. The water was removed from the bottom of the bottle via a vacuum probe, and the solvent sample transferred to a pre-cleaned 40 ml vial, reduced in volume with a stream of nitrogen, and placed in a freezer pending GC/MS analysis.

Biota: Intertidal byssal mussels *(4-6 cm long)* were collected at the water sampling sites in the three inner channels during the October and February surveys. Epibenthic invertebrates and benthic and pelagic fishes also were collected by trawling along a fixed transect in each channel (Figure 1). All specimens were wrapped in pre-cleaned aluminum foil and frozen pending sample preparation, which was conducted

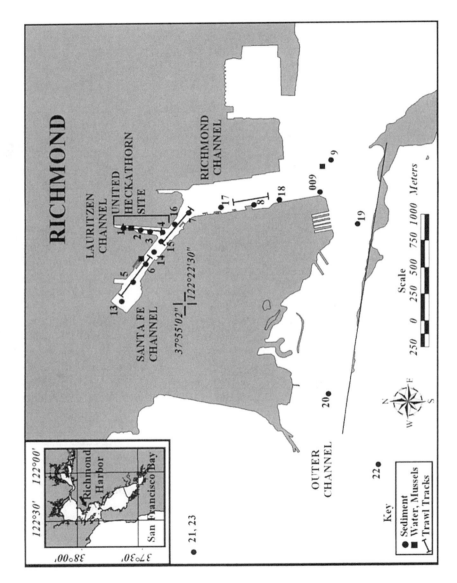

Figure 1. Collection sites in Richmond Harbor of San Francisco Bay, California.

within a few months of collection. In the laboratory, ten mussels (*Mytilus galloprovincialis*) were randomly selected, the byssus removed, and the soft tissues excised and composited. Composite samples of the epibenthic bay shrimp (*Crangon franciscorum*), briefly rinsed free of attached sediment in a stream of filtered seawater, also were obtained when possible. These samples, as well as whole body samples of an epibenthic Gobiid fish (probably the bay goby *Lepidogobius lepidus*), the pelagic shiner surfperch (*Cymatogaster aggregatus*), and a pelagic Anchoveta fish (probably the northern anchovy *Engraulis mordax*), also were collected and rinsed of attached sediment. The composite or single-specimen tissue samples were frozen in liquid nitrogen and pulverized with mortar and pestle. Subsamples were distributed for pesticide, percent solids, and lipid analyses. Subsamples were spiked with internal surrogate standards (4,4' DDE-d8 and ^{13}C-heptachlor epoxide), and pesticide extracts were obtained following the procedures of Ozretich and Schroeder (2) using C-18 and aminopropyl solid phase cartridges for the initial cleanup, followed by passage through a column of silica gel to further remove biogenic material in the extracts (1). Percent solids of most tissues were determined following freeze-drying; lipid content then was measured with a methanol-chloroform micro-volume technique (3).

GC/MS Analysis

Quantitation was by capillary gas chromatography – mass spectrometry selected ion monitoring using response factors relative to surrogate internal standards added prior to sample extraction. Pesticide identity was confirmed by retention time and use of characteristic ion pairs. Compound identities were considered confirmed if their peaks were found at expected retention times with ratios of ions within 20% of those expected from authentic standards. The level of degradation of DDT to DDE and DDD during the analysis was determined daily and corrections were applied to the extract concentrations of these compounds. Less than 10% degradation of DDT was observed during the course of this study.

Results and Discussion

Surficial Sediment

The magnitude of contamination of Richmond Harbor by DDT and dieldrin residues is indicated by the surficial sediment concentrations (Table I). The median of the four ΣDDT values for Lauritzen Channel was about 37,000 μg/kg (dry wt.); medians for Santa Fe Channel, Richmond Channel, and Outer Channel were 630, 60, and 15 μg/kg, respectively. Corresponding values for dieldrin were 480, 14, 1.2, and 1.9 μg/kg. For each pesticide there was a clear gradient away from the head of Lauritzen Channel, near the former site of the pesticide formulation plant. In addition, the distribution of ΣDDT changed substantially along this gradient. Within Lauritzen Channel, the two primary residues were 4,4'-DDT and 4,4'-DDD, which together constituted 85% - 90% of the

Table I. ΣDDT (with Percent Composition) and Dieldrin in Surface[a] Sediment (μg/kg dry wt.)

Station[b]	TOC[c]	ΣDDT	2,4'-DDE	4,4'-DDE	2,4'-DDD	4,4'-DDD	2,4'-DDT	4,4'-DDT	Dieldrin
1	2.38	77,700	0.1	2.0	8.2	42.1	3.3	44.4	748
2	1.78	47,800	0.1	1.8	5.3	32.8	2.3	57.6	528
3	1.73	26,000	0.1	2.5	9.6	58.9	0.6	28.4	442
4	1.46	2,740	0.5	4.0	7.9	55.5	0.0	32.1	36
13	1.55	556	0.8	8.7	11.1	67.5	0.0	11.9	17
5	1.49	420	0.6	11.0	9.6	57.5	1.2	20.0	5.5
6	2.98	2,340	1.5	24.8	10.2	40.6	2.8	20.1	78
14	1.51	730	0.4	7.3	10.8	69.4	0.0	12.1	17
15	1.46	522	0.4	5.9	9.8	65.4	0.9	17.6	9.4
16	1.35	696	0.4	5.7	6.9	49.2	0.5	37.3	11
7	1.08	368	0.6	7.4	9.5	59.6	0.5	22.4	9.5
17	1.34	132	0.5	7.3	8.7	54.7	0.0	28.9	3.0
8	1.18	82	0.7	11.5	9.9	56.0	0.5	21.3	1.7
18	1.25	38	1.6	12.2	9.6	63.5	0.0	13.3	0.0
009	1.22	24	1.2	15.4	11.7	58.3	0.0	14.2	0.0
9	0.87	12	0.9	17.2	11.2	45.7	0.0	25.9	0.6
19	1.14	17	2.4	18.1	9.6	41.0	9.0	21.1	2.4
20	1.10	15	2.7	17.8	9.6	41.8	0.0	28.8	1.9
22	0.76	11	2.7	10.7	14.3	43.8	0.0	28.6	0.0
21	1.05	18	3.2	15.1	12.4	52.4	0.0	18.4	0.0
23	1.02	15	2.7	17.1	10.3	34.2	0.0	35.6	3.4

a. Sediment layer: 0-10 cm
b. Lauritzen Ch.: Sta. 1-4; Santa Fe Ch.: Sta. 13, 5, 6, 14, 15, 16; Richmond Ch.: Sta. 7, 17, 8, 18, 009, 9; Outer Ch.: Sta. 19-23 (Figure 1)
c. Percent dry wt.

ΣDDT, while 4,4'-DDE constituted only about 2%. However, within Richmond Channel, the median concentration of 4,4'-DDT was lower by about a factor of two, while that of 4,4'-DDE was higher by a factor of four. Over the entire study area, the median ratio of 4,4'-DDD to 4,4'-DDT was about 2:1. This relative high ratio is consistent with the suggestion by Pereira et al. *(4)* and Venkatesan et al. *(5)* that much of the DDD measured in Richmond Harbor is attributable to the facility's processing and loss of technical DDD, which also was used as an agricultural insecticide, rather than to reductive dechlorination of DDT.

Near-Surface Water

Concentrations of ΣDDT and dieldrin measured in the near-surface waters collected from the three channels in October 1991 and February 1992 were remarkably similar (Table IIA). For Lauritzen Channel, average values for ΣDDT were 50 and 50 ng/liter, respectively. Corresponding values for dieldrin were 21 and 15 ng/liter. Similar agreement between seasonal values was observed for the other two inner channels. There is some indication that the water concentrations obtained from the October survey varied with tidal height; in four of the five cases where a comparison is possible, the average pesticide concentration at the lowest tide level exceeded that at the highest tide level. The three values obtained from the replicates collected sequentially at a station during the February survey showed a high degree of precision; on average, the relative standard error obtained was less than 10%. As was the case for the surficial sediment samples, somewhat more than half of the ΣDDT in the water samples from the three channels was 4,4'-DDD (Table IIB). In addition, the approximate two-fold decrease in the percentage composition of 4,4'-DDT from Lauritzen to Richmond Channel observed for surficial sediment also was obtained for the near-surface water samples. However, in contrast to the pattern observed for sediment, there was no substantial increase in the percentage composition of 4,4'-DDE measured in the near-surface water samples.

Intertidal Mussels

As was the case for sediment and water, large spatial gradients also were observed for ΣDDT and dieldrin in whole soft tissue of the intertidal mussel *Mytilus galloprovincialis* (Table IIIA). However, in contrast to the case for water, substantial differences were obtained between average (wet weight) concentrations in the Lauritzen and Santa Fe Channel samples from the October and February survey. These differences generally correspond to those measured in the lipid content of the mussel tissues, where, on average, the February values were lower than those for October by about a factor of three. As was the case for sediment and water, 4,4'-DDD constituted the largest percent of ΣDDT in the mussel tissue (Table IIIB). Also, percent values for 4,4'-DDT were highest, and percent values of 4,4'-DDE were lowest, in the Lauritzen Channel mussels. The percent composition results listed in Table IIIB contrast highly with those reported for the coastal mussel *M. californianus* collected in 1974 inshore of the municipal

210

Table IIA: ΣDDT and Dieldrin in Near-surface Water[a] (ng/liter)

Channel	Oct Tide[b]	Repl.	Oct.	Feb.	Combined	Oct.	Feb.	Combined
			ΣDDT			Dieldrin		
Lauritzen	6.0	1	60	54		19	16	
	4.9	2	22	50		10	15	
	2.1	3	68	47		34	13	
Ave.			50	50	50	21	15	18
S.E.			15	2	6	7	1	3
Santa Fe	5.7	1	9.3	9.8		2.8	1.7	
	5.5	2	5.1	6.6		3.1	1.7	
	2.0	3	14	6.7		<1.0	1.7	
Ave.			9.4	7.7	8.6	2.0	1.7	1.8
S.E.			2.5	1.1	1.3	1.0	0.0	0.4
Richmond	6.1	1	<0.2	1.5		<1.0	<1.0	
	2.9	2	0.3	1.5		<1.0	<1.0	
	2.2	3	1.5	1.3		<1.0	<1.0	
Ave.			0.6	1.4	1.0	<1.0	<1.0	<1.0
S.E.			0.5	0.1	0.3	- -	- -	- - -
Proc. Blank[c]			0.7	0.2		<1.0	<1.0	

Table IIB. Average Percent Composition of ΣDDT in Water[a]

Channel	ΣDDT[d]	2,4'-DDE	4,4'-DDE	2,4'-DDD	4,4'-DDD	2,4'-DDT	4,4'-DDT
Lauritzen	50	0.4	3.7	16.4	54.7	8.0	16.8
Santa Fe	8.6	1.6	5.1	20.2	57.8	5.8	9.5
Richmond	1.0	0.0	4.6	29.1	57.4	0.0	8.8

a. Unfiltered samples: water depth 0.3 m
b. Surface water elevation (ft.) above Mean Lower Low Water
c. Concentrations not corrected for procedural blank values
d. Units: ng/liter

Table IIIA. ΣDDT and Dieldrin in Intertidal Mussels[a] (μg/kg wet wt.)

Channel	Oct.	% Lipid	Feb.	% Lipid	Combined	Oct.	Feb.	Combined
			ΣDDT				Dieldrin	
Lauritzen								
Med.	2,600		670		1,600	100	23	57
Ave.	5,100	1.42	680	0.48	2,900	170	25	97
S.E.	1,700		150		1,100	56	3	36
(n)	(5)		(5)		(10)	(5)	(5)	(10)
Santa Fe								
Med.	480		110		440	32	6	18
Ave.	520	2.27	180	0.86	350	30	8	19
S.E.	42		67		69	3	2	4
(n)	(5)		(5)		(10)	(5)	(5)	(10)
Richmond								
Med.	40		36		38	<2	3	3
Ave.	40	1.71	40	0.95	40	2	5	4
S.E.	12		6		6	1	2	1
(n)	(5)		(5)		(10)	(5)	(5)	(10)
Control[b] (OR Coast)								
Med.			1.5				<1.7	
Ave.			1.4	1.05			<1.7	
S.E.			0.1				---	
(n)			(5)				(5)	
Anal. Blank			<3.8				<1.7	

Table IIIB. Average Percent Composition of ΣDDT in Mussels[a]

Channel	ΣDDT [c]	2,4'-DDE	4,4'-DDE	2,4'-DDD	4,4'-DDD	2,4'-DDT	4,4'-DDT
Lauritzen	2,900	0.8	11.1	11.4	42.4	11.6	22.8
Santa Fe	350	1.0	14.8	11.3	54.9	4.7	13.8
Richmond	40	1.5	21.2	10.4	43.9	4.2	18.8

a. *Mytilus galloprovincialis*; whole soft tissue
b. *Mytilus californianus*; whole soft tissue
c. Units: μg/kg wet wt.

wastewater outfall system off Palos Verdes Peninsula, a major source of DDT wastes. In that study, 4,4'-DDD and 4,4'-DDT constituted only about 3% and 1% , respectively, of the ΣDDT measured in the whole soft tissue, while 4,4'-DDE constituted about 85% of the ΣDDT (6).

Epibenthic and Pelagic Biota

Distinct spatial gradients of the target pesticides also were measured in the whole body concentrations of two epibenthic organisms, the bay shrimp *Crangon franciscorum* and a Gobiid fish (Table IV). This also was observed for the pelagic shiner surfperch (*Cymatogaster aggregatus*), but not for the Anchoveta (Table V). Available percent composition data yielded an average value of about 60 percent 4,4'-DDD in the whole body samples of the Gobiid fish (Table IVB), similar to the values presented above for sediment and water. However, a much lower average value for 4,4'-DDT (7 percent), and a much higher average value for 4,4'-DDE (29 percent) was measured in this epibenthic fish compared to the surficial sediment (Table I). Also, the percent composition data obtained for the whole body samples of the epibenthic Gobiid fish contrast with corresponding results for muscle tissue of shiner surfperch collected during 1997 from other parts of the Bay; average values reported for 4,4'-DDE, 4,4'-DDD, and 4,4'-DDT were 72%, 21%, and <4%, respectively (7). Even greater contrasts exist between the average percent composition values for the Gobid fish and muscle tissue of another epibenthic fish, the flatfish Dover sole (*Microstomus pacificus*), collected from the municipal wastewater discharge zone off Palos Verdes Peninsula, CA, in 1971-72. In those samples corresponding values for the three compounds were 84%, 6%, and 0.7% respectively (8). This high percentage of 4,4'-DDE found in the 1971-72 study of Dover Sole (and other organisms) is consistent with recent reports that 4,4'-DDE constitutes >80% of ΣDDT in sediments and biota off southern California (9, 10). Such agreement between percent composition values obtained over two to three decades argues against the hypothesis that the earlier reports of relatively high DDE metabolites in the region might be due principally to measurement error caused by degradation of DDT during analysis.

Comparison of Spatial Gradients

A listing of the comparable average concentrations of ΣDDT and dieldrin obtained for the samples discussed above is presented in Table VI. (The sediment averages are based on values for those stations included in, or immediately bracketing, the trawl tracks). This comparison illustrates the large spatial gradients measured in all sample types except the highly motile Anchoveta fish. The largest gradient was observed in the surficial sediment (associated with the trawl tracks), where the respective average ΣDDT and dieldrin values for Lauritzen Channel were 600 and 360 times those for Richmond Channel. In comparison, results from a similar sediment contamination survey of the area conducted in 1993 yielded site concentration ratios of 250:1 for ΣDDT and 270:1 for

Table IVA. ΣDDT and Dieldrin in Epibenthic Shrimp[a] and Gobiid Fish[b]
(μg/kg wet wt.)

| Channel | Shrimp | | Gobiid | | |
	ΣDDT[c]	Dieldrin[c]	ΣDDT[c]	Dieldrin[c]	% Lipid
Lauritzen					
Med.	310	12			
Ave.	310	12	5,400	200	2.2
S.E.	4	12			
(n)	(2)	(2)	(1)	(1)	
Santa Fe					
Med.	99	3	730	28	
Ave.	110	3	680	28	2.9
S.E.	10	1	57	3	
(n)	(3)	(3)	(5)	(5)	
Richmond					
Med.	24	2	120	6	
Ave.	24	2	130	6	1.1
S.E.	3	2	19	3	
(n)	(2)	(2)	(3)	(3)	

Table IVB. Average Percent Composition of ΣDDT in Gobiid Fish[b]

Channel	ΣDDT[d]	2,4'-DDE	4,4'-DDE	2,4'-DDD	4,4'-DDD	2,4'-DDT	4,4'-DDT
Lauritzen	5,400	0.1	44.0	0.6	45.9	0.4	9.0
Santa Fe	680	0.3	26.3	1.9	65.6	0.3	5.6
Richmond	130	1.3	16.8	2.8	72.7	0.0	7.4

a. *Crangon franciscorum*; whole body (no lipid value)
b. Probable *Lepidogobius lepidus*; whole body
c. Values for October 1991 and February 1992 surveys combined
d. Units: μg/kg wet wt.

214

Table VA. ΣDDT and Dieldrin in Shiner Surfperch[a] and Anchoveta Fish (µg/kg) wet wt.)

| Channel | Surfperch | | Anchoveta | | |
	ΣDDT[b]	Dieldrin[b]	ΣDDT[b]	Dieldrin[b]	% Lipid
Lauritzen					
Med.	8,300	340			
Ave.	7,500	390	98	15	1.8
S.E.	1,500	90			
(n)	(n)	(5)	(1)	(1)	
Santa Fe					
(n=1)	920	40	670	17	1.9
Richmond					
Med.	91	---			
Ave.	100	2	170	4	1.8
S.E.	21	2			
(n)	(5)	(5)	(1)	(1)	

Table VB. Average Percent Composition of ΣDDT in Anchoveta Fish

Channel	ΣDDT[c]	2,4'-DDE	4,4'-DDE	2,4'-DDD	4,4'-DDD	2,4'-DDT	4,4'-DDT
Lauritzen	7,500	0.0	20.1	9.1	49.4	5.9	15.5
Santa Fe	920	0.5	26.4	6.1	51.4	4.0	11.6
Richmond	100	1.0	38.6	5.2	42.2	3.0	10.0

a. *Cymatogaster aggregatus*; whole body (no lipid value)
b. Values for October 1991 and February 1992 surveys combined
c. Units: µg/kg wet wt.

Table VI. Spatial Comparison of ΣDDT and Dieldrin Average Concentrations[a,b]

Channel	Sediment	Water	Mussel	Shrimp	Gobiid	Surfperch	Anchoveta
				ΣDDT			
Lauritzen	50,500	50	2,900	310	5,400	7,500	98
Santa Fe	1,010	8.6	350	110	680	920	670
Richmond	84	1.0	40	24	130	100	170
				Dieldrin			
Lauritzen	570	18	97	12	200	390	15
Santa Fe	29	1.8	19	3	28	40	17
Richmond	1.6	<1	4	2	6	2	4

a. Sediment concentration average values based on stations included within or immediately bracketing trawl tracks (Figure 1)
b. Water: ng/liter; sediment: μg/kg dry wt.; tissue: μg/kg wet wt.

Table VII. BCFs[a] for ΣDDT and Dieldrin in Biota

Channel	ΣDDT				Dieldrin			
	Mussel	Gobiid	Surfperch	Anchoveta	Mussel	Gobiid	Surfperch	Anchoveta
Lauritzen	58,000	108,000	150,000	1,960	5,390	11,100	21,700	830
Santa Fe	40,700	79,100	107,000	77,900	10,600	15,600	22,200	9,440
Richmond	40,000	130,000	100,000	170,000	---	---	---	---
Ave.	46,200	106,000	119,000	83,300	8,000	13,400	22,000	5,140
S.E.	5,890	14,700	15,600	48,600	2,600	2,250	250	4,300

a. Bioconcentration Factor: Ratio of combined average concentration in tissue to combined average concentration in unfiltered water.

dieldrin *(4)*. Large ratios also were obtained between average contamination levels at these two sites (Table VI) for ΣDDT concentrations in the water (50:1), mussel (72:1), Gobiid (42:1), and surfperch (75:1) samples. Corresponding ratios for dieldrin were >18:1, 24:1, 33:1, and 195:1. These results indicate a relatively small degree of mixing of water or specimens of these two fishes from Lauritzen and Richmond channels.

Bioconcentration Factors.

Tissue/water bioconcentration factors (BCF) for both ΣDDT and dieldrin in the intertidal byssal mussel, the epibenthic Gobiid fish, and the pelagic surfperch (TableVII) were remarkably similar for the three channel areas. Relative standard error values for ΣDDT in the three estuarine organisms were 13 %, 14%, and 13%, respectively. Corresponding RSE values for dieldrin were 32%, 17%, and 1%. These results indicate that, in addition to the byssal mussel, the two fishes were relatively non-motile, and came to quasi-equilibrium in each of the exposure environments of the three channels. Thus, all three organisms would appear to be useful bioindicators of organochlorine contaminated sites in estuaries.

Acknowledgments

We thank David Specht, Kathy Sercu, and other scientists of the U.S. EPA laboratory at Newport, OR, and the on-site technical contractor AScI, for assistance in this research. We also thank Dr. Jay Davis, San Francisco Estuary Institute (CA), and Dr. Renee Falconer, Youngstown State University (OH) for their guidance. Finally, the support of Dr. Andrew Lincoff, U.S. EPA Region IX (San Francisco, CA) throughout this study is gratefully acknowledged.

Literature Cited

1. "Ecological Risk Assessment of the Marine Sediments at the United Heckathorn Superfund Site"; Lee II, H., Ed.; Final Report to U.S. EPA Region IX , ERL-N-269 (May 20, 1994); U.S. Environmental Protection Agency. Office of Research and Development. Environmental Research Laboratory: Newport, OR, 1994; EPA-600/X-94/029.
2. Ozretich, R.J.; Schroeder, W. P. *Anal. Chem.* **1986**, *58*, 2041-2048.
3. "United Heckathorn Superfund Site Study"; Lee II, H., Ed.; Planning Document No. ERL-N -199; U.S. Environmental Protection Agency. Office of Research and Development. Environmental Research Laboratory: Newport, OR, 1991; EPA-600/X-91/121.
4. Pereira, W.E.; Hostettler, F.D.; Rapp, J.B. *Mar. Environ. Res.* **1996**, *41*, 299-314.
5. Venkatesan, M.I.; deLeon, R.P.; van Geer, A.; Luoma, S.N. *Mar. Chem.* **1999**, *64*, 85-97.

6. Young, D.R.; Heesen, T.C.; McDermott, D.J. *Mar. Pollut. Bull.* **1976**, *7*, 156-159.
7. "Contaminant Concentrations in Fish from San Francisco Bay – 1997"; Technical Report of the San Francisco Estuary Regional Monitoring Program for Trace Substances, RMP Contribution No. 35 (May 1999); San Francisco Estuary Institute: Richmond, CA, 1999.
8. Young, D.R.; McDermott, D.J.; Heesen, T.C. *J. Wat. Pollut. Contr. Fed.* **1976**, *48*, 1919-1928.
9. Stull, J.K.; Swift, D.J.P.; Niedoroda, A.W. *Sci. Tot. Environ.* **1996**, *179*,73-90.
10. Zeng, E.Y.; Venkatesen, M.I. *Sci. Tot. Environ.* **1999**, *229*, 195-208.

Chapter 16

Deposition of Atmospheric Semivolatile Organic Compounds to Vegetation

Martine I. Bakker[1], Johannes Tolls[2], and Chris Kollöffel[1]

[1]Faculty of Biology, Transport Physiology Group, Utrecht University, P.O. Box 80084, NL 3508 TB Utrecht, The Netherlands
[2]Research Institute of Toxicology (RITOX), Environmental Toxicology and Chemistry, Utrecht University, P.O. Box 80058, NL 3508 TB Utrecht, The Netherlands

Introduction

In the last decades, there has been growing attention for the deposition of semivolatile organic compounds (SOCs) to vegetation [1]. SOCs, such as polychlorinated biphenyls (PCBs), polycyclic aromatic hydrocarbons (PAHs) and polychlorinated dibenzo-p-dioxins and dibenzofurans (PCDD/Fs), have vapor pressures roughly between 10^1 and 10^{-6} Pa [2]. They originate from a variety of anthropogenic (and also natural) sources and are of special interest because of their global distribution, persistence, tendency to bioaccumulate and known or suspected toxicity.

From the early 1980s, the cuticles of leaves and pine needles have been found to accumulate gaseous SOCs from ambient air [3-7]. Also, mosses [8, 9], tree bark [10] and lichens [11] are shown to accumulate these compounds.

The contamination of vegetation with SOCs causes concern, as vegetation is a starting point for food chain transfer to animals and humans. This is of major importance for the exposure of the human population to SOCs in industrialized countries, particularly via the pathway atmosphere → vegetation → cattle → milk/ dairy/ beef → humans [12-16].

A second point of interest is that vegetation serves as a sink for SOCs and in this way cleanses the air [16-18]. It was estimated that 44% of the PAH emissions in the northeastern United States was removed from the atmosphere by vegetation [17]. However, after sampling at more locations, this estimate was corrected to 4% [18]. Forests are shown to be effective air filters for organic chemicals [19, 20], at least for those with $\log K_{oa}$ (octanol-air partition coefficients) values between 7 and 11. This range includes a large number of PAHs, PCBs and PCDD/Fs [21].

© 2001 American Chemical Society

Naturally, plants can only act as a temporary sink, since they live only for a limited period of time. Litterfall [19, 22] and erosion of wax layers [22, 23] serve as an important vector for SOCs to the soil. Furthermore, there is some evidence that, after deposition to vegetative surfaces, some SOCs (PCDD/Fs) are photodegraded by sunlight [24, 25], but this could not be confirmed [26].

Finally, the deposition of SOCs to vegetation gets attention because plants are being used as biomonitors for air pollution, as vegetation integrates contamination over time and plant samples are much easier to collect than air samples. Plants can be used to identify point sources and to qualitatively determine regional and global contamination levels of organic pollutants [1].

The scope of this paper is to cover the mechanisms and kinetics of atmospheric deposition of SOCs to plants, and to discuss the factors influencing the deposition process. Uptake from soil via roots will not be addressed, because the contamination of aerial plant parts with SOCs occurs primarily via the atmosphere. As SOCs are very hydrophobic and therefore have low water solubilities, they partition to the epidermis of the root, and uptake via the roots followed by translocation to the shoots is negligible for most plants [27-31]. The only exceptions known are plants of the family of *Cucurbitaceae* (zucchini, pumpkin) [32, 33].

Little research has been done on toxicity of SOCs to or degradation in plants. Most of these studies investigated these processes with cell cultures. Hence, little is known of the relevance of these processes in intact plants. For these reasons, neither toxicity nor degradation will be discussed with in this paper.

Deposition Pathways

Introduction

Deposition of atmospheric SOCs to plants occurs via several pathways. Because SOCs partition between the gas phase and atmospheric particles, a major division can be made between gaseous and particle-bound deposition. Compounds bound to atmospheric particles can reach the plant surface by both dry and wet deposition. Since the solubility of the hydrophobic SOCs is very low in rain droplets or other precipitation, wet deposition of gases is of minor importance [21, 34].

Also, contaminated soil particles can be transported directly to the plant surface by wind or splash [35, 36]. Uptake of compounds that are volatilized from highly contaminated soil is another pathway, particularly in more basal aerial plant parts [35].

The deposition of airborne compounds, whether they are gaseous or particle-bound, to plant leaves involves three steps (Figure 1). First, the contaminant is transported from the (turbulent) atmosphere to the laminar air boundary layer surrounding the leaf. Subsequently, the contaminant has to be carried across the boundary layer. In this layer the airflow is parallel to the leaf surface, and the wind speed is highly reduced but increases with distance from the surface [37]. The final step in the uptake process is the interaction of the compound with the leaf surface. Deposited particles may simply adhere to the leaf surface, react with it chemically, or bounce off [38]. For gases, the third step comprises the adsorption to the surface or diffusion into the cuticle [39].

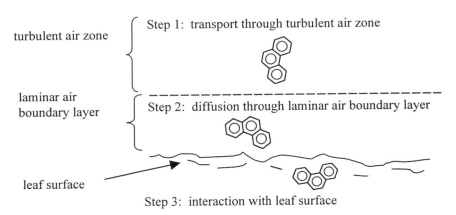

Figure 1. Three steps in the uptake process. (Reproduced with permission from reference 40. Copyright 1999 Environmental Toxicology and Chemistry.)

The different deposition pathways are all a function of 1) the physicochemical properties of the compound (such as vapor pressure, hydrophobicity, molecular weight), 2) environmental characteristics (temperature, wind) and 3) plant characteristics (surface area, lipid content and composition, architecture of the plant) [27].

Particle-bound Deposition

The particle-bound fraction of SOCs in the atmosphere depends on the ambient temperature, the available particle surface and the compound's volatility [41]. Recently, Finizio *et al.* [42] proposed the compound's K_{oa} (octanol-air partition coefficient) to govern the gas-particle distribution, with high K_{oa} values favoring high particle-bound fractions.

The particle size distribution of PAHs has been intensively studied and shows that PAHs are largely bound to particles < 1-2 μm [43-48]. Data for other SOCs of interest are scarce. Nevertheless, also PCDD/Fs were shown to be primarily associated with small particles [48]. However, although the SOC loading of the large particles is low, dry particle deposition fluxes to the earth's surface appear to be dominated by large particles, since their deposition velocity is relatively high [48, 49]. In contrast, wet deposition of particle-bound PCDD/Fs was reported to be dominated by fine particles [48].

Whether dry or wet deposition is the most important route of particles to plant surfaces is primarily dependent on the amount of precipitation. In a coastal environment, wet deposition is likely to be of more importance in supplying SOCs than at an inland site [36, 50]. Nonetheless, the role of wet deposition of SOCs to plant surfaces is not very clear. On one hand, results indicate that rain and hail increased plant concentrations [51], but on the other hand precipitation may cause a wash-off of already deposited material [52]. In the remainder of this chapter we will consider wet and dry particle deposition as a whole.

SOC transfer from the deposited particles to the cuticle has been studied with wash-off experiments. Washing of lettuce with water removed a considerable amount of the high molecular weight (MW) PAHs, but little of the small PAH phenantrene, indicating that only the latter, gaseous PAH was sorbed in the cuticle [53]. In contrast, rinsing maize leaves with aqueous solutions could only extract a minor part of the high MW PCDD/Fs and PAHs, suggesting that the compounds were desorbed from the particles or that the particles were encapsulated in the cuticle [51]. While this remains inconclusive, we will use the term uptake only to refer to the dry deposition of gases, whereas this term will not be used for the deposition of particle-bound SOCs.

Dry Gaseous Deposition

Mechanism

The uptake of gaseous SOCs involves passive diffusion of the compounds between the atmosphere and the cuticle of the plant. The cuticle is an extracellular, non-living, lipid layer which forms the interface between the atmosphere and the plant and protects the plant from desiccation and from fungal and insect attack [54]. It is usually characterized by the presence of two specific classes of lipids: soluble waxes and insoluble polyester cutins [55]. Recently, it has been proposed that also another major lipid polymer, cutan, is present in the cuticular membrane [56]. Waxes occur embedded in the cuticular membrane (intracuticular waxes) and lying across the surface of the membrane (epicuticular waxes). They consist of complex mixtures of long-chain aliphatic and cyclic components [54, 57]. The cuticle is not homogeneous, but composed of a number of layers of which the characteristics vary according to species, age of the plant [57] and environmental conditions [58].

Uptake of SOCs via stomata is probably not important. They will prefer to enter the cuticle, hence the contribution of the stomatal pathway will be negligible to that of the cuticle [59]. Moreover, there is some evidence that the cuticle extends farther along the cell walls that form the epidermal boundary of the substomatal cavity [60]. So, even after entering the stomata, SOCs can still be sorbed to these "internal" cuticular waxes.

Partitioning process

The diffusion of gaseous SOCs into the cuticle can be viewed as a chemical partitioning process between the gas phase and the vegetation, in which a compound is transferred from the atmosphere to the plant until an equilibrium has been reached [27, 50]. Relationships have been found between experimentally determined plant-air concentration factors (K_{pa}) and the octanol-air partition coefficient (K_{oa}) [1, 61-65]. The K_{oa} of a compound can be estimated by dividing K_{ow} by the dimensionless Henry's Law constant. In addition, direct measurements have been performed and for an increasing number of SOCs (PCB's, chlorinated benzenes, PAHs), measured K_{oa} values at different temperatures are now available [66-68].

In many experiments a high correlation between $\log K_{pa}$ and $\log K_{oa}$ is found (Table I) [63-65, 69], which means that the air-to-vegetation and air-to-octanol free energies of phase change are similar. When the slope of a plot of $\log K_{pa}$ versus $\log K_{oa}$ is equal to one (as in [63], see Table I), K_{pa} and K_{oa} are linearly related and the lipid fraction of the plant behaves as octanol. As a consequence, under equilibrium conditions, accumulation of SOCs in vegetation can be described as

$$K_{pa} = L \cdot K_{oa} \qquad (1)$$

in which L is the lipid fraction of the plant [27, 70].

Table I. Slopes and correlation coefficients of plots of $\log K_{pa}$ (or $C_{vegetation}/C_{air}$) versus $\log K_{oa}$ from several studies

Plant	Compounds	Study	Slope	r^2	Reference
azalea	a, b, c	lab	0.91[e]	0.85	[69]
ryegrass	b, c, d	lab	1.00[e]	0.95	[63]
needles, leaves seeds, tree bark	d	field	0.48[e]	0.98	[64]
pasture	b	field	0.32-0.47	0.66-0.96	[71]
ryegrass	b	lab	1.15	0.98	[65]
clover	b	lab	0.70	0.86	[65]
plantain	b	lab	0.87	0.98	[65]
hawk's beard	b	lab	0.74	0.97	[65]
yarrow	b	lab	0.57	0.93	[65]
ryegrass	b, c, d	field	0.60[f]	0.93	[72]
creeping thistle	b, c, d	field	0.62[f]	0.96	[72]
dandelion	b, c, d	field	0.78[f]	0.98	[72]
plantain	b, c, d	field	0.65[f]	0.96	[72]
yarrow	b, c, d	field	0.35[f]	0.72	[72]
lady's mantle	b, c, d	field	0.53[f]	0.95	[72]
sunflower	b, c, d	field	0.39[f]	0.85	[72]
autumn hawkbit	b, c, d	field	0.77[f]	0.97	[72]
white clover	b, c, d	field	0.66[f]	0.90	[72]
corn	b, c, d	field	0.57[f]	0.90	[72]

a: pesticides, b: PCBs, c: chlorobenzenes, d: PAHs e: K_{oa}'s calculated instead of measured, f: calculated with the nine compounds from reference [72] with the lowest K_{oa} values (for which equilibrium was approached).

However, $\log K_{pa}$-$\log K_{oa}$ plots for many plant species give slopes different from one. In a field experiment, a slope of the log-log plot for PAHs in needles, leaves and tree bark of 0.48 was observed [64], while the slope for PCBs in a field pasture amounted to ~0.4 [71] (Table I). Kömp and McLachlan [65] and Böhme et al. [72], studied uptake of SOCs in different plant species and found slopes varying considerably between species (Table I). The large differences between the slopes were attributed to the differences in lipid quality [65]. Another explanation for shallow slopes of the log-log plots is a deviation from equilibrium, leading to greater underestimations of K_{pa}'s of the compounds with higher K_{oa} values. However, equilibrium conditions were checked in all experiments, except for reference [64].

A slope smaller than one indicates that the lipophilicity of octanol is higher than that of the vegetation and vice versa. Hence, when slopes deviate from one, octanol is not a good descriptor for plant lipids [65].

One- or two-compartment model

Bacci and coworkers [61, 73, 74] performed controlled uptake and clearance experiments with azalea and a series of organic compounds. The results were interpreted using a simple one-compartment linear first order model of diffusive exchange, in which the leaf behaves as a well-mixed homogeneous compartment. This model is described by the following equation:

$$\frac{dC_{leaf}}{dt} = k_1 \cdot C_{air} - k_2 \cdot C_{leaf} \qquad (2)$$

in which C_{leaf} (ng/g) and C_{air} (ng/g) are the concentrations in the leaf and in the air, t is time (h), and k_1 (1/h) and k_2 (1/h) are the uptake and elimination rate constant, respectively. When $t \rightarrow \infty$, equilibrium is approached and the ratio of C_{leaf} and C_{air} is equal to K_{pa} (= k_1/k_2). This model was used to describe uptake of SOCs in spruce needles [75], grass [24], and in a number of different other plant species (both foliage and fruits) [76].

Several other researchers tried to fit their uptake and clearance data into a model. They concluded that a one-compartment model failed to describe the data and that a second compartment should be added to the model leaf [39, 63, 77-79]. In the two-compartment model proposed by Tolls and McLachlan [63], the leaf consists of a small surface compartment, reacting fast to changing air concentrations, connected to a larger reservoir compartment. The model described the experimental data well [63, 78].

Whether a model with one or two compartments is preferable to describe plant uptake, depends on the timescale of the experiment. For most purposes the one-compartment model will suffice to describe the data, although it is clear that the assumption that a leaf is one homogeneous compartment is not fulfilled. In experiments in which short-term kinetics are studied, an additional surface compartment may be needed to accurately describe the concentration changes in the plant.

Transport through the Cuticle

Designation of the compartments

Although the model fits showed that a leaf consists of more than one compartment, it is not clear which parts of the leaf actually represent these compartments. It has been proposed that, in needles, compounds first adsorb to the surface and subsequently partition into the cuticle [39]. Also, parts of the interior of conifer needles, namely membrane lipids and essential oil (mainly composed of lipophilic terpenoids), were thought to act as a reservoir compartment [79].

In experiments in which needles and leaves were separated in a leaf wax and inner compartment by means of fractionated extractions, the presence of small SOCs in the inner compartment was demonstrated [5, 78]. As compounds can move through the cuticle exclusively by molecular diffusion, their mobility is inversely proportional to their molar volume [80-84]. Consequently, only the small compounds could reach the inner compartments. Reischl [5] proposed that the compounds diffused through the cuticular waxes (surface compartment) to accumulate in the cutin or possibly the interior of the needle (reservoir compartment). Riederer and coworkers also pointed out the high sorption capacity of cutin [85, 86].

Tortuosity of cuticular waxes

The thickness of the cuticle can vary from tenths of micrometers to more than ten micrometers [87], while the thickness of the epicuticular wax layer is estimated on 10% of the cuticle thickness [88]. According to Fick's Law, the time needed to diffuse through a layer depends on the thickness of the layer. However, diffusion through cuticles seems independent of its thickness [87, 89]. Because extraction of the cuticular waxes from the cuticle increased the mobility of the compounds by one to three orders of magnitude, waxes are considered the main transport barriers in the cuticle [86, 88, 90]. Therefore, the composition and arrangement of cuticular waxes is of more importance for the permeability of the cuticle than the cuticle thickness.

The structure of cuticular waxes consists of two distinct phases: an amorphous and a crystalline phase [86, 91]. Diffusion (and also partitioning) of compounds only takes place in the former [82, 83, 86]. The impermeable wax crystals reduce the mobility of compounds in the wax in two ways: by decreasing the available volume and by creating a tortuous pathway. Consequently, the density of the crystals (the crystallinity) and the arrangement of the crystalline waxes determine the transport rate through the waxes [83].

The crystallinity of wax layers is relatively constant in the temperature range 5-40°C [84]. Although this range is similar to the range of ambient temperatures, temperatures of leaves exposed to sunlight can be much higher (up to 24°C) than the ambient temperature [92]. It appears that at higher temperatures the aliphatic crystallinity of the waxes decreases [93]. This leads to a loss of size selectivity of the waxes, and hence, to a higher mobility of large compounds.

The tortuosity of cuticular waxes may vary greatly between different plant species, e.g. from a factor of 28 (*Citrus grandis*) to 759 (*Ilex paraguariensis*). This leads to estimated path lengths (tortuosity * 10% of cuticle thickness) ranging from 7 to 410 μm for these plants [84]. An average SOC has a molar volume of ±150 cm^3/mol (calculated with McGowan's characteristic volumes [94]). This leads to a diffusion coefficient of ± 10^{-16}-10^{-17}cm^2/s (extrapolated from [80, 81, 83]), which is around the magnitude of the diffusion coefficients of solid materials [82, 86, 88, 95]. With Fick's Law it can be estimated that it takes this SOC 24-240 s to cross the *Citrus* and 58–580 days to cross the 410 μm *Ilex* cuticular waxes to reach the cutin polymer. So, although cutin is known to be an effective storage compartment for SOCs, in the case of plant leaves having highly crystalline waxes, compounds may never reach this compartment because the lifetime of the leaf may be too short.

Factors Influencing Atmospheric Deposition to Vegetation

Influence of K$_{oa}$ on Deposition Pathway

Dry gaseous versus particle-bound deposition
First estimations showed that each of the three deposition pathways: dry gaseous, dry particle and wet deposition, contributed to the concentrations of SOCs measured in spruce [96]. Later, SOC concentrations in plants in outdoor locations and in greenhouses, which contained filtered or non-filtered air, were compared, which led to more detailed information. For a number of chlorinated pesticides and a range of PCBs in spruce [97] and for lower chlorinated PCDD/Fs in ryegrass [26], dry gaseous deposition was found the principal deposition pathway. Results indicated that for ryegrass, for hepta- and octachlorinated PCDDs (having high K$_{oa}$-values) the contribution of large particles may be important. This was confirmed by a similar experiment with native grassland cultures [98]. In this experiment, dry particle bound deposition clearly played an important role for PCDD/F congeners with 6 and more chlorine atoms, while for lower chlorinated congeners dry gaseous deposition was dominant.

Nakajima [99], Kaupp [51] and Simonich and Hites [64] concluded from field experiments that also for high K$_{oa}$ PAHs (and also high K$_{oa}$ PCDD/Fs [51]) the dry deposition of particles is a significant pathway for deposition to vegetation.

Three dominant deposition pathways
Although a number of models have been presented in the literature, validation studies for these models are scarce [1]. The fugacity model for dry gaseous deposition to ryegrass, developed by Tolls and McLachlan [63], which was modified into a one-compartment model, was validated with a field experiment. The predicted and measured values were in very good agreement [70]. In a recent publication, McLachlan [59] used the results of this field validation to propose a mathematical framework which helps to interpret SOC measurements in plants by identifying the

dominant deposition process for a given compound. In the framework (as in the field validation) three dominant processes are distinguished. The first process, equilibrium partitioning, is dominating for compounds with low K_{oa} values. Uptake of these substances is therefore linearly related with K_{oa} (Figure 2) [59]. In the second process, the uptake is still dominated by dry gaseous deposition, but it is independent of the compound's K_{oa} (Figure 2). This can be explained by the extremely high storage capacity of the vegetation for these chemicals. As a consequence, the time needed to approach equilibrium is longer than the exposure time, so uptake of these compounds is kinetically limited. [70]. Particle-bound deposition, the third process, is controlling for high K_{oa} compounds. A higher K_{oa} leads to a higher the particle-bound fraction in the air and this leads to a higher contribution of particle-bound deposition (Figure 2).

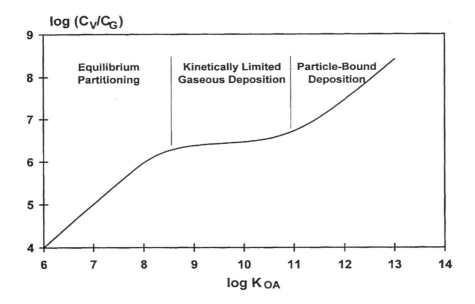

Figure 2. Illustrative plot of log $C_{vegetation}/C_{gaseous}$ phase vs log K_{oa}. (Reproduced from reference 59. Copyright 1999 American Chemical Society.)

This framework was applied by Böhme *et al.* [72], who analyzed SOCs in ten different plant species collected simultaneously at a semirural site in Central Europe, to identify which of the three processes caused the accumulation of a given compound. The compounds from 5 different families fit nicely on the framework curves for the ten plant species, which demonstrates that the interpretive framework is the first model to successfully cover the whole scope of atmospheric deposition of SOCs to plants. The behavior of the PAHs was different in some cases, probably due to their different gas-particle distribution from that of other compound classes [72].

Compounds of which the deposition is kinetically limited, will probably never reach an equilibrium with the plant, since most leaves do not live long enough [70]. This is in agreement with a field study, in which concentrations of chlorinated organic compounds in spruce needles were still increasing after 5 years [100]. This information is relevant for food studies, where food crops are investigated, which are only present on the field for a relatively short time period. In these cases, equilibrium may not be reached for the high K_{oa} compounds.

However, Thomas et al. have shown that PCBs do reach equilibrium with grass within two weeks, because concentrations in two-week, six-week and three-months old mixed grass swards were similar [71]. The fact that these measurements were performed on windy sites near the UK coast (i.e. higher turbulence, smaller boundary layers), in contrast with the inland site used by McLachlan and coworkers, may explain the difference between the studies.

In another field study in the UK all PCDD/F congeners were found to be transferred with the same efficiency to grass [36]. This may mean that the deposition of all congeners was kinetically limited, but since large PCDD/Fs are very hydrophobic, a contribution of particle deposition is to be expected. The authors considered the wet deposition possibly important in the maritime UK environment. Another explanation may be that PCDD/Fs sorb to ultrafine particles, and may behave as gases [36].

In conclusion, the compound's K_{oa} determines the dominating deposition pathway, being equilibrium partitioning, kinetically limited dry gaseous or particle-bound deposition. However, the environmental conditions, together with the plant species, determine the K_{oa} values which mark the transition from one process to the other.

Influence of Temperature

The partitioning process between air and vegetation is expected to be dependent on the ambient temperature. According to the Clausius-Clapeyron relationship, a plot of $\ln K_{pa}$ vs $1/T$ gives a straight line with a positive slope. This means, as the ambient temperature decreases, partitioning to vegetation increases.

Kömp and McLachlan [101] studied the temperature dependence of K_{pa} of PCBs in ryegrass under controlled laboratory conditions in a fugacity meter. A very strong temperature dependence was found: K_{pa} decreased by a factor of 30 (some dichlorobiphenyls) to 2000 (some octachlorobiphenyls) in a temperature range between 5 and 50°C, indicating that under environmental conditions K_{pa} values are highly variable. However, simulations with the fugacity model [70] showed that for most SOCs, the ambient temperature has little effect on the concentrations in a field vegetation. As the dry gaseous deposition of SOCs with relatively high K_{oa} values is kinetically limited and equilibrium is not approached, changes in temperature will only affect concentrations in plants after longer time periods. Only for compounds with lower K_{oa} values, the plant air concentration ratio can respond fairly quickly to a change in temperature [101].

In contrast, in a few field studies, in which PAHs or PCDD/Fs were determined in the atmosphere and in vegetation in different seasons, the semilogarithmic Clausius-Clapeyron relationship has been confirmed [64, 99, 102]. Although the plant-air ratios in these studies may not have been genuine equilibrium ratios, they apparently react to temperature changes fast enough to observe a seasonal trend.

Some studies even report diurnal variation of plant concentrations. The concentrations of volatile organic compounds (VOCs) in different plants varied during the day, because of the large variation in VOC concentrations in air, due to different temperatures and traffic flows during the day [103]. As VOCs are not very hydrophobic and have low K_{oa} values, they can rapidly adjust to varying air concentrations. In a study in a remote forested bog, the logarithm of the SOC concentrations in air was related to the reciprocal temperature (r^2 up to 0.9) [104]. The authors attributed the variation to strong diurnal variations in temperature, resulting in adsorption and subsequent volatilization of the compounds from the plant surfaces [104], but the compounds sorbed in the surface compartment of the leaves may also react fast to changes in temperature.

Influence of Plant Characteristics

SOC concentrations in different plant species

In several studies the SOC concentrations of different plant species were compared. Buckley [3] found an eightfold range in PCB-concentrations (ΣPCBs varying from 32 ng/g dry weight to 250 ng/g dry weight) in foliage from 18 plant species in New York State. Concentrations of PCBs and some organochlorine pesticides in 12 woodland species near Siena (Italy) differed by a factor 2-6 [4], while the differences in moss, lichen, beech leaves and spruce needles in south Germany were less than a factor of 3 [105]. For VOCs, large differences in concentrations (factor 100-400) were found between foliage from different trees and plants [103, 106]. In a study in Norway, lichens were compared to spruce needles. Differences in SOC concentrations ranged from a factor of 3 up to >50 [107].

This synopsis shows that the variation between different plant species can be considerable. Consequently, it is important to know which plant characteristics cause the differences in SOC concentrations in different species.

Lipid content

Differences in accumulation between different plants can be partly explained by the lipid content of the species. Normalizing to the extractable lipid content decreased the differences in PAH concentrations between leaf, needle, seed and tree bark samples collected on the same site from a factor of 5 to 2 [64]. However, the lipid content of plants is only important in (near-)equilibrium situations, since uptake kinetics seem independent of it [76]. Besides, the extractable lipid content may not represent the actual storage volume, as SOCs may also accumulate in non-extractable lipid material, such as cutin. On the other hand, SOCs may not have reached the internal lipids, but these are included in the total lipid content. Therefore, the fractional volume of the cuticle to the leaf volume may be a better normalizing factor [72]. In practice, however, this factor is difficult to estimate.

The interspecies variation in the study of Böhme *et al.* [72] was a factor of ~30 (on a dry weight basis), for the compounds with low K_{oa} values which had approached equilibrium. However, leaving out two of the ten species (yarrow and sunflower), reduces the variation to a factor of ~5. In addition, although the authors [72] argued that the variability in K_{pa}'s was not related to the lipid content, the extractable lipid content for the eight species (without yarrow and sunflower) correlate well with K_{pa} (r^2's between 0.5 an 0.8).

Lipid composition

In the study of Böhme *et al.* [72], the SOC-concentrations in yarrow and sunflower leaves were much higher than those in the other plants, even after normalization to lipid content and volume fraction of the cuticle. This was probably caused by a very different cuticular wax composition, which was also indicated by the relatively shallow slopes of the $\log K_{pa}$-$\log K_{oa}$ plots of these plants [72]. These results agree with the study of Kömp and McLachlan [65], in which the slope of the $\log K_{pa}$-$\log K_{oa}$ plot of yarrow was also the lowest (see Table I) and the K_{pa}'s of the lower chlorinated PCBs were the highest [72].

In Kömp and McLachlan's study [65], K_{pa} values were found to vary a factor of 20 for the five species. The differences in both the slopes (see Table I) and the K_{pa}'s were attributed to differences in the composition of the plant lipids.

Hiatt [103] attempted to relate the composition of the leaf cuticular wax to the concentrations of VOCs in plants. The presence of monoterpenes, a class of wax components, was held responsible for the high concentrations of VOCs found in some species [103]. However, the monoterpenes were the only class of wax components determined in the waxes, hence a causal relationship can not be shown.

Leaf age

The large difference found between SOC concentrations in pine needles and lichens in Norway could not be explained by the lipid contents of the plants. However, it was suggested that the differences in plant ages (2 years for the pine needles and 25-50 years for the lichens) may play a role [107]. SOC concentrations in air of a few decades ago were likely higher than those in contemporary air. Hence, the older lichens may have been approaching higher equilibrium concentrations of PCBs, and are now releasing the compounds very slowly [107].

Leaf surface area

Dry gaseous deposition occurs to all surfaces of the plant and therefore the total surface area (two sides of the leaves) influences the uptake rate. The variability in uptake rate constants of 2,3,7,8-TCDD in four plant species was reduced from a factor of 50 to 4 when the constants were normalized to the surface area of the plant [76]. Similarly, in the study of Böhme *et al.* [72], a significant relationship was found (except for yarrow, sunflower and creeping thistle) between uptake and the surface area to volume ratios of the different plants, for those gaseous compounds that had not reached equilibrium. In theory, the best results will be obtained when the true total surface area (i.e. taking into account the relief of the leaf surface and cuticular waxes) and not the superficial surface area would be used [39]. In practice, however, the true surface area is difficult to determine.

Particle deposition on different types of vegetation was found to increase with increasing leaf area index (surface area of leaves -one side- divided by m² ground) [108]. In contrast with dry gaseous deposition, only (a part of) the upper side of the leaf will be receiving the particles. Researchers have tried to define this receiving leaf area as being the horizontal leaf area [72], since the horizontal orientation of the leaves (together with the rough hairy surface) was found to be of importance [98]. A linear relationship between the horizontal area to volume ratio and the deposition of particle-bound compounds (except for two species) was found in the study of Böhme *et al.* [72]. Also, the projected leaf surface area (one side of the leaves), representing the horizontal surface area that is not covered by other leaves has been used to estimate the receiving leaf area [40].

Plant architecture

In dry gaseous deposition, the rate limiting step in the uptake process is in the air [59]. As a consequence, the plant architecture is an important plant characteristic for the uptake rate. It determines the aerodynamic surface roughness (and therefore the turbulence) and the thickness of the laminar boundary layer, which influence step 1 and 2 of the uptake process, respectively (Figure 1) [40]. The aerodynamic surface roughness is influenced by the height of the plant and the size and orientation of the leaves (higher roughness leading to higher turbulence). The thickness of the laminar boundary layer is determined by the size and shape of the leaves, the roughness of the leaf surface and the presence of leaf hairs. It is also influenced by wind speed and irradiation. In a field study, where PAH concentrations in three *Plantago* (plantain) species were compared, it is reported that *Plantago media* contained the lowest concentrations of low-K_{oa}, mainly gaseous, PAHs. This was explained by the lower surface roughness and larger thickness of laminar boundary layer of this densely hairy plant, which leaves are spreading low to the ground [40].

As for gases, a high aerodynamic surface roughness leads to efficient turbulent transport of small particles [109]. Wind tunnel experiments with radio-labeled particles with sizes ranging from 0.03–44 µm have shown that hairy leaves are better particle collectors than glabrous leaves (e.g. [52, 110-112]). Hairy leaves act as an efficient particle trap: because of the thick boundary layer, wind eddies can not penetrate down to blow off the particles from the leaf surface [37]. Another explanation for the increased particle collection efficiency of hairy leaves is that leaf hairs may cushion the impact and therefore reduce the bounce-off of particles [113].

The increased particle collection efficiency of hairy leaves was confirmed in the *Plantago* study, in which the concentrations of the high K_{oa}, mainly particle-bound, PAHs were highest in the densely hairy *Plantago media* leaves [40].

Summary and Concluding Remarks

Deposition pathways

In the 1980's, it was demonstrated that atmospheric deposition of SOCs, rather than uptake via roots, is the dominant pathway for contamination of aerial plant parts. In recent years, important progress has been made to elucidate the relevant pathways of atmospheric deposition. Both dry gaseous deposition and particle bound deposition (wet and dry) have been shown to contribute to the deposition of SOCs. Which one is actually controlling the deposition, depends on the K_{oa} of the compound, the plant species and the prevailing temperature and wind. Under given environmental conditions, the deposition pathway to a given plant species is governed by the K_{oa}, in which the controlling process goes from equilibrium partitioning to kinetically limited dry gaseous deposition to particle bound deposition with increasing K_{oa}.

Compartments

Uptake of atmospheric SOCs in vegetation can be described with a one- or two-compartment model. One-compartment models, in which all lipids are lumped together, are mostly used to describe long-term atmospheric deposition, while two-compartment models give a more detailed view of the short-term diffusion behavior of the chemical within the leaf. The first (or surface) compartment may consist of the cuticular waxes and the second (or reservoir) compartment of the cutin, although the time to reach the cutin may be very long, due to the high tortuosity of the diffusion pathway through the cuticular waxes.

Factors influencing deposition

It is clear that the K_{oa} is the most important compound property, as it determines the controlling deposition process and can predict the K_{pa} well. However, as slopes of $\log K_{pa}$-$\log K_{oa}$ plots are generally different from one, lipids of most plant species have a different lipophilicity than octanol.

Temperature and wind are important environmental parameters. Under equilibrium conditions, temperature has a large influence on K_{pa}. The effect of temperature in non-equilibrium situations is less pronounced, because it takes time for concentrations in plants to change, particularly for compounds with high K_{oa}'s. Because of different wind speeds, the rate at which plant concentrations adjust to new situations can be different for different sites. Wind can increase the transfer of the chemical from the air to the plant surface, by increasing the turbulence and decreasing the thickness of the laminar boundary layer. This illustrates the important role of the wind, although the effect of wind on uptake rate has never been studied directly.

In many studies only small differences (< a factor of ~8) between SOC concentrations in plants have been found, but sometimes also large differences (up to a factor of >50) are measured. Under equilibrium conditions, normalizing SOC concentrations to extractable lipid content may decrease the variation, although the insoluble lipid polymer cutin may contribute to the total storage capacity of the plant. The large differences between equilibrium concentrations in plants may be explained by the composition of the lipids. However, it is difficult to relate the presence of certain wax components to the storage capacity of leaves. In cases where equilibrium has not been approached, the age of the leaves, the leaf area, the aerodynamic surface roughness of the plant, the orientation of the leaves and the roughness of the leaf surface are key factors in determining the SOC concentrations in plants.

Concluding remarks

The (sometimes) large differences between concentrations in different plant species may have consequences for the precision of predictions of SOC concentrations in food crops. Most predictive models do not take species differences into account, although some include lipid content and/or surface area to volume ratio [59, 114, 115]. As we are just starting to begin to grasp the effect of lipid composition and plant architecture on kinetically limited and particle-bound deposition, it is not yet possible to refine the existing models with respect to plant characteristics.

The same holds for estimates of the importance of vegetation as a pollutant sink. Also for biomonitoring purposes, the large variation in concentrations between species may have consequences. Special attention should be given to the choice of the plant species, which should be ubiquitous. In addition, the age of the sample leaves should be taken into account. However, the various environmental conditions under which the leaves are sampled may have a larger influence on the concentrations in plants than the plant characteristics themselves. Hence, it is difficult to draw conclusions from comparison of samples from different locations. This may not be of major influence when biomonitoring is used for identifying point sources, but it seriously complicates the use of plants as indicators of regional or global contamination levels.

Acknowledgement

We are grateful to the Inspectorate of Health Protection of the Dutch Ministry of Health, Welfare and Sports for funding our research on the deposition of SOCs in plants.

References

1. Simonich, S. L.; Hites, R. A. *Environ. Sci. Technol.* **1995**, *29,* 2905-2914.
2. Bidleman, T. F. *Environ. Sci. Technol.* **1988**, *22,* 361-367.
3. Buckley, E. H. *Science* **1982**, *216,* 520-522.
4. Gaggi, C.; Bacci, E.; Calamari, D.; Fanelli, R. *Chemosphere* **1985**, *14,* 1673-1686.
5. Reischl, A.; Reissinger, M.; Hutzinger, O. *Chemosphere* **1987**, *16,* 2647-2652.

6. Frank, H.; Frank, W. *Environ. Sci. Technol.* **1989**, *23*, 365-367.
7. Eriksson, G.; Jensen, S.; Kylin, H.; Strachan, W. *Nature* **1989**, *341*, 42-44.
8. Thomas, W.; Herrmann, R. *Staub-Reinhalt. Luft* **1980**, *40*, 440-444.
9. Thomas, W. *Ecotox. Environ. Saf.* **1986**, *11*, 339-346.
10. Hermanson, M. H.; Hites, R. A. *Environ. Sci. Technol.* **1990**, *24*, 666-671.
11. Muir, D. C. G.; Segstro, M. D.; Welbourn, P. M.; Toom, D.; Eisenreich, S. J.; Macdonald, C. R.; Whelpdale, D. M. *Environ. Sci. Technol.* **1993**, *27*, 1201-1210.
12. Fürst, P.; Fürst, C.; Groebel, W. *Chemosphere* **1990**, *20*, 787-792.
13. Theelen, R. M. C.; Liem, A. K. D.; Slob, W.; Van Wijnen, J. H. *Chemosphere* **1993**, *27*, 1625-1635.
14. Fries, G. F. *J. Anim. Sci.* **1995**, *73*, 1639-1650.
15. McLachlan, M. S. *Environ. Sci. Technol.* **1996**, *30*, 252-259.
16. Duarte-Davidson, R.; Sewart, A.; Alcock, R. E.; Cousins, I. T.; Jones, K. C. *Environ. Sci. Technol.* **1997**, *31*, 1-11.
17. Simonich, S. L.; Hites, R. A. *Nature* **1994**, *370*, 49-51.
18. Wagrowski, D. M.; Hites, R. A. *Environ. Toxicol. Chem.* **1997**, *31*, 279-2882.
19. Matzner, E. *Water Air Soil Pollut.* **1984**, *21*, 425-434.
20. Horstmann, M.; McLachlan, M. S. *Atmos. Environ.* **1998**, *32*, 1799-1809.
21. McLachlan, M. S.; Horstmann, M. *Environ. Sci. Technol.* **1998**, *32*, 413-420.
22. Horstmann, M.; McLachlan, M. S. *Environ. Sci. Technol.* **1996**, *30*, 1794-1796.
23. Chamberlain, A. C. *Atmos. Environ.* **1970**, *4*, 57-78.
24. McCrady, J.; Maggard, S. P. *Environ. Sci. Technol.* **1993**, *27*, 343-350.
25. Schuler, F.; Schmid, P.; Schlatter, C. *Chemosphere* **1998**, *36*, 21-34.
26. Welsch-Pausch, K.; McLachlan, M. S.; Umlauf, G. *Environ. Sci. Technol.* **1995**, *29*, 1090-1098.
27. Paterson, S.; Mackay, D.; Tam, D.; Shiu, W. Y. *Chemosphere* **1990**, *21*, 297-331.
28. Wang, M.-J.; Jones, K. C. *Environ. Sci. Technol.* **1994**, *28*, 1260-1267.
29. Schroll, R.; Scheunert, I. *Chemosphere* **1992**, *24*, 97-108.
30. McCrady, J.; McFarlane, C.; Gander, L. K. *Chemosphere* **1990**, *21*, 359-376.
31. Wild, S. R.; Jones, K. C. *Atmos. Environ.* **1992**, *26a*, 1299-1307.
32. Hülster, A.; Müller, J. F.; Marschner, H. *Environ. Sci. Technol.* **1994**, *28*, 1110-1115.
33. Ecker, S.; Horak, O. *Chemosphere* **1994**, *29*, 2135-2145.
34. Duinker, J. C.; Bouchertall, F. *Environ. Sci. Technol.* **1989**, *23*, 57-62.
35. Trapp, S.; Matthies, M. *Environ. Sci. Technol.* **1997**, *331*, 71-74.
36. Jones, K. C.; Duarte-Davidson, R. *Environ. Sci. Technol.* **1997**, *31*, 2937-2943.
37. Gregory, P. H. In *The microbiology of the atmosphere;* Polunin, N., Eds.; Plant Science Monographs; Leonard Hill: London, 1961, pp. 22-27.
38. Davidson, C. I.; Wu, Y.-L. In *Control and fate of atmospheric trace metals;* Pacyna, J. M.; Ottar, B., Eds.; NATO-ASO Series C, Mathematical & Physical Sciences; Kluwer: Dordrecht, 1989, 268, pp. 147-202.

234

39. Schreiber, L.; Schönherr, J. *Environ. Sci. Technol.* **1992**, *26*, 153-159.
40. Bakker, M. I.; Vorenhout, M.; Sijm, D. T. H. M.; Kollöffel, C. *Environ. Toxicol. Chem.* **1999**, *18*, 2289-2294.
41. Pankow, J. F. *Atmos. Environ.* **1987**, *21*, 2275-2283.
42. Finizio, A.; Mackay, D.; Bidleman, T. F.; Harner, T. *Atmos. Environ.* **1997**, *31*, 2289-2296.
43. Venkataraman, C.; Friedlander, S. K. *Environ. Sci. Technol.* **1994**, *28*, 583-572.
44. Poster, D. L.; Hoff, R. M.; Baker, J. E. *Environ. Sci. Technol.* **1995**, *29*, 1990-1997.
45. Schnelle, J.; Wolf, K.; Frank, G.; Hietel, B.; Gebefügi, I.; Kettrup, A. *Analyst* **1996**, *121*, 1301-1304.
46. Chen, S. J.; Liao, S. H.; Jian, W. J.; Lin, C. C. *Environ. Internat.* **1997**, *23*, 475-488.
47. Chen, S.; Hwang, W.; Chiu, S.; Hung, M.; Lin, C. *J. Environ. Sci. Health* **1997**, *A32*, 1781-1805.
48. Kaupp, H.; McLachlan, M. S. *Atmos. Environ.* **1999**, *33*, 85-95.
49. Holsen, T. M.; Noll, K. E.; Liu, S.-P.; Lee, W.-J. *Environ. Sci. Technol.* **1991**, *25*, 1075-1081.
50. McLachlan, M. S. *Organohal. Comp.* **1995**, *22*, 81-103.
51. Kaupp, H. Ph.D. thesis, University of Bayreuth , Bayreuth, Germany, 1996
52. Wedding, J. B.; Carlson, R. W.; Stukel, J. J.; Bazzaz, F. A. *Environ. Sci. Technol.* **1975**, *9*, 151-153.
53. Larsson, B.; Sahlberg, G. In *Polynuclear aromatic hydrocarbons: Physical and biological chemistry;* Cooke, M.; Dennis, A. J.; Fisher, G. L., Eds.; Sixth Int. Symposium; Springer-Verlag: New York, 1982, pp. 417-426.
54. Eglinton, G.; Hamilton, R. *Science* **1967**, *156*, 1322-1335.
55. Holloway, P. J. In *Air poluttants and the leaf cuticle;* Percy, K. E.; Cape, J. N.; Jagels, R.; Simpson, C. J., Eds.; NATO-ASI Series G, Ecological Sciences,; Springer-Verlag: Berlin, Germany, 1994, 36, pp. 1-14.
56. Jeffree, C. E. In *Plant cuticles; an integrated functional approach;* Kerstiens, G., Eds.; Environmental Plant Biology; Bios Scientific: Lancaster, UK, 1996, pp. 33-82.
57. Kirkwood, R. C. In *Pesticides on plant surfaces;* Cottrell, H. J., Eds.; Critical reports on applied chemistry; J. Wiley and Sons: New York, 1987, 28, pp. 1-25.
58. Cape, J. N.; Percy, K. E. *New. Phytol.* **1993**, *125*, 787-799.
59. McLachlan, M. S. *Environ. Sci. Technol.* **1999**, *33*, 1799-1804.
60. *The cuticles of plants;* Martin, J. T.; Juniper, B. E., Eds.; Edward Arnold: London, 1970; Vol. 95-107.
61. Bacci, E.; Cerejeira, M. J.; Gaggi, C.; Chemello, G.; Calamari, D.; Vighi, M. *Chemosphere* **1990**, *21*, 525-535.
62. Travis, C. C.; Hattemer-Frey, H. A. *Chemosphere* **1988**, *17*, 277-283.
63. Tolls, J.; McLachlan, M. S. *Environ. Sci. Technol.* **1994**, *28*, 159-166.
64. Simonich, S. L.; Hites, R. A. *Environ. Sci. Technol.* **1994**, *28*, 939-943.

65. Kömp, P.; McLachlan, M. S. *Environ. Sci. Technol.* **1997**, *31*, 2944-2948.
66. Harner, T.; Mackay, D. *Environ. Sci. Technol.* **1995**, *29*, 1599-1606.
67. Kömp, P.; McLachlan, M. S. *Environ. Toxicol. Chem.* **1997**, *16*, 2433-2437.
68. Harner, T.; Bidleman, T. F. *J. Chem. Eng. Data* **1998**, *43*, 40-46.
69. Paterson, S.; Mackay, D.; Bacci, E.; Calamari, D. *Environ. Sci. Technol.* **1991**, *25*, 866-871.
70. McLachlan, M. S.; Welsch-Pausch, K.; Tolls, J. *Environ. Sci. Technol.* **1995**, *29*, 1998-2004.
71. Thomas, G.; Sweetman, A. J.; Ockenden, W. A.; Mackay, D.; Jones, K. C. *Environ. Sci. Technol.* **1998**, *32*, 936-942.
72. Böhme, F.; Welsch-Pausch, K.; McLachlan, M. S. *Environ. Toxicol. Chem.* **1999**, *33*, 1805-1813.
73. Bacci, E.; Gaggi, C. *Chemosphere* **1987**, *16*, 2515-2522.
74. Bacci, E.; Calamari, D.; Gaggi, C.; Vighi, M. *Environ. Sci. Technol.* **1990**, *24*, 885-889.
75. Reischl, A.; Schramm, K.-W.; Beard, A.; Reissinger, M.; Hutzinger, O. *VDI Berichte* **1989**, *745*, 511-523.
76. McCrady, J. *Chemosphere* **1994**, *28*, 207-216.
77. Schreiber, L.; Schönherr, J. *New. Phytol.* **1993**, *123*, 547-554.
78. Hauk, H.; Umlauf, G.; McLachlan, M. S. *Environ. Sci. Technol.* **1994**, *28*, 2372-2379.
79. Keymeulen, R.; Schamp, N.; Van Langenhove, H. *Chemosphere* **1995**, *31*, 3961-3975.
80. Schreiber, L.; Schönherr, J. *Pestic. Sci.* **1993**, *38*, 353-361.
81. Schreiber, L. *Pestic. Sci.* **1995**, *45*, 1-11.
82. Baur, P.; Marzouk, H.; Schönherr, J.; Bauer, H. *Planta* **1996**, *199*, 404-412.
83. Schreiber, L.; Kirsch, T.; Riederer, M. *Planta* **1996**, *198*, 104-109.
84. Baur, P.; Marzouk, H.; Schönherr, J. *Plant Cell Environ.* **1999**, *22*, 291-299.
85. Riederer, M.; Schönherr, J. *Ecotox. Environ. Saf.* **1984**, *88*, 236-247.
86. Riederer, M.; Schreiber, L. In *Waxes, chemistry, molecular biology and functions;* Hamilton, R. J., Eds.; The Oily Press Ltd.: Dundee, 1995, pp. 131-156.
87. Price, C. E. In *The plant cuticle;* Cutler, D. F.; Alvin, K. L.; Price, C. E., Eds.; Linnaeus Society Symposium Series; Academic Press: New York, 1982, 10, pp. 237-252.
88. Bauer, H.; Schönherr, J. *Pestic. Sci.* **1992**, *35*, 1-11.
89. Norris, R. F. *Am. J. Bot.* **1974**, *61*, 74-79.
90. Kerler, F.; Schönherr, J. *Arch. Environ. Contam. Toxicol.* **1988**, *17*, 7-12.
91. Kirkwood, R. C. *Pestic. Sci.* **1999**, *55*, 69-77.
92. Baur, P.; Buchholz, A.; Schönherr, J. *Plant Cell Environ.* **1997**, *20*, 982-994.
93. Merk, S.; Blume, A.; Riederer, M. *Planta* **1998**, *204*, 44-53.
94. Abraham, M. H.; McGowan, J. C. *Chromatographia* **1987**, *23*, 243-246.
95. Schönherr, J.; Riederer, M. *Rev. Environ. Contam. Toxicol.* **1989**, *108*, 1-70.
96. Umlauf, G.; McLachlan, M. S. *Environ. Sci. Poll. Res.* **1994**, *1*, 146-150.

97. Umlauf, G.; Hauk, H.; Reissinger, M. *Environ. Sci. Poll. Res.* **1994**, *1*, 209-222.

98. Welsch-Pausch, K.; McLachlan, M. S. *Organohal. Comp.* **1996**, *28*, 72-75.

99. Nakajima, D.; Yoshida, Y.; Suzuki, J.; Suzuki, S. *Chemosphere* **1995**, *30*, 409-418.

100. Jensen, S.; Eriksson, G.; Kylin, H.; Strachan, W. M. J. *Chemosphere* **1992**, *24*, 229-245.

101. Kömp, P.; McLachlan, M. S. *Environ. Sci. Technol.* **1997**, *31*, 886-890.

102. Wagrowski, D. M.; Hites, R. A. *Environ. Toxicol. Chem.* **1998**, *32*, 2389-2393.

103. Hiatt, M. *Analytical Chemistry* **1998**, *70*, 851-856.

104. Hornbuckle, K. C.; Eisenreich, S. J. *Atmos. Environ.* **1996**, *30*, 3935-3945.

105. Morosini, M.; Schreitmüller, J.; Reuter, U.; Ballschmiter, K. *Environ. Sci. Technol.* **1993**, *27*, 1517-1523.

106. Keymeulen, R.; Schamp, N.; Van Langenhove, H. *Atmos. Environ.* **1993**, *27A*, 175-180.

107. Ockenden, W. A.; Steinnes, E.; Parker, C.; Jones, K. C. *Environ. Sci. Technol.* **1998**, *32*, 2721-2726.

108. Heil, G. W. In *Vegetation structure in relation to carbon and nutrient economy;* Verhoeven, J. T. A.; Heil, G. W.; Werger, M. J. A., Eds.; SPB Academic: The Hague, 1988, pp. 149-155.

109. Burghardt, M.; Schreiber, L.; Riederer, M. *J. Agric. Food Chem.* **1998**, *46*, 1593-1602.

110. Romney, E. M.; Lindberg, R. G.; Hawthorne, H. A.; Bystrom, B. G.; Larson, K. H. *Ecology* **1963**, *44*, 343-349.

111. Chamberlain, A. C. *Proc. Roy. Soc., Ser. A* **1967**, *296*, 45-70.

112. Little, P.; Wiffen, R. D. *Atmos. Environ.* **1977**, *11*, 437-447.

113. Chamberlain, A. C.; Little, P. In *Plants and their atmospheric environment;* Grace, J.; Ford, E.; Jarvis, P., Eds.; The 21st symposium of the British Ecological Society; Blackwell Scientific: Oxford, 1981, pp. 147-173.

114. Riederer, M. *Environ. Sci. Technol.* **1990**, *24*, 829-837.

115. Trapp, S.; Matthies, M. *Environ. Sci. Technol.* **1995**, *29*, 2333-2338.

Chapter 17

Hair Mercury Analysis and Its Application to Exposure Studies in NHEXAS and NHANES IV

G. M. Cramer[1], K. Bangerter[2], R. Fernando[3], G. M. Meaburn[4], and E. Pellizzari[3]

[1]Office of Seafood, U.S. Food Administration, Washington, DC 20204
[2]Medical University of South Carolina, Charleston, SC 29425
[3]Research Triangle Institute, Research Triangle Park, NC 27709
[4]National Ocean Service, Charleston, SC 29412

The results of a scoping probability-based population study utilizing biological measurements of methylmercury exposure suggests that exposure from fish in six Great Lakes states may be considerably lower among the general population than historic public health limits. Due to concerns that historic limits may not be adequately protective of the developing fetus, the federal government has begun a national survey using methods from the Great Lake study to characterize the distribution of methylmercury exposure for U.S. population sub-groups that are considered most sensitive to the effects of methylmercury.

Introduction

The massive methylmercury (MeHg) poisoning which occurred in Minimata Japan in the 1950s and 1960s alerted the world to the presence of MeHg in fish and to the hazards of consuming large quantities of fish that are heavily contaminated with MeHg (*1*). Subsequent studies have demonstrated that MeHg occurs in essentially all fishery products and for most people, the relatively low levels in fish represent the predominant source of exposure to all forms of mercury (*2*).

© 2001 American Chemical Society

Over the past several years, the extent of MeHg exposure among U.S. fish consumers as well as the public health consequences of such exposure has been the subject of considerable debate within the Federal government and at state level. This debate has been fueled by concern about the sensitivity of the developing fetus to the effects of prenatal MeHg exposure, the emergence of new public health studies on the effects of MeHg exposure, and concern that mercury emissions may have a potential impact on MeHg exposure among fish consumers.

This paper does not deal with these latter points. Instead, it reports on the results of a pilot MeHg exposure study and describes the application of analytical methods from that study to the development of a national data base that is intended to resolve some of the MeHg exposure debate and provide a framework for more accurate characterization of the risks to U.S. fish consumers.

In the U.S., fish consumption is highly variable, reflecting the influence of geographic location, season of the year, ethnicity, and personal food preferences. The extent of MeHg exposure from fish meals is also variable since the concentration of MeHg in fish varies by species, size of fish, age of fish, and location of harvest. Although dietary modeling approaches are routinely used to estimate MeHg exposure, the outcomes of such efforts are model dependent and their accuracy in predicting MeHg exposure within the context of a complex diet which includes fish with different MeHg content has been the subject of considerable debate.

Another approach for characterizing dietary exposure to MeHg is to use biomarker data, such as blood or hair mercury levels, which provide a direct measure of dietary MeHg intake. Hair and blood mercury levels are routinely used as a biomarker of MeHg exposure in health effect studies of populations who are dependent on fish consumption (3, 4, 5, 6, 7). Studies with radiolabeled MeHg have shown that once it is ingested, it is quantitatively absorbed from the diet, distributed throughout the body within a few hours, and after a 20-30 hour clearance phase, blood levels decrease exponentially with a half life of about 50 days (8). By allowing for the rapid clearance phase, changes in blood mercury levels have been shown to be directly proportional to the levels of MeHg ingested, with nearly all of the mercury found in blood in the form of MeHg (9).

In the case of scalp hair, the concentration is approximately 250 times the blood concentration at the time the hair is formed. Once formed, hair strands grow at the rate of approximately 1 cm per month, providing a record of previous mercury exposures that remains unchanged for periods of up to 11 years (10). Although blood and hair can both be used to document MeHg exposure, hair-strand analysis is preferred because it provides a simple, non-invasive sampling procedure that allows monitoring of MeHg intake. Since MeHg accounts for about 80% of the total mercury levels found in hair (11, 12), total mercury determination is typically used to characterize MeHg exposure from fish.

In 1995, at the EPA sponsored National Forum on Mercury in Fish, it was suggested that a national data base be developed on total hair mercury levels. Such a data base would provide direct measure of the extent of MeHg exposure that occurs among U.S. fish consumers. Importantly, such information would help resolve the MeHg exposure debate and allow direct comparison of MeHg exposure measurements in the U.S. with populations who were the subject of MeHg human health studies.

Experimental

In 1996, FDA and the National Marine Fisheries Service (NMFS) had the opportunity to collaborate with Research Triangle Institute (RTI) to develop a hair mercury data set in conjunction with the National Human Exposure Assessment Survey (NHEXAS), an EPA sponsored project headed by RTI (*13*). One component of this survey involved a 300 home probability-based sampling in EPA Region V (Illinois, Indiana, Michigan, Minnesota, Ohio, and Wisconsin). The primary focus of the study was to document multi-media exposure to various metals, pesticides, and other organics, and characterize the exposure distribution for the general population of a relatively large area. The target population of the study had racial, ethnic, socioeconomic, and other demographic characteristics similar to the national profile. Since dietary records for a four day period along with duplicate diet portions were collected from each participant providing a hair sample, the NHEXAS study provided the opportunity to not only characterize the distribution of hair mercury levels among NHEXAS participants, but the opportunity to examine the relationship of hair mercury levels to short term fish consumption information.

Hair samples for the NHEXAS study were obtained by cutting a bundle of hair approximately the size of an eraser in diameter from the occipital region of the head. The bundle was cut with clean stainless steel scissors as close to the scalp as possible and the bundle was kept intact during cutting and storage using a plastic clip. The 3-cm segment of hair closest to the scalp, representing approximately 3 months of integrated exposure to MeHg, was used for analysis. Hair strands less than 3-cm were analyzed in their entirety.

In order to describe the entire range of total mercury levels in NHEXAS hair samples, a sensitive analytical procedure which would permit the analysis of 5 mg samples of hair was developed by RTI based on atomic fluorescence spectrometry. Hair samples, washed with acetone to remove surface contamination, were digested in sulfuric acid in airtight vials at 90 °C for 6-8 hours. Following the conversion of all forms of Hg to Hg (ll) using a bromide/bromate solution, hydroxylamine addition to remove excess bromide/bromate, and tin chloride addition to reduce Hg (ll) to Hg (0), the samples were analyzed by cold vapor atomic fluorescence spectrometry (CVAFS) using a Hg specific fluorescence detector.

Hair samples were analyzed in batches of 25-30 samples. Each batch included 4-5 matrix standards (hair composite spiked with Hg reference standards), triplicate reagent blanks, and triplicate performance evaluation samples (a certified hair mercury sample). In addition, duplicate analyses were performed on a subset of 12% of the hair samples as well as a subset of 12% of the sample extracts. A QC check standard was analyzed every 10th sample.

The method performance was evaluated for total Hg measurement in human hair. The correlation coefficient for standard curves in the hair matrix was 0.9983, while the precision and the bias were <1.6%, and < 8%, respectively. Over the 7 to 8 months of analysis, the method detection limit (MDL) averaged 12 ppb (range 4-22), recovery of Hg in a certified hair mercury sample (NIES-13) averaged $100 \pm 3\%$, precision for duplicate extract analysis was 4.6%, and precision for duplicate sample analysis was 12.5%. The low MDL yielded 95% of the 182 hair samples with measurable values, permitting the entire distribution of Hg levels to be described.

Results and Discussion

Weighted population estimates of means, median and other percentiles were generated using demographic variables such as age, race, income, to adjust for non-response bias and seasonality of the hair collection period (14). In general, the demographics indicate that participants providing hair samples were representative of the general population of EPA Region V.

Total Mercury Levels Observed in NHEXAS Hair Samples

Statistic	Levels (ppb)
5th percentile	36
10th percentile	54
25th percentile	93
50th percentile	204
75th percentile	335
90th percentile	531
95th percentile	893
Minimum	14
Mean	287
Maximum	3,505

The concentrations observed were all lower than 5000 ppb, a figure used by the U.S. FDA and the World Health Organization (15) as a measure of exposure that would result from ingesting MeHg over extended periods at levels corresponding to the tolerable daily intake (0.43 ug/kg/day). The mean and median NHEXAS hair mercury levels are comparable to the averages observed in a 1972 EPA study in Birmingham, Alabama and Charlotte, North Carolina (*16*). They are also similar to the median levels of methylmercury (360 ppb) observed in a study of U.S. women of child-bearing age who ate fish at least once a month (*17*). In contrast, the levels observed in this study are much lower than levels observed among populations who depend on fish as a principal component of their diet. For example, in a multi-year study of pregnant women in the Seychelles Islands who ate an average of 12 fish meals per week, the mean total Hg levels for each years hair collection fell in the range of 5900 to 8200 ppb (*12*).

Previous studies have noted that organic mercury levels in hair decrease when individuals use hair treatments containing thioglycolic acid, such as hair relaxers and cold wave preparations (*3, 18*). Reduced total mercury levels are usually observed in the hair of women since they are the primary users of such products. Although the NHEXAS participants were not questioned about the use of such products, the finding that males had mean levels about 20% less than women (260 ppb vs. 315 ppb) suggests that women in this study did not make extensive use of such hair treatments.

To investigate the relationship between fish ingestion and hair mercury levels, dietary records for a 4-day period from each NHEXAS participant were reviewed for reported fish meals. Out of a total of 182 participants, 39 (21%) reported eating a fish meal at least once during the 4-day survey period. The mean hair level for all fish consumers was 418 (median = 268) ppb while the mean for persons not eating fish during the survey period was about 297 (median = 183) ppb. For women of child-bearing age (16-49 years), the mean level was 416 (median = 296) ppb among fish consumers and 326 (median = 201) ppb for those not eating fish. Among children 9 years and younger, none reported eating fish during the 4-day survey and the mean hair mercury level was 177 (median = 133) ppb. Interestingly, the person whose hair had the highest level in this study (3505 ppb) did not report eating fish during the 4-day survey period.

Estimates of total MeHg intake from fish meals over the 4-day survey period were derived for each participant using the weight of fish meals obtained from duplicate diet portions, USDA recipe files on the percent of fish in given fish meals, and FDA and NOAA data on the MeHg content of different fish species (*19, 20*). In order to compare estimated intakes of MeHg to hair mercury levels, average daily intakes for the four day survey period were normalized to the body weight of the consumer. Estimates of the average daily intake ranged from 0.0008 ug/kg/day to 0.55 ug/kg/day. The average for all fish eaters was 0.078 ug/kg/day and the 90th percentile was 0.153 ug/kg/day.

Estimates of MeHg intake were also derived from the observed hair mercury levels using an average figure of 0.084 ug/kg/day derived from three studies (21, 22, 23) as a measure of MeHg intake that will result in 1 ppm MeHg in hair. MeHg intakes corresponding to the mean, 90th percentile, and maximum hair levels were estimated to be 0.035, 0.080, and 0.150 ug/kg/day, respectively. Although these values are 2-3 times lower than the corresponding mean, 90th percentile, and maximum intake estimates derived from the 4-day survey results, regression analysis did not reveal a significant correlation between the normalized daily intakes and hair mercury levels. One explanation may the small sample size in this scoping study. Another may be that the 4-day survey interval is too short for modeling purposes to yield reliable predictions of intake that occur as part of a complex diet and over time frames consistent with the accumulation of mercury in hair. This finding is not surprising since about 85% of the U.S. population eats fish over 2-3 months and only 21% of the NHEXAS participants ate fish during the 4-day survey.

The findings from the NHEXAS study demonstrate that the development of sensitive analytical methods for mercury in appropriate biological matrices will permit broad population studies to be conducted to address MeHg exposure issues now being considered by a number of U.S. government agencies. The findings also suggest that efforts to characterize exposure using dietary models as well as the relationship between biomarker measurements and dietary records may be more successful if dietary information is available for the extended time intervals reflected in biological measurements.

Such a study has just begun. The Center for Disease Control and Prevention's (CDC) fourth National Health and Nutrition Examination Survey (NHANES IV) is a 6 year study whose primary purpose is to produce descriptive statistics that can be used to measure and monitor the health and nutritional status of the general population in the civilian non-institutional U.S. The Blood-Hair Mercury Component of the NHANES IV (which is planned for FY 1999, 2000, and 2001) represents a collaborative effort of the CDC and the FDA that is co-sponsored by CDC, FDA, the Environmental Protection Agency (EPA), the National Institutes for Environmental Health Sciences (NIEHS), the National Oceanic and Atmospheric Administration (NOAA), the Department of Energy (DOE), and the Department of Health and Human Services (DHHS).

The primary aim of the study will be the development of a nationally representative biomarker data set that will allow the distribution of MeHg exposures to be characterized for U.S. population sub-groups that are considered most sensitive to the effects of MeHg, i.e., women (16-49 years) and young children (aged 1-5 years). The study of nearly 4000 participants will involve the determination of total mercury levels in hair strands ~ 1.5 cm long (approximately 1 months growth) and total, inorganic, and organic mercury concentrations in blood. The sensitivity of cold vapor atomic fluorescence spectrometry should permit the entire range of mercury concentrations in blood and hair to be described. Hair strands will be analyzed by RTI and the blood samples will be analyzed by the CDC's National Center for Environmental Health.

Secondary aims of the Blood-Hair Mercury Component will be to characterize the quantitative relationship between mercury concentrations found in matched blood and hair samples and develop predictive equations that relate mercury blood and hair concentrations to estimates of exposure developed from dietary records. Although the NHEXAS scoping study did not reveal any quantitative relationship, the likelihood of finding correlations is expected to be greater with the NHANES IVstudy, particularly among high-end fish consumers. The size of the study population will be 20-fold larger and the dietary questionnaire has been improved based on the NHEXAS experience by adding questions that will permit month long fish consumption patterns to be more reliably characterized. The length of hair strand that will be analyzed has been reduced to more closely reflect the time interval described in the fish consumption questionnaire. Questions about the use of hair treatment formulations that could alter the organic mercury content of hair have also been added.

The results of the mercury exposure component of the NHANES study, which will be available in 2002, should provide a nationally representative biomarker data set that will allow MeHg exposure and health risks among U.S. fish consumers to be more reliably characterized and permit the role of blood and hair biomarker data sets and dietary modeling for characterizing MeHg exposure to be better defined.

Literature Cited

1. Harada, M. *Crit. Rev. Toxicol.* **1991**, *25*, 1-24.
2. Simpson, R. E.; Horwitz, W.; and Roy, C. A. *Pestic. Monit J.* **1974**, *7*, 127-138 .
3. Suzuki, T. In *Biological Monitoring of Toxic Metals* ; Clarkson, T.W.; Friberg, L.; Nordberg, G. F.; and Sager, P. R. Eds; Plenum Press, New York,1988; pp. 623-640.
4 Airey, D. *Sci. Total Environ.* **1983**, *31*, 157-180.
5 Haxton, J.; Lindsay, D. G.; Hilsop, J. S.; Salmon, L.; Dixon, E. J.; Evans, W. H.; Reid, J. R.; Hewitt, C. J.; and Jeffires, D. S. *Environ. Res.* **1979**, *18*, 351-368.
6 Oskarsson, A.; Ohlin, B.; Ohlander, E.;M.; and Albanus, L. *Food Addit. Contam.* **1990**, *7*, 555-562.
7 Sherlock, J. C.; Lindsay, D. G.; Evans, W. H.; and Collier, T. R. 19 *Arch Environ. Health* **1982**, *37*, 271-278.
8 Miettinen, J. K.; Rahola, T.; Hattula, T.; Rissanen, K.; and Tillander, M. *Annals Clin.. Res.* **1971**, *3*, 116-122.
9 Sherlock, J.; Hislop, J.; Newton, D.; Topping, G.; and Whittle, K. *Human Toxicol.* **1984**, *3*, 117-131.
10 Suzuki, T. In *Advances in Mercury Toxicology*; Suzuki, T.; Imura, N.; Clarkson, T. W. Eds. Plenum Press, New York, 1991; pp 459-483.

11 Phelps, R. W.; Kershaw, T. G.; Clarkson, T. W.; and Wheatley, B. *Arch. Environ. Health* **1989**, *35*, 161-168.

12 Cernichiari, E.; Toribara, T. Y.; Liang, L.; Marsh, D. O.; Berlin, M. W.; Myers, G. J.; Cox, C.; Shamlaye, C. F.; Choisy, O.; Davidson, P.; and Clarkson, T. W. *Neurotoxicology* **1995**, *16*, 613-628.

13 Pellizzari, E.; Lioy, P.; Quackenboss, J.; Whitmore, R.; Clayton, A.; Freeman, N.; Waldman, J.;Thomas, K.; Rodes, C.; and Wilcosky, T. *J. Exposure Anal. Environ. Epidemiol.* **1995**, *5*, 327-358.

14 Clayton, C. A.; Pellizzari, E. D.; Whitmore, R.W.; and Perritt, R. L. *J. Exposure Anal. Environ. Epidemiol.* **1999**, *9,* 381-392.

15 Tollefson,L. and Cordle, F. *Envir. Health Perspect.* **1986**, *68*, 203-208.

16 U.S. EPA. Human Scalp Hair: An environmental exposure index for trace elements , EPA-600/1-78-037c, U.S. Environmental Protection Agency, Research Triangle Park, NC, 1978.

17 Smith, J.C.; Von Burg, R.; and Allen, P. V. *Arch. Environ. Health* **1997**, *54*, 476-480.

18 Yamamoto, R. and Suzuki, T. *Int. Arch.. Occup.. Health* **1978**, *42*, 1-9.

19 Cramer, G. U.S. Food and Drug Administration. Unpublished results.

20 Hall, R.A.; Zook, E. G.; and Meaburn, G. M. *NOAA Technical Report NMFS WWRF-721*, U.S. Department of Commerce, Washington DC 1978.

21 Skerfving, S. *Toxicology* **1974**, *2*, 3-23.

22 Gearhart, J. M.; Clewell, H. J.; Crump, K. S.; Shipp, A. M., and Silvers, A *Water, Air, and Soil Pollution* **1995**, *80*, 49-58.

23 Stern, A. H. *Risk Anal.* **1993**, *13*, 355-364.

Chapter 18

Dioxins in Food from Animal Sources

V. J. Feil and G. L. Larsen

Biosciences Research Laboratory, Agriculture Research Service, U.S. Department of Agriculture, P.O. Box 5674 University Station, Fargo, ND 58105–5674

Pentachlorophenol treated wood in production facilities has been found to correlate with chlorinated dibenzo-*p*-dioxins and chlorinated dibenzofurans in animal fats.   This source of dioxin/furan contamination has been observed in beef animals from a feeding study, in beef animals from several state and federal production-type research facilities, and in elk from North Dakota. Elk from near the Yellowstone National Park burn areas showed no evidence of this type of contamination.

Introduction

Dioxins are ubiquitous environmental pollutants that are formed during the manufacture of some industrial chemicals (i.e. herbicides, antibacterials, fungicides) and during combustion processes (i.e. forest fires, incinerators, fireplaces, combustion engines). The term "dioxin" became known to the general public because of issues dealing with Agent Orange and represented 2,3,7,8-tetrachlorodibenzo-*p*-dioxin. The environmental problem, however, consists of 75 different congeners containing from one to eight chlorine atoms generating very complex chemical, health, and regulatory issues. Adding to the complexity is the fact that some polychlorinated dibenzofurans (135 congeners) and polychlorinated biphenyls (209 congeners) have similar toxicities. The chlorinated furans are also formed in burning processes while the chlorinated biphenyls were manufactured in large quantities for industrial applications. A common hypothesis for dioxin exposure to the general population is that chlorinated dioxins and furans are carried on particulate matter generated in burning processes, deposited on forage, eaten by cattle, stored in fat, and finally consumed by humans in milk and meat. We review here our research that has established another source of dioxins in food from animal sources; namely, the exposure of animals to wood components in feeding facilities that were treated with pentachlorophenol (PCP).

U.S. government work. Published 2001 American Chemical Society

Discussion

Our research has shown that dioxin levels in domestic beef are generally low, but are much above average, and are highly variable at some locations. The analyses in our studies were done by EPA method 1613 (1). A feeding study in which small amounts of some dioxin and furan congeners were fed to animals yielded very useful information in this regard. In that study some of the control animals had higher levels of some dioxin congeners than the dosed animals, suggesting an unknown source of contamination. The source of contamination was subsequently traced to PCP treated wood in the feeding facility (2). The animals used in the feeding experiment were born, raised, and fed feed grown on the research facility. Serum dioxin levels taken at the beginning of the experiment were below the limits of detection for nearly all of the congeners (the animals entered the feeding experiment at weights of 220 to 262 kg and were fed for 120 days). Congener patterns in back fat and kidney fat were similar to those found in samples from treated walls and supporting posts of the pole barn taken after the analyses suggested a source of contamination (Figure 1). A major difference

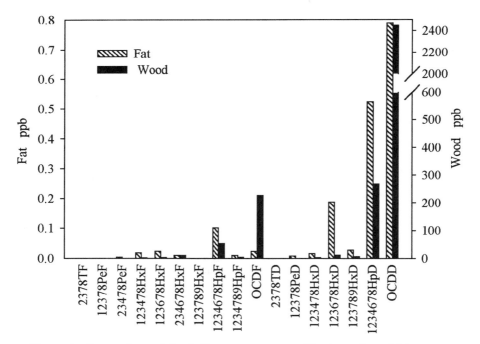

Figure 1. Comparison of dioxin/furan congener profiles in perirenal fat of control animals and wood with the highest dioxin/furan concentrations found at the Carrington, ND feeding facility.

in congener patterns between wood and beef fat is a reduction of several congeners in fat relative to wood, most notably 1,2,3,4,6,7,8,9-octachlorodibenzo-*p*-dioxin (OCDD) and to a lesser extent, 1,2,3,4,6,7,8,9-octachlorodibenzofuran and 1,2,3,4,6,7,8-heptachlorodibenzo-*p*-dioxin. This is likely due to poor absorption and different rates of metabolism among the different congeners (3,4). This alteration of congener patterns in beef versus wood is in agreement with relative bioconcentration factors (BCF) reported by Fries and coworkers in a feeding study with lactating Holstein cows (5). They found BCF values of 3.7, 0.68 and 0.08 for 1,2,3,6,7,8-HxCDD, 1,2,3,4,6,7,8-HpCDD and OCDD respectively. Stephens and coworkers have reported BCF values in chicken adipose tissue (chickens fed contaminated soil from a PCP manufacturing plant) of 6.84, 1.61 and 0.36, respectively, for the same congeners (6). Although the congener patterns in wood samples vary to a fairly large extent (likely due to variability in production lots and weathering), the pattern of the most highly contaminated wood sample at the feeding site was dominated by the following: OCDD (15 ppm); 1,2,3,4,6,7,8-HpCDD (1.7 ppm); 1,2,3,6,7,8-HxCDD (0.1 ppm); OCDF (1.1 ppm); and 1,2,3,4,6,7,8-HpCDF (0.36 ppm). In contrast, concentrations of 2,3,7,8-TCDD, 1,2,3,4,7,8-HxCDD and 1,2,3,7,8,9-HxCDD were 0.0015, 0.02 and 0.05 ppm, respectively. Other wood samples from the feeding facility had lower dioxin/furan concentrations but retained the pattern. These results suggest that large variations in dioxin/furan concentrations in the fat of animals at a given production facility are likely to exist because of differences in animal tendencies to rub, lick and chew, and because of variable opportunities for contact with treated wood components.

We have since associated high concentrations of dioxins in animals with PCP treated wood in thirteen production facilities at state and federal research locations around the nation. Every site that yielded animals with significant dioxin levels was subsequently found to contain components made of PCP treated wood. Our study yielded I-TEQ (7) values of 0.37-30.8 (ave. 5.63) ppt in fat from 11 bulls and values of 0.33-7.8 (ave. 2.29) ppt in fat from 9 beef cows (8). A nationwide beef survey conducted jointly by the U. S. Food Safety Inspection Service and the U. S. Environmental Protection Agency yielded values of 2.52 and 4.10 ppt for 2 bulls and 0.52-2.0 (ave. 0.81) ppt for 6 beef cows (9). Animals in each study showed PCP congener patterns for many of the samples, especially those with the higher concentrations. Purdue University and Pennsylvania State University each have two off-campus production type research facilities. At each university one site produced animals with low dioxin levels in kidney fat and one site produced animals with elevated levels. Subsequent investigations revealed PCP treated wood (i.e. ppm levels of OCDD) components at the site where animals with high dioxin levels were raised. For example, dioxin congener patterns in wood samples taken from one of the Purdue University feeding facilities were similar to those found in perirenal fat from animals raised at that site (Figure 2). No PCP components were found at the sites where animals with low dioxin levels were raised; however, the PCP dioxin/furan congener pattern was often recognizable in spite of levels being near the limits of detection, suggesting that we overlooked some PCP treated wood components or that the dioxins/furans were air borne, probably due to burning of treated wood.

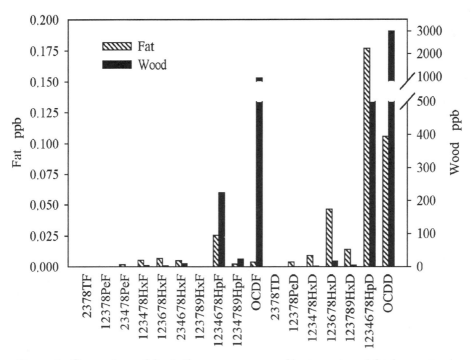

Figure 2. Comparison of dioxin/furan congener profiles in perirenal fat from animals raised at one of the Purdue University facilities and from wood samples from that site.

Figure 3 compares dioxin/furan congener profiles in perirenal fat from animals at the two Purdue University facilities, one group with TEQ values of less than 2, the other with TEQ values greater than 5. Animals that were exposed only or primarily to non treated wood (no PCP treated wood was found at the facility where animals with low TEQ values were raised) showed some characteristics of the PCP dioxin/furan pattern; however, their TEQ values received considerably more contribution from tetra and penta congeners. Some samples of supermarket foods (*10*) and of human blood and milk (*11*) also showed PCP dioxin congener patterns.

We have also measured dioxin levels in elk from two sites in North Dakota and from forest fire areas near Yellowstone National Park. Elk and bison have been observed to forage on charred debris in Yellowstone Park (*12*) and may provide a measure of forest fire contributions to the environmental dioxin burden. Elk were chosen for study because they have feeding habits similar to cattle and because they are not raised in confined areas. Elk from the Yellowstone areas had I-TEQ values in kidney fat ranging from 0.16 to 0.58 ppt while elk from the North Dakota sites had values ranging from 0.09 to 10.79 ppt (*13*). The only North Dakota elk with a TEQ greater than 2.0 ppt was a one half year old calf that likely had been nursed by a female with a dioxin level of approximately 2 ppt. Lactation is a known means of excreting dioxins and related compounds. An investigation of the North Dakota areas inhabited

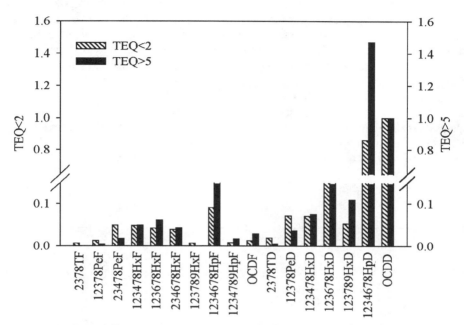

Figure 3. Dioxin/furan congener patterns in fat from animals raised at the two facilities at Purdue University. Patterns were determined from five animals with the lowest TEQ values raised at the site where no PCP treated wood was found and from five animals with the highest TEQ values raised at the site where PCP treated wood was found. The patterns are normalized to OCDD.

by the elk revealed that local wildlife clubs had constructed feeding facilities from old utility poles and from wood that had apparently been treated with PCP. The feeding facilities were used to support animal survival during severe winter weather. The utility posts used in construction of one of the facilities had an OCDD level of 89 ppm, while wood in another facility had an OCDD level of 35 ppm, providing ample opportunity for ingestion of the dioxins/furans contained in the PCP treated wood. The dioxin/furan congener patterns in elk fat from animals with the higher levels and from beef animals raised in facilities containing PCP treated wood are shown in Figure 4. Other than absolute values, the patterns are quite similar.

The use of PCP for treating fence posts and barns was heavily promoted through agricultural extension service personnel in the 50s and 60s. Many of these facilities remain in use because of the long life of treated wood. The ubiquitous nature of PCP and its impurities is not surprising as U.S. production has been reported to be 24,000 tons in 1974 (*14*) and annual world production has been reported to be approximately 100 million kg per year for several years (*15*). Most of the toxic effects found in food producing animals have been associated with impurities found in technical grade PCP (*14,15,16,17,18*). Modern production protocols for dairy and beef involve the feeding of hay and grain (our analyses of hay and grain have shown dioxin levels below the detection limits for most congeners) in large scale feeding facilities that contain little or no wood. This practice likely dilutes the dioxin concentration in foods from animal sources, reducing human exposure. The ubiquity of the PCP generated dioxins/furans is emphasized by the results of our geographical survey where the presence of the PCP dioxin congener pattern in fats from animals with TEQ levels below 1.0 is often recognizable. This may be because many of the animals being fed in large feedlots spend the early portion of their lives in small cow-calf operations where exposure to PCP treated wood components is quite likely or because the animals are exposed to air borne dioxins/furans due to the burning of PCP treated wood in incinerators, fire places, and open fires.

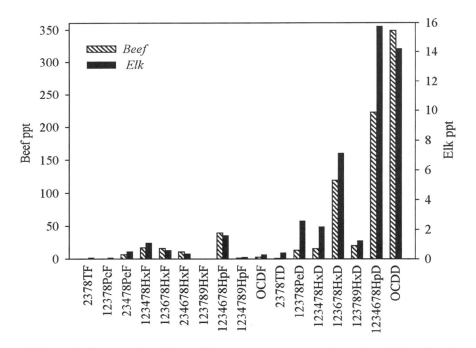

Figure 4. Comparison of dioxin/furan congener patterns in perirenal fat of beef with the five highest TEQ values from the geographical survey and in elk with the five highest TEQ values.

Conclusions

Our research shows a strong correlation between concentrations of polychlorinated dibenzo-*p*-dioxins and dibenzofurans in perirenal fat of beef cattle and elk, and exposure to wood treated with pentachlorophenol. Elimination of PCP treated wood components from production facilities, especially components that present a high opportunity for contact, would likely bring about a significant reduction in dioxin levels in beef.

References

1. United States Environmental Protection Agency. Office of Water Regulations and Standards. Industrial Technology Division. April, 1990.
2. Feil,V. J.; Larsen, G.L.; Fries, G. F. *Livestock Environment V*, **1997,** Vol. II, 1004-9, American Society of Agricultural Engineers, St. Joseph, MI.
3. Birnbaum, L. S.; Couture, L. A. *Toxicol. Appl. Pharmacol.* **1988,** *93*, 22-30.
4. Feil, V. J. *unpublished studies.*
5. Fries, G. F.; Paustenbach, D. J.; Mather, D. B.; Luksemburg, W. J. *Environ. Sci. Technol.* **1999,** *33*, 1165-1170.
6. Stephens, R. D.; Petreas, M. X.; Hayward, D. G. *Sci. Total Environ.* **1995,** *175*, 253-273.
7. NATO/CCMS, *North Atlantic Treaty Organization Committee on the Challenges of Modern Society* **1988,** North Atlantic Treaty Organization, Brussels, report no. 176.
8. Feil,V. J.; Davison, K. L.; Tiernan, T. O. *Organohalogen Compounds* **1995,** *26*, 117-9.
9. Winters, D.; Cleverly, D.; Meier, K.; Dupuy, A.; Byrne, C.; Deyrup, C.; Ellis, R.; Ferrario, J.; Harless, R.; Leese, W.; Lorber, M.; McDaniel, D.; Schaum, J.; Walcott, J. *Chemosphere* **1996,** *32*, 469-78.
10. Schecter, A.; Startin, J.; Wright, C.; Kelly, M.; Papke, O.; Lis, S.; Ball, M.; Olson, J. *Environmental Health Perspectives* **1998,** *102*, 962-6.
11. Schecter, A.; Papke, O.; Furst, P. *Organohalogen Compounds* **1996,** *30*, 57-60.
12. Stone, R. *Science* **1998,** *280*, 1527-28.
13. Feil,V. J.; Larsen, G.L. *Organohalogen Compounds* **1998,** *39*, 113-15.
14. Ahlborg, U.; Thunberg, T. *CRC Critical Reviews in Toxicology* **1980,** 1-35.
15. McConnell, E.; Moore, J.; Gupta, B.; Rakes, A.; Luster, M.; Goldstein, J.; Haseman, J.; Parker, C. *Toxicol. Appl. Pharmocol.* **1980,** *52*, 468-90.
16. Hughes, B.; Forsell, J.; Sleight, S.; Kuo, C.; Shull, L. *J. Animal Science* **1985,** *61*, 1587-1603.
17. Parker, C.; Jones, W.; Matthews, H.; McConnell, E.; Hass, J. *Toxicol. Appl. Pharmacol.* **1980,** *55*, 359-69.
18. Kerkvliet, N.; Brauner, J.; Matlock J. *Toxicology* **1985,** *36*, 307-24.

Chapter 19

Dioxin Body Burdens in California Populations

Myrto Petreas, Jianwen She, Michael McKinney, Pat Visita,
Jennifer Winkler, Mandy Mok, and Kim Hooper

Hazardous Materials Laboratory, Cal/EPA, 2151 Berkeley Way, Berkeley, CA 94704

Dioxins bioaccumulate in the food web and are measurable in fatty tissues of biota. The first systematic studies of dioxins in human adipose and breast milk from California revealed congener patterns and concentrations similar to those from the rest of the world. Comparison with limited adipose measurements conducted in the late 1980's indicates small but statistically significant decreases in concentrations in California populations.

INTRODUCTION

Polychlorinated Dibenzo-p-Dioxins and Polychlorinated Dibenzofurans (PCDD/PCDFs or Dioxins) are byproducts of chemical and thermal processes. Atmospheric deposition is currently the primary route of dioxins into the environment. Due to their lipophilic properties, dioxins bioaccumulate in biota in the food chain and, therefore, the diet (primarily dairy, meat, poultry and fish) accounts for over 90% of non-occupational human exposure (1). It is well known (2) that specific congeners of PCDD/PCDFs and Polychlorinated Biphenyls (PCBs) have significantly different potency in inducing diverse enzymes, modulating hormone receptor-binding activities, altering levels of thyroid hormone and vitamin A, and resulting in immunotoxicity, teratogenicity, hepatotoxicity, cancer and acute toxicity in various cell systems and

252 © 2001 American Chemical Society

animals. Of the over two hundred dioxin and furan congeners, seventeen are chlorinated in the 2,3,7,8 positions. The most extensively studied congener of this group is 2,3,7,8-tetra dioxin (TCDD). All seventeen 2,3,7,8-substituted congeners have a planar structure, exhibit the highest affinity for the Ah receptor *(2)* and bioaccumulate in human tissues *(3)*. Of the 209 Polychlorinated Biphenyls (PCBs), those substituted on both para- and at least two meta- positions are approximate isostereomers of 2,3,7,8 TCDD and exhibit high affinity for the Ah receptor *(2)*. Recently, the anti-estrogenic potential of a number of these PCDD/PCDF and PCB congeners has been shown *(4) (5) (6)*. In general, their order of potency paralleled their binding affinities for the Ah receptor *(4)*.

A conventional way to capture the toxicity of PCDD/PCDF mixtures is to use Toxic Equivalency Factors (TEFs) developed for individual congeners. The use of International Toxic Equivalents (I-TEQs) is widely accepted to express total dioxin toxicity as a weighted sum of the seventeen 2,3,7,8-substituted toxic congeners *(7)*.

The best way to monitor body burdens of persistent, bioaccumulative contaminants is to systematically sample a targeted population. Human milk, serum, and adipose tissues have been used successfully in many studies to monitor lipophilic chemicals. Caution should be exercised, however, in interpreting findings and in generalizing and extrapolating from such monitoring studies. Accessibility to tissues, limits the population that can be characterized (human milk may be used to assess body burdens in women of a certain age range; adipose can, usually, only come from subjects undergoing some type of surgery or from cadavers). In addition, comparisons of measurements in different types of specimens are not always straight forward, even when concentrations are expressed on a lipid basis *(8,9)*.

Many industrialized countries have on-going monitoring programs where body burdens of dioxins and dioxin-like compounds are being measured. Surveys of targeted populations with consistent protocols enable researchers to examine trends and to compare groups *(10)*. Downward trends in dioxin body burdens have been shown in Sweden *(11)*, The Netherlands *(12)* and Germany *(13)*. No such trends can be established for the USA because there are no systematic monitoring programs. The National Human Adipose Tissue Survey (NHATS) *(14)*, a systematic nation-wide study of pesticide residues in cadavers, included dioxin analyses only in 1982 and 1987, and has since been discontinued. In California, serum levels of PCDD/PCDF were measured in the late 1980s in a small group of residents of Oroville, during an investigation of a Pentachlorophenol contaminated site. These serum levels were higher than PCDD/PCDF levels measured in a small group of controls *(15)*. The lack of any systematic body burden data from California impedes the work of public health and regulatory agencies and fuels concerns from citizens.

This paper presents the first systematically collected data on dioxin body burdens from California. The data include dioxin measurements in adipose tissues of women undergoing breast surgery (biopsies or lumpectomies) but who are free of cancer.

These women comprise the control group in a breast cancer case-control study centered in the San Francisco Bay Area. In addition, this paper presents measurements of dioxins in human milk from a group of low-income women participating in the Women with Infants and Children (WIC) support program in the Stockton area of the Greater San Francisco Bay Area.

EXPERIMENTAL METHODS

Study Populations.

It is important to define the populations sampled in order to make meaningful inferences and comparisons. In this study, two very distinct populations of women were sampled: The first group was comprised of middle aged women participating in a breast cancer study as controls (cancer-free). The second group was comprised of young first-time mothers. Each group represents a distinct demographic population and the two groups combined encompass a wide range of age, race, social and economic backgrounds (Table 1).

Breast cancer study:
Women (28-65 years old) undergoing breast surgery for suspected breast cancer at Stanford University, California, were recruited in the study. Small amounts of breast adipose tissue were collected during surgery and women were interviewed regarding demographics, environmental and occupational exposures, and medical and reproductive histories. Only women with non-malignant pathologies (controls) are included in this comparison. Whereas the study was open to all races the recruited population was over 80% Non-Hispanic White. This population was highly educated with over 85% college graduates, of whom over 40% had graduate degrees. Similarly, over 80% had family incomes exceeding $50,000, of whom over 40% had family incomes exceeding $100,000. This population can, therefore, be characterized as having a high socioeconomic status (SES).

Milk study:
Women participating in the Stockton area Women, Infants and Children (WIC) program were recruited in the study. Stockton is a city of about 200,000 located approximately 50 miles east of San Francisco, within the Greater San Francisco Bay Area. The WIC program supports low-income mothers and their young children by providing perinatal check ups and nutritional supplements. Milk was collected

Table 1. Demographic Characteristics of Populations Studied

		ADIPOSE SF-98 N=45	MILK 1998 N=40	P*	ADIPOSE SF-88 N=17	P**
Age				0.001		0.89
	Mean	45.2	22.7		46.4	
	SD	8.5	5.1		9.0	
	Range	28-65	16-35		32-65	
	Median	45	22		43	
Race						
	Non-Hispanic White	82%	8%		94%	
	Hispanic	6%	45%		0%	
	African American	2%	18%		6%	
	Asian	8%	18%		0%	
	Other	4%	12%		0%	
SES		High	Low		Unknown	

*p-value for Wilcoxon Test for difference in age of women in the Adipose SF-98 vs. Milk groups;
** p-value for Wilcoxon Test for difference in age of women in the Adipose SF-98 vs. Adipose SF-88 groups

according to the WHO protocol (*16*) which targets primaparae, with 2 to 10 week-old infants, and where both mother and infant are in good health. The women provided approximately 100 mL of milk in a precleaned jar, and were interviewed regarding demographics, exposures, residential history, and medical and reproductive histories.

Comparison Populations:

Data from two populations are used for comparison with the current study. The first are adipose tissue samples from 17 women undergoing surgery for reasons other than cancer. These samples were collected in 1988 from hospitals in the San Francisco Bay Area as part of a larger survey that included men and women from San Francisco and Los Angeles (*17*). These unpublished data are the only comparable data from the region. The second data set consists of nationwide average concentrations from 48 composite cadaver adipose tissue samples from the 1987 NHATS survey (*14*).

Sample Analysis

Adipose sample preparation. Samples were stored at -20 °C until analysis. Samples were thawed, weighed, mixed with Na_2SO_4, homogenized with 1:1 dichloromethane:hexane, and spiked with internal standards. Following centrifugation, 1/10 of the extract was analysed for organochlorine pesticides and PCBs and the remaining 9/10 analysed for PCDD/PCDFs and coplanar PCBs.

Milk sample preparation. Samples were stored at -20 °C until analysis. Samples of milk (100 g) were thawed, spiked with internal standards, and extracted with 2:1:1 ethanol:hexane:ethyl ether in the presence of sodium oxalate.

PCDD/PCDF Analysis. A modified Smith-Stalling procedure was used for sample clean up (*18*). Labeled internal standards were used for all seventeen 2,3,7,8-substituted PCDD/PCDF congeners. Briefly, samples were serially processed through columns containing Na_2SO_4 and AX21 Carbon. PCDD/PCDFs were eluted from the carbon column with toluene and the eluate cleaned up with alumina and acid silica columns; recovery standards were added and the sample concentrated to 10 μL. Lipid content was determined gravimetrically in an aliquot of the extract.

HRGC/HRMS Analysis. PCDD/PCDFs and coplanar PCBs were analyzed by HRGC/HRMS (Finnigan MAT 90) with a 60m, 0.25 mm ID, 0.25 μm film thickness, DB-5ms column. The samples were introduced through a Septum Programmable Injector (SPI). The temperature was held initially at 220 °C for 2 minutes, increased to 260 °C at 5 °C/min, and then to 300 °C at 1 °C/min. PFK was used for the lock masses and the MS was operated in an EI mode with multiple ion monitoring.

RESULTS AND DISCUSSION

An important finding of the adipose study was the high variability in lipid content observed in the breast adipose specimens. Lipid content ranged from less than 10% to over 90% with a mean of 72%. This variability may be explained by breast tissue physiology, where adipose is interspersed within non-fatty connective tissue. It may also reflect differential presence of blood or other non-lipid tissues in the sample submitted for analysis. Given the small size of these samples (often less than 1 g), these non-fatty tissues may impact the lipid content. To compensate for these differences, all results were expressed on a lipid basis, making all measurements comparable. As shown in Figure 1, the % lipid content of the adipose specimens correlated with the age of the patient (R^2=0.159, p<0.001). Given that age is a known risk factor for breast cancer, non lipid-adjusted concentrations may lead to misclassifications and significantly confound Odds Ratios for disease. It should be noted that some of the studies examining links between organochlorines and breast cancer did not use lipid adjusted measurements (19-21), which may explain, in part, the contradictory findings of these studies (19-29).

Because of the small size of the adipose sample and, often, its low lipid content, many congeners were below the detection limit. In such cases, half the detection limit was used to calculate I-TEQs. Nevertheless, congeners who were consistently non-detected (congener below detection in over 50% of the samples) were not used for statistical analysis. In addition, a new summary measure (Adjusted TEQ, or Adj-TEQ) was devised incorporating only those congeners that were consistently measured above the detection limit. The eight congeners that comprised the Adj-TEQ are shown in Table 2 in bold print. The conventional I-TEQs correlated well with the Adj-TEQs (R^2=0.94, p<0.001) and, therefore, Adj-TEQs were used in statistical analyses to minimize uncertainties. In addition, Adj-TEQs allowed comparison with other data sets where most congeners were reported above detection, mainly because larger sample weights had been analyzed.

Table 2 shows summary statistics for the adipose and milk measurements from the two California populations. Overall, whereas the congener pattern is consistent, milk concentrations are significantly lower than adipose concentrations. Figure 2 shows Adj-TEQs (pg/g fat) for both milk and adipose samples as a function of age. The figure shows little overlap between the two populations, with the younger milk donors having lower Adj-TEQs than the older adipose donors. Regression analysis showed an R^2 of 0.451 with a slope indicating an increase of 0.5 pg/g fat for each year of age. This positive correlation of TEQs with age has been reported extensively (30).

Major PCDD/PCDF congeners in adipose of 45 cancer-free women (breast cancer study controls) are shown in Figure 3 as SF-98, along with measurements from the two (14, 17) comparison populations. In 1988, adipose tissue samples from 17 women undergoing surgery for reasons other than cancer were collected from

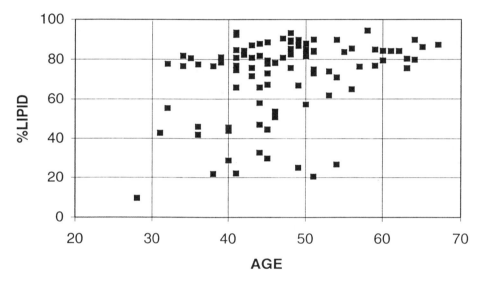

Figure 1. Lipid content of breast adipose samples as a function of age.

Table 2. PCDD/PCDF Concentrations (pg/g fat) In Human Adipose And Milk

	ADIPOSE, SF-98, n=45			MILK, 1998, n=40				ADIPOSE, SF-88, n=17			
	% >DL	MEDIAN	RANGE	% >DL	MEDIAN	RANGE	p *	% >DL	MEDIAN	RANGE	p **
2,3,7,8-TCDD	55	3	0.4-20	58	1.3	0.2-5.3	0.0001	100	5.9	2.3-12.5	0.002
1,2,3,7,8-PeCDD	44	4	0.4-24	68	3.7	0.3-12.5	0.25	100	12.6	4.4-25	0.001
1,2,3,4,7,8-HxCDD	67	6	1-35	58	2.1	0.3-9.2		88	11	6.8-25	
1,2,3,6,7,8- HxCDD	100	54	21-180	100	16	3-59	0.0001	100	65	32-124	0.03
1,2,3,7,8,9- HxCDD	64	5.5	1.6-26	73	4	1-11		94	11	5-32	
1,2,3,4,6,7,8- HpCDD	100	60	23-200	100	28	7-112	0.0001	100	108	34-334	0.003
OCDD	100	396	169-3230	100	118	42-438	0.0001	100	760	137-1230	0.004
2,3,7,8-TCDF	50	2.7	0.3-16	48	1	0.3-3.9		100	2.7	1-5.4	
1,2,3,7,8-PeCDF	13	1.3	0.3-16	18	0.7	0.3-2.4		38			
2,3,4,7,8- PeCDF	89	9	2-26	83	3.7	0.4-17	0.0001	94	4	0.6-24	0.02 #
1,2,3,4,7,8-HxCDF	80	4	0.4-48	73	1.9	0.4-42	0.0001	100	7.9	4.3-17	0.002
1,2,3,6,7,8- HxCDF	76	3.5	0.4-13	55	1.8	0.4-28	0.0001	100	5.6	2.3-12	0.03
1,2,3,7,8,9- HxCDF	0			0				18			
2,3,4,6,7,8- HxCDF	53	3.3	1.4-23	38	1.5	0.6-8.5		94	1.6	0.5-10	
1,2,3,4,6,7,8- HpCDF	84	9.2	3-29	75	4.6	1.5-29		100	11	4-20	
1,2,3,4,7,8,9- HpCDF	0			0				35			
OCDF	16	3.4	0.2-64	10	1.4	0.2-16		100	0.8	0.4-4.6	
I-TEQ	100	19	10-60	100	8	1.6-29	0.0001	100	27.2	13-63	0.01
Adj-TEQ	96	16	8-51	100	7	1.4-27	0.0001	100	23.9	11-56	0.009

* p-value for Wilcoxon Test for Adipose SF-98 vs. Milk; ** p-value for Wilcoxon Test for Adipose SF-98 vs. Adipose SF-88;
# Adipose SF- 98 greater than Adipose SF-88

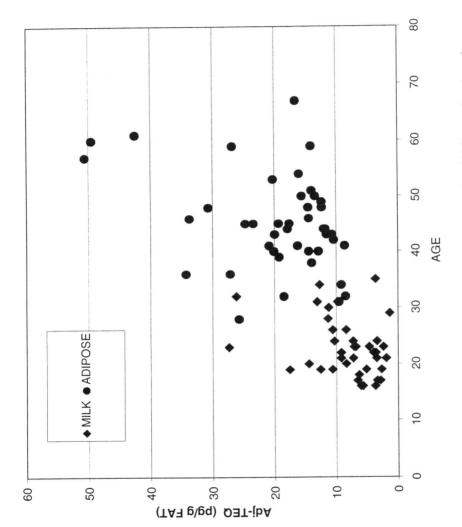

Figure 2. Adj-TEQ (pg/g fat) of milk and adipose samples (1998) as a function of age. Adj-TEQ=-3.879+0.54*(Age), $R^2=0.451$, N=83.

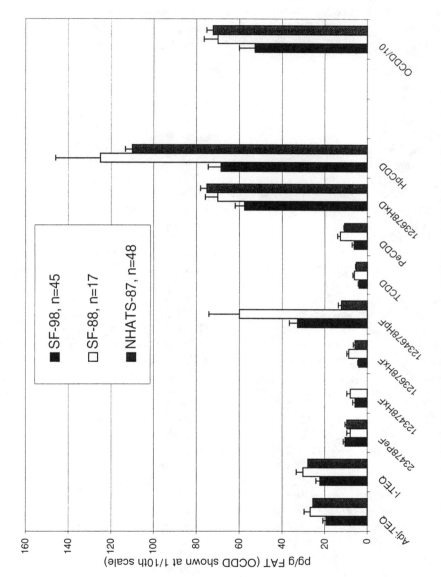

Figure 3. Major PCDD/PCDF congeners, I-TEQ and Adj-TEQ in adipose tissues from the San Francisco Bay Area in 1998 and in 1988, and from the NHATS survey in 1987.

hospitals in the San Francisco Bay Area as part of a larger survey that included men and women from San Francisco and Los Angeles (17). These data are shown as SF-88 in Figure 3. Nationwide average estimated concentrations from 48 composite cadaver adipose tissue samples from the 1987 NHATS survey (14) are also shown in Figure 3. Because of the small sample weights available from the breast cancer study (often less than 1 g), many PCDD/PCDF congeners were below detection in the adipose tissues. An adjusted TEQ, (Adj-TEQ) based on only the congeners consistently measured in all 3 data sets is also shown in addition to the I-TEQ, to facilitate comparisons.

The pattern of the prevalent congeners is consistent across all three data sets (Figure 3) and also consistent with congener patterns reported in other non-occupationally exposed populations (30). The congener distributions were tested for normality by the Shapiro-Wilk test and found non-normally distributed. Most congeners remained non-normally distributed following logarithmic transformation and, therefore, non-parametric techniques were used for comparisons. For adipose samples, the Wilkoxon rank sum test revealed significant decreases over the past decade in the concentrations of Adj-TEQ, I-TEQ, and in all but one of the individual congeners examined (Table 2). The 1988 levels measured in the San Francisco Bay Area women may be considered representative of the whole State, since no statistically significant differences by sex, age or geographic area were reported (17). As the age distributions of the SF-98 and SF-88 women were similar (Table 1), the apparent decrease is not confounded by age. This first documented decrease in California body burdens is consistent with worldwide observed decreases (30). It is believed that body burden decreases reflect lower PCDD/PCDF emissions due to implementation of pollution controls and industrial process substitutions. The one exception was 23478-PeCDF, whose levels were significantly higher in the SF-98 than in the SF-88 group. Elevated body burdens of 23478-PeCDF have been linked to high fish intake (30). As no exposure information exists on the SF-88 data, it is not clear why 23478-PeCDF levels appear to have risen over the last decade.

Levels of PCDD/PCDFs, I-TEQs and Adj-TEQs measured in the 40 human milk samples from the WIC population were significantly lower than levels in adipose from the contemporary women (Table 2). The only congener whose levels were not statistically different between the two groups was 23478PeCDD, probably due to its high standard deviation in the adipose data set. The lower PCDD/PCDF levels can be explained in part by the statistically significant differences in age distributions among these two populations (Table 1). A significant correlation with age has been demonstrated in these data sets, as well as in other populations. However, these milk concentrations are lower than reported World averages (30). At this point it is unclear why this group of mostly non-white, low SES women has low PCDD/PCDF body burdens. Questionnaire information along with measured body burdens are examined to assess pathways of exposures (31).

ACKNOWLEDGMENT

We thank all participants, and the volunteers from the Community Breast Health Project. The breast cancer study is performed in collaboration with Drs. P Reynolds, D Smith and D Gilliss and Ms. S Hurley of the California Department of Health Services and Drs. S Jeffrey and ME Mahoney of Stanford University. The breast cancer study is partially funded by the US Army Medical Research and Materiel Command, Grant #DAMD17-94-J-4429 P6003. The WIC study is performed in collaboration with WIC staff and is partially funded by the Public Health Trust. Ms. Amber Cheng and Ms. Kimberly Walters assisted in data management and statistical analyses.

LITERATURE CITED

1. CC Travis and HA Hattemer-Frey. Human exposure to 2378-TCDD. Chemosphere 16: 2331-2342, 1987
2. Safe S. PCBs, PCDDs, PCDFs, and Related Compounds: Environmental and Mechanistic Considerations Which Support the Development of Toxic Equivalency Factors (TEFs). Crit Rev Tox 21(1):51-88, 1990
3. Rappe C, Andersson R, Bergquist PA, et al. Overview on environmental fate of chlorinated dioxins and dibenzofurans. Sources, levels, and isomeric pattern in various matrices. Chemosphere 16:1603, 1987
4. Gallo MA, Hesse EJ, MacDonald GJ, Umbreit TH. Interactive effects of estradiol and 2,3,7,8-TCDD on hepatic cytochrome P-450 and mouse uterus. Toxicol Let. 32:123-132, 1986
5. Safe S. ED: Environmental Toxin Series 1. Polychlorinated Biphenyls (PCBs): Mammalian and Environmental Toxicology. New York: Springer-Verlag, 1987
6. Krishnan V, Safe S. PCBs, PCDDs, and PCDFs as Antiestrogens in MCF-7 Human Breast Cancer Cells: Quantitative Structure-Activity Relationships. Tox Appl Pharmac. 120, 55-61, 1993
7. Ahlborg UG, A Brower, M Fingerhut, et al. Impact of PCDD/PCDFs/PCBs on human and environmental health, with special emphasis on application of the toxic equivalency factor concept. Europ J Pharmacol.-environ. Toxicol. Pharmacol Sect. 228:179-199, 1992
8. Greizerstein HB, C Stinson, P Mendola, GM Bucks, PJ Kostyniak, JE Vena. Comparison of PCB congeners and pesticide levels between serum and milk from lactating women. Environmental Resrearch Section A.80:280-286, 1999
9. Archibeque-Engle et al. Comparison of organochlorine pesticide and PCB residues in human breast adipose tissue and serum. J Toxicol Env Health, 52:285-293, 1997

10. Liem AKD, UG Ahlborg, H Beck at al. Levels of PCBs, PCDD/PCDFs in human milk. Results of the 2nd round of a WHO-coordinated exposure study. Organohalogen Compounds, 30:268273, 1996
11. Noren C and D Meironyte. Contaminants in Swedish human milk. Decreasing levels of organochlorine and increasing levels of organobromine compounds. Organohalogen Compounds, 38:1-4, 1998
12. Liem AKD, JMC Albers RA Baumann, et al. PCBs, PCDD/PCDFs and Organochlorine pesticides in human milk in The Netherlands. Levels and trends. Organohalogen Compounds, 26:69-74, 1995
13. Furst P, Furst C, Wilmers K. Human milk as a bioindicator for body burden of PCDDs, PCDFs, organochlorine pesticides and PCBs. Env Health Persp. 102 suppl 1: 187 1994
14. Orban JE, Stanley J, Schwemberger J, Remmers J. Dioxins and Dibenzofurans in adipose tissue of the general US population and selected subpopulations. Am J Public Health: 84:439-445, 1994
15. Goldman LR, M Harnly, J Flattery, DG Patterson, LL Needham. Serum PCDDs and PCDFs among people eating contaminated home-produced eggs and beef. Envir Health Persp, In Press, 1999
16. World Health Organization. WHO Regional Office for Europe. Levels of PCBs, PCDDs and PCDFs in Breast milk. Results of WHO-coordinated Interlaboratory Quality Control Studies and Analytical Field Studies. Yrjanheikki EJ, ed. Environmental Health Series Report #34. Copenhagen, 1989.
17. Midwest Research Institute. Determination of body burdens for PCDDs and PCDFs in California residents. Final Report. For the California Air Resources Board. ARB Contarct No. A6-195-33. 1989
18. HML. Hazardous Materials Laboratory. Analysis of PCDD/PCDFs. Method 880, 1991
19. Wolff MS, Toniolo PG, Lee EW, Rivera M, Dubin N. Blood levels of organochlorine residues and risk of breast cancer. J Natl Cancer Inst 85:648-652, 1993
20. Krieger N, MS Wolff, RA Hiatt, et al. Breast Cancer and Serum Organochlorines: a Prospective Study among, White, Black and Asian Women. JNCI. 86:589-599, 1994
21. Dewailly E, S Dodin, R Verrealt, et al. High Organochlorine Body Burden in Women with Estrogen Receptor-Positive Breast Cancer. JNCI. 86: 232-234, 1994
22. van't Veer P, I Lobbezoo, J M Martin-Moreno, et al. DDT (dicophane) and postmenopausal breast cancer in Europe: case-control study. Br. Med J. 315:81-85, 1997
23. Unger M, Kiaer H, Blichett-Toft M, Olsen J, Clausen J. Organochlorine compounds in human breast fat from deceased with and without breast cancer and in a biopsy material from newly diagnosed patients undergoing breast surgery. Environ Res 34:24-28, 1984

24. Mussalo-Rauhamaa H, Hasanen E, Pyysalo H, Kauppila AR, Pantzar P. Occurrence of Beta-Hexachlorocyclohexane in Breast Cancer Patients. Cancer 66:2124-2128, 1990

25. Falck F Jr, Ricci A Jr, Wolff MS, et al. Pesticides and polychlorinated biphenyl residues in human breast lipids and their relation to breast cancer. Arch Envir Hlth 47:143-146, 1992

26. Hardell L, G Lindstrom, Liijegren, et al. Increased concentrations of OCDD in cases with breast cancer-results from a case control study. Eur J Cancer Prevention, 4:351-57, 1996

27. Lopez-Carillo L, A Blair, M Lopez-Cervantes, et al. DDE serum levels and breast cancer risk: a case-control study from Mexico. Cancer Research 57:3728-32, 1997

28. Hunter D, S Hankinson, F Laden et al. Plasma organochlorine levels and the risk of breast cancer. NEJM. 337:1253-58, 1997

29. Moysich K, C Ambrosone, J Vena, et al. Environmental Organochlorine exposure and postmenopausal breast cancer risk. Cancer Epid Biomarkers and Prevention 7:181-188, 1998

30. World Health Organization. International Agency for Research on Cancer. IARC Monographs on the Evaluation of Carcinogenic Risks to Humans. Vol 69, Polychlorinated Dibenzo-p-Dioxins and Polychlorinated Dibenzofurans. Lyon, France.1997

31. K Hooper, M Petreas, J She, P Visita, M Mok, J Winkler, M McKinney, Y Cheng, B Reisberg, K Ruhstaller. PCDD/PCDF. Levels in Breast Milk from an Ethnically Diverse Population Near A Hazardous Waste Site in Stockton, California. Proceedings of the 19th International Dioxin Conference, Venice, Italy, 1999

Chapter 20

Biomagnification of Toxic PCB Congeners in the Lake Michigan Foodweb

A. G. Trowbidge and D. L. Swackhamer[1]

Environmental and Occupational Health, School of Public Health,
University of Minnesota, Minneapolis, MN 55455
[1]Corresponding Author

This study examined whether AHH-inducing PCBs preferentially biomagnify relative to total PCBs in aquatic foodwebs. Organisms from the Lake Michigan lower trophic level foodweb were collected at two locations over the course of two years. This study determined that not only do AHH-inducing PCBs biomagnify, but they biomagnify to a greater extent than do total PCBs. TEQs were calculated and were also found to preferentially biomagnify.

INTRODUCTION

Polychlorinated biphenyls (PCBs) have been investigated by researchers since 1966 when they were first identified in fish and eagle tissue and human hair samples (*1*). Today, these chemical compounds still pose a serious environmental threat even though their production in the United States was formally banned in 1979. They have been identified in a multitude of biotic and abiotic samples, including samples from remote pristine locations (*2-9*)

Concerns attributed to PCBs stem from their environmental persistence, ability to bioaccumulate and their potential toxicity. These stable compounds are of particular concern in aquatic systems where their concentrations are usually low due to their hydrophobic nature, but reach much higher concentrations in top predator fish because of their ability to bioaccumulate and biomagnify in the aquatic foodweb. This poses a serious health threat to humans since the primary source of PCBs in the human diet is due to the consumption of fish from contaminated water (*10-12*).

© 2001 American Chemical Society

Not all PCBs invoke the same toxic responses in all species. The PCB family of 209 congeners exhibit different magnitudes of toxicity and manifest their toxicity under different modes of action (*13-16*). One small subset of individual PCB congeners has been the focus of much toxicological research. These congeners include PCBs with no ortho, two para and two or more meta chlorines, their mono-ortho analogs and some di-ortho congeners (*14*). They are structurally similar to highly toxic 2,3,7,8-tetrachlorodibenzo-*p*-dioxin (TCDD) and exhibit similar toxic responses (*13, 15, 17*). These toxic compounds promote induction of the aryl hydrocarbon hydroxylase (AHH) enzyme system which elicits such symptoms as body weight loss, dermal toxicity, immunotoxicity, and adverse effects on reproduction and development (*15*). While these PCB compounds are less toxic than TCDD, they often contribute more to total TCDD-type toxicity due to their higher concentrations in environmental matrices (*17-19*). Smith et al. found that more than 90% of the TCDD-type toxicity in chinook salmon was attributed to 3 of the AHH-inducing PCBs (*20*). Even with this toxicity potential, there are only a handful of reports concerning their distribution or dynamics in aquatic or terrestrial ecosystems (*20-25*).

Because of the lack of exposure data for AHH-inducing PCBs, risk assessments are most often performed using total PCB measurements. An assumption inherent in this approach is that the exposure to AHH-inducing congeners is proportional to the total PCBs throughout the risk assessment model. This study examines congener specific concentrations for 16 individual AHH-inducing PCBs found in lower trophic level organisms in the Lake Michigan foodweb. It compares the proportion of AHH-inducing PCB congeners relative to total PCBs, biomagnification factors for AHH-inducing PCBs and total PCBs and toxic equivalency (TEQs) values among the foodweb trophic levels.

LAKE MICHIGAN SAMPLING

Lake Michigan is one of the Great Lakes that has been severely affected by the discovery of PCBs in top predator fish. Lake trout and coho salmon exceed the U.S. Food and Drug Administration concentration guidelines for commercial sale. This discovery created devastating effects on a multibillion-dollar fishing and tourism industry (*26*). As a result, PCBs in Lake Michigan's top predator fish have been researched quite extensively, but little is known about PCB concentrations in the components that make up their food source. The primary components of the pelagic foodchain include phytoplankton at the base of the foodweb, herbivorous zooplankton, *Mysis*, forage fish such as alewife and rainbow smelt and top predators such as lake trout and coho salmon. Components of the benthic foodchain include detritus, *Diporeia sp*, forage fish such as bloater chub and sculpin, and the top predator lake trout and coho salmon (*27*). Both foodweb branches obtain PCBs from water, and the benthic branch includes sediments as a PCB source.

Based on this understanding of the Lake Michigan foodweb, major components occupying the lower trophic levels of the pelagic foodweb were chosen for sample collection along with a major benthic component. Sample collection was executed as part of the Lake Michigan Mass Balance Study (LMMBS) funded by the U.S. EPA Great Lakes National Program Office. Samples for the LMMBS were collected from eleven sites over seven cruises and analyzed on a congener specific basis for 110 PCB congeners. A subset of the LMMBS samples were re-analyzed for concentrations of AHH-inducing PCBs which often co-elute with other PCB congeners. Figure 1 illustrates site locations, dates of collection and sample media which were used for this study. Briefly, two site locations were chosen for re-analysis, sites 180 and 380. Samples were collected during 1994 and 1995. Sample media collected included water (dissolved phase), phytoplankton (10-100 μm), bulk net zooplankton (>100 μm), *Mysis relicta*, and *Diporeia* sp.

Water samples were collected at the same depth as phytoplankton samples. Dissolved phase PCBs were collected on XAD-2 resin. Approximately 300L of water was processed for each dissolved phase sample. Phytoplankton samples were collected via "phytovibes", a cone-shaped frame with a $1m^2$ opening which held a stationary 10 μm Nitex net with a detachable collection cup at the bottom. A motor vibrated the structure to facilitate water removal. Samples were filtered onto GF/F fiber filters and ranged from 0.14 to 3.93 g dry weight. Microscopic analysis indicated a pure phytoplankton sample for this size fraction (10 - 100 μm).

Organisms collected under the heading of zooplankton included bulk zooplankton (>100 μm), *Mysis relicta* and *Diporeia sp.* Bulk zooplankton was collected via vertical net tows using 100 μm nets that produced a sample dominated by zooplankton, but occasionally containing some colonial diatoms. *Mysis relicta* were collected either by vertical net tows using 500 μm nets or by benthic sled tows. *Diporeia* were collected by benthic sled tows. Organisms were hand picked to get a suitable and pure sample. Subsamples of bulk phytoplankton, zooplankton, *Mysis* and *Diporeia* were collected for particulate organic carbon (POC) analysis.

ANALYTICAL PROCEDURES

All samples (whole organisms) used in this project were extracted in a Soxhlet apparatus. Extracts were subsequently batch extracted to remove additional water, lipids were removed using 6% deactivated alumina column, and compound separation was performed using a combined column of 0% deactivated silica and 1% deactivated alumina. Prior to extraction, PCB congeners #14 (3,5-dichlorobiphenyl), #65 (2,3,5,6-tetrachlorobiphenyl) and #166 (2,3,4,4',5,6-hexachlorobiphenyl) were added as surrogates. Media-appropriate volumes of two internal standards,

Field Sampling

Biota Samples

Phytoplankton   10 - 100 μm
Zooplankton   > 100 μm bulk
Mysis   Hand picked
Diporeia   Hand picked
Dissolved   XAD

Seven cruises:
June, August, October, 1994
Jan., March, Aug., Sept. 1995

Analytes:
16 AHH inducing PCBs

Samples analyzed for AHH-inducing PCBs
were collected as part of the EPA Lake
Michigan Mass Balance Study. (Sites 180 & 380)

FIGURE 1. Sampling sites, collection dates and media collected from Lake Michigan.

congeners #30 (2,4,6-trichlorobiphenyl) and #204 (2,2',3,4,4',5,6,6'-octachlorobiphenyl) were added prior to instrumental analysis.

Extracts were re-analyzed for specific AHH-inducing PCBs (Table I) using gas chromatographic (Hewlett Packard 5890A) mass spectrometry (Hewlett-Packard 5988A) with selected ion monitoring (SIM) and the mass spectrometer operated in the negative ion mode (GC/MS-NI). A method developed by Schmidt and Hesselberg (28) was adapted for use in this project. This method utilizes the fact that GC/MS in the negative ion mode is very selective and sensitive to highly chlorinated compounds. AHH-inducing PCB congeners often co-elute with other congeners having a different number of chlorine atoms which allows them to be differentiated by negative ion mass spectrometry.

To ensure high quality data a strict set of QA criteria were invoked for this research project. A full Quality Assurance Project Plan was prepared and approved by EPA (29). This plan included criteria for accuracy, precision, blank values, and detection limits.

RESULTS AND DISCUSSION

Of the 16 AHH-inducing PCB congeners analyzed, only two were not used in quantitation of total toxic PCB congeners. PCB 126 was removed from the data set due to interference from surrogate standard PCB 166. An attempt was made to account for this interference, however the resulting PCB 126 concentrations were not reliable. PCB 170 was removed from the data set since it could not be separated from congener 190 and has the same molecular weight.

The first step in analyzing the data was to determine if concentration normalization to either lipid or organic carbon was appropriate. Normalizing data is only appropriate when contaminant concentration varies in direct proportion to the normalizing factor (30). If these criteria are not met and data normalization is performed, incorrect conclusions can result. Although lipid content varied among trophic levels (see Table II), the relationship between both lipid and organic carbon to contaminant concentration was weak or insignificant for each sample type analyzed. It was not deemed appropriate to normalize congener concentrations to either lipid or organic carbon. Therefore these values remain on a dry weight basis.

Concentration data was evaluated for spatial and temporal differences using ANOVA (SPSS Inc.). No significant differences were discovered and the data were averaged for further analysis.

Table II. Average total AHH-inducing PCB concentration and ΣTEQ for each sample media collected.

	N	% Lipid ±std	MEAN (ng/g)	STD DEV	ΣTEQ (pg/g) (Safe)	ΣTEQ (pg/g) (WHO human)	ΣTEQ (pg/g) (WHO fish)
Diss.	10	--	1.0E-05	9.37E-06	2.7E-06	1.1E-06	3.4E-08
Phyto	12	5±2	5.1	2.2	2.1	0.8	0.02
Zoopl	12	20±9	30.7	23.0	5.5	1.9	0.10
Mysis	11	15±8	95.2	41.4	14.3	6.7	0.32
Dip.	10	16±11	134.1	33.6	30.9	9.1	0.48

Table I. Analyzed AHH- Inducing PCB congeners

CONGENER NUMBER	CO-ELUTING PCB CONGENERS	SUBSTITUTION PATTERN	ORTHO SUBSTITUTION
81		3,4,4',5	non
77	110	3,3',4,4'	non
123	149	2',3,4,4',5	mono
118		2,3',4,4',5	mono
114	131	2,3,4,4',5	mono
105		2,3,3',4,4'	mono
138	163	2,2',3,4,4',5'	di
158		2,3,3',4,4',6	di
126	166	3,3',4,4',5	non
128		2,2',3,3',4,4'	di
167		2,3',4,4',5,5'	mono
156		2,3,3',4,4',5	mono
157		2,3,3',4,4',5'	mono
169		3,3',4,4',5,5'	non
170	190	2,2',3,3',4,4',5	di
189		2,3,3',4,4',5,5'	mono

Concentrations

Average concentrations of total AHH-inducing PCBs (sum of 14 congeners) and the sum of the toxic equivalent concentration (ΣTEQ) for each environmental media are listed in Table II. For summation purposes values less than the method detection limit were set to zero.

Total AHH-inducing PCB concentrations for the dissolved phase ranged from 2 to 30 pg/L, which corresponds to 2e-06 to 3e-05 ng/g in equivalent units to the foodweb. Phytoplankton concentrations ranged from 2 to 9 ng/g, and bulk zooplankton concentrations ranged from 3 to 71 ng/g. The concentrations for *Mysis* and *Diporeia* ranged from 34 to 167 ng/g and 79 to 196 ng/g.

Patterns of individual AHH-inducing PCB congeners in each media are shown in Figure 2. A similar pattern is apparent for each media with congeners 118 and 138 dominating in all cases. There is an overall trend for AHH-inducing PCBs to increase with each step of the foodchain.

Preferential Biomagnification

Before examining the dynamics of AHH-inducing congeners relative to total PCB, we first determined if the AHH-inducing congeners biomagnified between trophic levels. The top graph of Figure 3 represents average trophic level contaminant concentrations that were determined by summing the 14 AHH-inducing PCBs analyzed in this study. Bioaccumulation of AHH-inducing PCBs is clearly indicated. Significant increases are observed between dissolved phase and *Mysis* and *Diporeia* ($p<0.05$, $p<0.05$), phytoplankton and *Mysis* and *Diporeia* ($p<0.05$, $p<0.05$) and zooplankton and *Mysis* and *Diporeia* ($p<0.05$, $p<0.05$). Thus there is evidence for biomagnification between phytoplankton and *Mysis* and bulk zooplankton and Mysis in the pelagic foodweb, and phytoplankton and *Diporeia* in the benthic foodweb (assuming that the detrital phytoplankton that serve as the *Diporeia* food source have the same concentration as live organisms). The bottom graph of Figure 3 shows a similar bioaccumulation pattern for total PCBs (sum of 110 congeners). In this case significant differences between all levels are observed ($p<0.05$) except between the dissolved phase and phytoplankton.

Biomagnification of AHH-inducing PCBs does occur, but are they preferentially accumulated proportionally to all other PCBs? To determine this, the ratio of AHH-inducing PCBs to total PCBs was calculated. If the toxic congeners bioaccumulate similarly to other congeners, the ratios would be similar across all media. However, this is not what is observed. Instead, Figure 4 shows significant increases ($p<0.05$) between the dissolved phase and zooplankton, dissolved phase and *Mysis*, phytoplankton and *Mysis*, zooplankton and *Mysis*, but no significant increase

AHH-Inducing PCB Congener Patterns

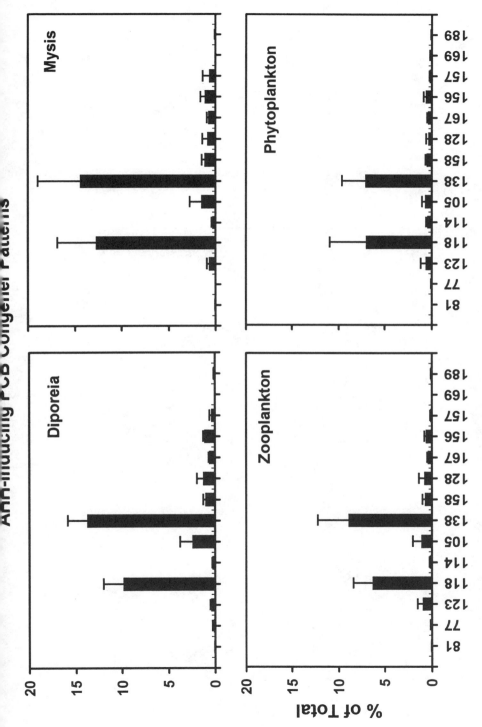

FIGURE 2. Proportion (%) of individual AHH-inducing PCB congeners relative to total PCBs in water, phytoplankton, zooplankton, Mysis and Diporeia. Continued on next page.

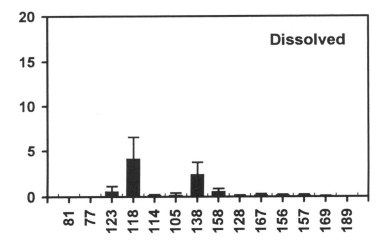

Figure 2. *Continued.*

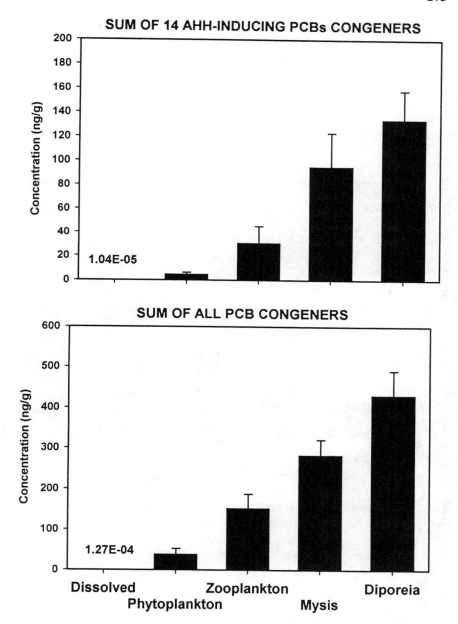

FIGURE 3. Concentrations of the sum of 14 AHH-inducing PCBs (top graph) and total PCBs (sum of 110 congeners; bottom graph) in trophic levels of the Lake Michigan foodweb in ng/g dry weight (dissolved phase is ng/g water for comparison) with 95% confidence intervals.

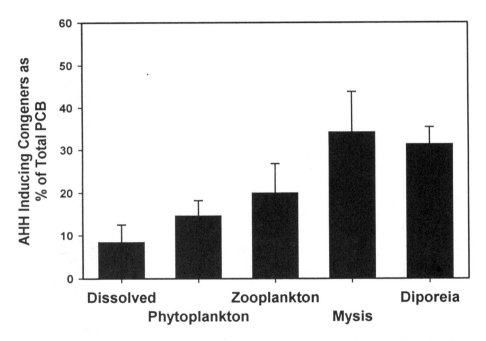

FIGURE 4. The ratios of the sum of AHH-inducing PCBs to the sum of total PCBs for each trophic level with 95% confidence intervals.

between phytoplankton and zooplankton. Significant differences ($p<0.05$) occur between dissolved phase and *Diporeia* and phytoplankton and *Diporeia*, but no significant differences are observed between *Diporeia* and zooplankton or *Mysis*. Another interesting observation is that while the *Diporeia* contaminant concentrations were greater than *Mysis* for both the sum of AHH-inducing PCBs and total PCBs, the *Diporeia* ratio is less than the *Mysis* ratio. The smaller ratio for *Diporeia* compared to *Mysis* may reflect the different contaminant exposure routes which these two organisms experience. *Mysis* represent a tertiary trophic level in the pelagic foodweb, feeding on smaller zooplankton. *Diporeia* are a secondary trophic level in the benthic foodweb, residing in sediments and feeding largely on settled phytoplankton detritus. Thus this result is consistent with the theory that preferential bioaccumulation of AHH-inducing PCBs occurs as food chain levels increase.

Biomagnification factors (BMFs) were calculated to further explore the concept of preferential biomagnification:

$$BMF = \frac{\text{Mean concentration in predator}}{\text{Mean concentration in prey}}$$

Figure 5 provides a comparison of BMFs for the phytoplankton-to-zooplankton, zooplankton-to-*Mysis*, and phytoplankton-to-*Diporeia* food chain steps. The first bar in each group is the average BMF for the AHH-inducing congeners. The second bar in each group represents the average BMF for all 110 PCB congeners analyzed. The third bar is the average BMF for AHH-inducing congeners that have been normalized to the concentration of congener 153, following the method of Boon et al. (*31, 32*). Congener 153 represents the maximum bioaccumulation potential for recalcitrant PCBs and has been documented as being resistant to biotransformation in many organisms (*33, 34*). Thus the 153-normalized BMF indicates the degree to which the AHH-inducing congeners are conserved in each trophic level.

This comparison shows that the BMFs for AHH-inducing PCBs and total PCBs are all significantly greater than 1 at the 95% confidence level. Furthermore, the BMFs for AHH-inducing PCBs are greater than the BMFs for total PCBs by a factor of two for all three trophic steps. Thus there is additional evidence for biomagnification between trophic steps for the AHH-inducing congeners, and the degree of biomagnification is significantly greater than the biomagnification of total PCBs. The average BMF for AHH-inducing congeners normalized to 153 for all food chain steps is not significantly different from 1. This indicates that AHH-inducing PCB congeners bioaccumulate in the food chain similarly to congener 153, and are resistant to degradation and metabolism (*3*). This suggests that the biomagnification of the AHH-inducing congeners is not a result of enhanced uptake due to the co-planar configuration, but rather a result of their recalcitrance in biological systems.

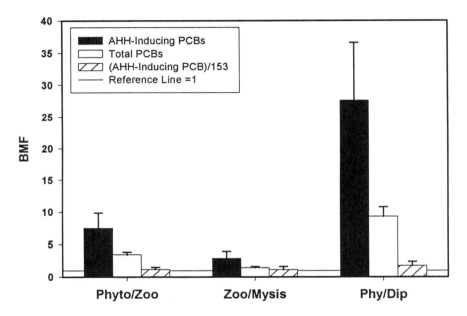

FIGURE 5. AHH-inducing PCB BMFs, total PCB BMFs and 153-normalized BMFs with 95% confidence intervals for three trophic steps.

TEQs

The interest in the bioaccumulation of these compounds lies in the potential toxic effects in both top predator fish and in humans who consume fish. To further interpret these data, we examined the concentrations expressed as toxic equivalents (TEQs) relative to 2,3,7,8-TCDD. This was accomplished by multiplying the concentrations of individual AHH-inducing congeners by weighting factors called toxic equivalency factors (TEFs). Two sources were used for the TEF values; Safe (*13, 14*) and the World Health Organization (WHO) (*35*). Different TEF values will result in differences in the TEQs (*24*). The resulting TEQ values are listed in Table II.

There are several important distinctions between the two sets of TEF values used to determine TEQs. WHO provides separate TEF values for human and fish toxicity while the Safe values do not make a distinction between species. The top graph of Figure 6 provides a comparison between the three sets of TEQ values in pg/g resulting from the three different TEF sources. Regardless of the source of TEFs used, the TEQ increases with increasing trophic level. This graph shows that the Safe TEQ values are greater than either the WHO human or fish TEQ values. Safe TEQ values include di-*ortho* PCBs while WHO concluded that there was insufficient evidence to continue using these congeners in toxicity calculations. In addition, toxicity studies examined by WHO indicated that fish were extremely insensitive to mono-*ortho* PCBs, and as a result the WHO fish TEF values assigned to these PCB congeners are much lower than Safe's TEF values or the WHO human TEF values. These factors explain the large discrepancies observed between the toxicity values. However, these data also clearly indicate that toxicity biomagnifies with increasing trophic level in the foodchain independent of TEF values utilized.

A better comparison of the TEQs using different TEFs can be made from bioaccumulation factors (BAFs). This is the ratio of the TEQ of each trophic level to the dissolved phase TEQ. The resulting log TEQ BAFs are shown in Figure 6. It is observed that the TEQ BAFs are quite similar regardless of the TEF value used in the calculation. Biomagnification is also indicated by these data, with the greatest biomagnification occurring for the WHO fish TEFs.

The individual congeners contributing the most to AHH-inducing concentrations are not the same congeners contributing the most to toxicity (TEQ). The three AHH-inducing congeners that contributed the most to concentration were congeners 138 (di-*ortho* hexa), 118 (mono-*ortho* penta), and 105 (mono-*ortho* penta). Their sum accounted for 83% of AHH-inducing PCB concentration in *Diporeia*, 84% in *Mysis*, 81% in bulk zooplankton, 82% in phytoplankton and 78% in the dissolved phase. This is consistent with what has been reported in Lake Michigan chinook salmon (*24*). These congeners were the top three contributors to concentration in all media except for the dissolved phase where congener 105 was the fifth greatest contributor. However, the congeners contributing to AHH toxicity are somewhat different. While

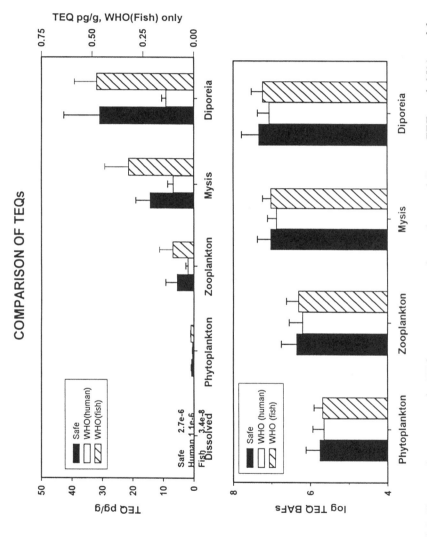

FIGURE 6. Top graph compares the TEQ concentrations obtained using different TEFs with 95% confidence intervals. The TEQs calculated by WHO fish TEFs are plotted on the right y-axis, and the other TEQs are plotted on the left y-axis. Bottom graph compares the log TEQs BAFs using different TEFs, with 95% confidence intervals.

congener 138 contributed the most to concentration, it contributes little to Safe's TEQ values due to its low TEF and is not considered in the WHO TEQ values. Congener 118 contributes second to concentration and is the greatest contributor to toxicity independent of TEF values used. PCB 105 is not a major toxicity contributor in either WHO human or fish TEQs, but contributes 15-30% of total PCB toxicity in Safe's TEQ value. It is unfortunate that congener 126 was not able to be determined in this study. This congener would likely contribute significantly to the TEQ due to its large TEF (*20, 25*).

CONCLUSIONS

The ability to estimate toxic effects resulting from chemical contamination is essential to both human and ecological protection. Understanding the environmental distribution of toxic chemicals provides a basis for assessing risk to humans and wildlife as a result of exposure to the chemical. This study provides insights into the dynamics and distribution of toxic AHH-inducing PCB congeners in the lower trophic levels of the Lake Michigan foodweb. Results from this study have determined that not only do these PCB congeners biomagnify, but they preferentially biomagnify relative to total PCBs. This biomagnification is not related to lipid or organic carbon content of the organism.

Examination of toxicity as a result of PCB bioaccumulation indicates that toxicity expressed as TEQs also biomagnifies. TEQ values were observed to be proportional to AHH-inducing PCB concentration, however the same congeners which account for the majority of the concentration are not the same congeners that account for the majority of the toxicity. Additionally, resulting TEQ values differ depending on which set of TEF values are used in the TEQ calculation. It is important to assess the species of interest and use the appropriate TEF values corresponding to that species when doing exposure assessments.

Current risk assessment practices often estimate exposure to AHH-inducing PCBs from total PCB concentration data. Since this study shows that AHH-inducing PCBs preferentially biomagnify relative to total PCBs and that the PCBs contributing most to the concentration are not always the most toxic compounds, this practice should be used with caution. Individual PCB congener concentrations provide a more reliable and accurate estimate of the true exposure to the species of interest.

LITERATURE CITED

1. Jensen, S. *New Sci* **1966**, *32*, 612.
2. DeWailly, E., Nantel, A., Weber, J.,Meyer, F. *Bull. Environ. Contam. Toxicol.* **1989**, *43*, 641-646.

3. Letcher, R.J., Norstrom, R. J., and Muir, D. C. G. *Environ. Sci. Technol.* **1998**, *32*, 1656-1661.
4. Norstrom, R.J., Simon, M., Muir, D.C.G.,Schweinsburg, R.E. *Environ. Sci. Technol.* **1988**, *22*, 1063-1071.
5. Risebrough, R.W., Walker III, W., Schmidt, T.T., de Lappe, B.W.,Connors, C.W. *Nature* **1976**, *264*, 738-739.
6. Swain, W.R. *J. Great Lakes Res.* **1978**, *4*, 398-407.
7. Harvey, G.R., and Steinhauer, W. G. *Atmos. Environ.* **1974**, *8*, 777-782.
8. Peel, D.A. *Nature* **1975**, *254*, 324-325.
9. AMAP, *AMAP Assessment Report: Arctic Pollution Issues.* Arctic Monitoring and Assessment Programme (AMAP): Oslo, Norway, 1998; 859.
10. Wantanabe, I., Yakushiji, T., Kuwabara, K., Yoshida, S., Maeda, K., Kashimoto, T., Loyama, K. and Kunita, H. *Arch. Environ. Contam. Toxicol.* **1979**, *8*, 67-75.
11. Schwartz, P.M., Jacobson, S. W., Fein, G. G., Jacobson, J. L. and Price, H. A. *Am. J. Public Health* **1983**, *73*, 293-296.
12. Sawhney, G.L., and Hankin, L. *J. Food Protection* **1985**, *48*, 442-448.
13. Safe, S.H. *CRC Crit. Rev. Toxicol.* **1990**, *21*, 51-88.
14. Safe, S.H. *CRC Crit. Rev. Toxicol.* **1994**, *24*, 87-149.
15. Safe, S.H. *CRC Crit. Rev. Toxicol.* **1984**, *13*, 319-395.
16. Shain, W., B. Bush, R. Seegal. *Toxicol. Appl. Pharmacol.* **1991**, *111*, 33-42.
17. Kannan, N., Tanabe, S., and Tatsukawa, R. *Bull. Environ. Contam. Toxicol.* **1988**, *41*, 267-276.
18. Kannan, N., Tanabe, S. , and Tatsukawa, R. *Arch. Environ. Health* **1988**, *43*, 11-14.
19. Tanabe, S., Kannan, N., Subramanian, A., Watanabe, S. and Tatsukawa, R. *Environ. Pollut.* **1987**, *47*, 147-163.
20. Smith, L.M., Schwartz, T.R., Fletz, K.,Kubiak, T.J. *Chemosphere* **1990**, *21*, 1063-1085.
21. Leonards, P.E.G., Y. Zierikzee, U. A. T. Brinkman, W. P. Cofino, N. M. van Straalen and B. Van Hattum. *Environ. Toxicol. Chem.* **1997**, *16*, 1807-1815.
22. Metcalfe, T.L.a.M., C. D. *Sci. Total Environ.* **1997**, *201*, 245-272.
23. Willman, E.J., J. B. Manchester-Neesvig, D. E. Armstrong. *Environ. Sci. Technol.* **1997**, *31*, 3712-3718.
24. Williams, L.L. ,Giesy, J.P. *J. Great Lakes Res.* **1992**, *18*, 108-124.
25. Williams, L.L., Giesy, J.P., DeGalan, N., Verbrugge, D.A., Tillitt, D.E.,Ankley, G.T. *Environ. Sci. Technol.* **1992**, *26*, 1151-1159.
26. Swackhamer, D.L. *Iss. Environ. Sci. Technol.* **1996**, *6*, 137-153.
27. Madenjian, C.P., DeSorcie, T.J.,Stedman, R.M. *Trans. Am. Fish. Soc.* **1998**, *127*, 236-252.
28. Schmidt, L.J., and Hesselberg, R. J. *Arch. Environ. Contam. Toxicol.* **1992**, *23*, 37-44.
29. Swackhamer, D.L., Trowbridge, A. Quality Assurance Project Plan for PCBs and Hg in the Lake Michigan Lower Pelagic Foodweb as part of the Lake

Michigan Mass Balance Study. US EPA Great Lakes National Program Office, Chicago, IL. 1995.

30. Hebert, C.E., and Keenleyside, K. A. *Environ. Toxicol. Chem.* **1995**, *14*, 801-807.

31. Boon, J.P., van Arnhem, E., Jansen, S., Kannan, N., Petrick, G., Schulz, D. E., Duinker, J. C., Reijnders, P. J. H., Goksoyr, A. in *Persistent Pollutants in Marine Ecosystems*, Livingstone, C.H.W.a.D.R., Editor Pergamon Press: Oxford. 1992; p. 119-159.

32. Boon, J.P., Oostingh, I., van der Meer, J., Hillebrand, J. T. J. *European Journal of Pharmacology - Environmental Toxicology and Pharmacology* **1994**, *270*, 237-251.

33. Muir, D.C.G., Norstrom, R. J., Simon, M. *Environ. Sci. Technol.* **1988**, *22*, 1071-1079.

34. Muir, D.C.G., Wagemann, R., Hargrave, B. T., Thomas, D. J., Peakall, D. B., Norstrom, R. J. *Sci. Total Environ.* **1992**, *122*, 75-134.

35. Van den Berg, M., Birnbaum, L., Bosveld, A.T.C., Brunstrom, B. Cook, P., Feeley, M., Biesy, J. P., Hanberg, A., Hasegawa, R. Kennedy, S. W., Kubiak, T., Larsen, J. C., van Leeuwen, F. X. R., Djien Liem, A. K., Nolt, C., Peterson, R. E., Poellinger, L., Safe, S., Schrenk. *Environ. Health Perspect.* **1998**, *106*, 775-792.

Chapter 21

Concentration of Persistent Organochlorine Compounds in the Placenta and Milk of the Same Women

Katarzyna Czaja[1], Jan K. Ludwicki[1], Mark G. Robson[2], Katarzyna Góralczyk[1], Pawel Struci|ski[1], and Brian Buckley[2]

[1]Department of Environmental Toxicology, National Institute of Hygiene, Chocimska 24, 00–791 Warsaw, Poland
[2]Environmental and Occupational Health Science Institute, 170 Frelinghuysen Road, P.O. Box 1179, Piscataway, NJ 08855–1179

Samples of breast milk collected on the 4th day after delivery, and placenta from the same donors, and breast milk samples from donors in different regions of Poland were tested for HCB, HCH isomers, DDT and its metabolites and total PCBs. There was negative relationship between the concentrations of individual DDT metabolites in the placentas and the corresponding concentrations in milk. p,p'-DDE was the dominant metabolite in milk, while p,p'-DDD - in the placentas. All of the milk samples have been found to have the highest quantities of p,p'-DDE and total PCBs, while all of the placenta samples had the highest quantities of p,p'-DDD and total PCBs.

© 2001 American Chemical Society

Introduction

Persistent organochlorine compounds are continually found in all of the elements of the environment. Due to their physical and chemical properties, they manage to bioaccumulate in the adipose tissue and biomagnify. Slow metabolism of these compounds results in their long half-life in human tissues, especially in the tissues of infants who do not have fully developed detoxication mechanisms yet. Humans are exposed to these compounds throughout their life, starting as early as in the fetal period. In all of their growth phases, the fetus and small children are particularly sensitive to toxic and teratogenic impact of chemical substances. This is related to the negative implications that might surface later during the child's physical and psychic development (1).

While sufficient information is available on the impact of organochlorine compounds on mammal organisms, little is known of how these compounds affect a child at such an early stage prenatal and infant lives. When predicting the possible impact of exposure to toxic compounds, one must bear in mind that small children have a different absorption, tissue distribution, biotransformation, elimination, and body response to environmental pollution. In addition, the fetus and infants have very low deposits of spare fat (2) which helps toxic compounds accumulate in high doses in such vital organs as the liver, adrenal gland, and brain (3). Excessive growth in the concentrations of these toxic compounds may pose a serious threat to the fetus, which has been confirmed when examining the cases of PCBs poisoning in Japan in 1968 and in Taiwan in 1979 as well as the cases of poisoning from the hexacholorobenzene-treated seed in Turkey (4). In all of these cases, the exposure of the children in *utero* resulted in indisputable fetus damage and drastic disorders in the physical and psychic development at a later time (5).

The toxic compounds accumulating in the mother's tissues are released during pregnancy and lactation (6). The identification of these compounds in the tissues of breast-fed children does not yet prove that the compounds have penetrated the placenta, as in such case one needs to consider also the children's exposure to these compounds *via* the mother's milk. The presence of organochlorine compounds is found in the fetuses (7), stillborn (8), as well as placental and cord blood (9). This shows that these compounds are released from the mother's body and carried to the fetus tissues (10). The placenta is not an effective barrier to the lipophilic organochlorine compounds (11). It can be assumed, however, that the possible accumulation of such compounds in the placenta can indirectly prove that the placenta indeed makes it more difficult for some compounds to penetrate to the fetus.

Much higher doses of organochlorine compounds (from 10 to 20 times higher) penetrate the infant's body *via* the milk than *via* the transplacental route (12). This is confirmed by the studies of experimental animals. For rats, only 2.7% of PCBs provided were found to be transported *via* the placenta and 39.2% *via* the milk. For mice, too, 75% of the PCBs provided to the mother were found in breast-fed descendants, while only 1% of PCBs were transported *via* the placenta (13).

286

Nevertheless many premises lead one to believe that the exposure of the child in *utero* to organochlorine compounds produces higher toxic effects than the exposure to these compounds *via* their presence in the mother's milk *(14)*.

This research was an attempt to examine how much and which compounds were retained in the placenta and what correlations there are between their concentrations in the milk and the placentas of the same donors.

Materials and Methods

The research material obtained from a maternity clinic included 27 samples of breast milk collected on the 4th day after delivery and 27 samples of the placenta from the same donors as well as 365 samples of breast milk from Poland's 7 regions of various industrial and agricultural development rate. All milk samples were collected after the 4th minute of breast-feeding, than a child was brought to the other breast, while approximately 20 ml of milk was collected from the first breast for analysis. The donors completed a detailed survey regarding their age, number of deliveries, health condition, diet, fluctuation in body weight, stress situations passed, etc. All the donors and their children were healthy. The content of HCB, HCH isomers, DDT and its metabolites: p,p'-DDE and p,p'-DDD and the total PCBs in the samples was determined using a gas chromatograph with an electron capture detector.

To assure the quality of results, the laboratory simultaneously used the same analytical method in the international proficiency testing organized by FAPAS (Food Analysis Performance Assessment Scheme) by the UK Ministry of Agriculture, Food, and Fisheries. In addition, the certified reference materials and own fortified samples were analyzed as part of a routine, internal analytical quality assurance procedure. The research was based on an analytical approach for which the following parameters were reported: limit of detection, linearity, precision, repeatability, and reproducibility. The above mentioned parameters differ across the compounds and amount to: the repeatability relative standard deviation from 1% to 5%, and the reproducibility relative standard deviation from 10% to 25%.

The resulting data were analyzed using t-Student test and one way analysis of variance (ANOVA).

Results and Discussion

Organochlorine compound levels in the breast milk samples collected in various regions of Poland are shown in Table I. The concentrations of the compounds found in the milk samples from the 4th day and from the placentas coming from the same donors are illustrated in Tables II and III.

Among eight chlorinated hydrocarbons from all milk samples examined, the concentrations of p,p'-DDE and total PCBs were always the highest. From all HCH

Table I. Concentrations of persistent organochlorine compounds in human breast milk in Poland (mg/L)

	HCB	α-HCH	β-HCH	γ-HCH	p,p'-DDT	P,p'-DDD	p,p'-DDE	ΣPCBs
LD	0.0002	0.0002	0.0002	0.0002	0.0008	0.0005	0.0004	0.0010
X	0.0020	0.0005	0.0033	0.0004	0.0034	0.0005	0.0282	0.0544
SD	0.0025	0.0018	0.0068	0.0014	0.0093	0.0025	0.0250	0.0814
Min	0.0002	0.0002	0.0002	0.0002	0.0008	0.0005	0.0009	0.0010
Max	0.0152	0.0155	0.0835	0.0156	0.1355	0.0381	0.1850	0.6590

LD - limit of detection; X - mean concentration; SD - standard deviation

Table II. Concentrations of persistent organochlorine compounds in 4th-day milk (mg/L)

	HCB	α-HCH	β-HCH	γ-HCH	p,p'-DDT	P,p'-DDD	p,p'-DDE	ΣPCBs
LD	0.0002	0.0002	0.0002	0.0002	0.0008	0.0005	0.0004	0.0010
X	0.0013	0.0002	0.0014	0.0002	0.0050	0.0009	0.0211	0.0076
SD	0.0008	0.0002	0.0008	0.0001	0.0024	0.0003	0.0140	0.0041
Min	0.0005	0.0002	0.0004	0.0002	0.0019	0.0008	0.0066	0.0017
Max	0.0040	0.0013	0.0036	0.0004	0.0094	0.0025	0.0528	0.0165

LD - limit of detection; X - mean concentration; SD - standard deviation

Table III. Concentrations of persistent organochlorine compounds in the placentas (mg/kg)

	HCB	α-HCH	β-HCH	γ-HCH	p,p'-DDT	p,p'-DDD	p,p'-DDE	ΣPCBs
LD	0.0010	0.0010	0.0010	0.0010	0.0050	0.0030	0.0020	0.0050
X	0.0061	0.0050	0.0128	0.0051	0.0155	0.0328	0.0120	0.0285
SD	0.0028	0.0002	0.0156	0.0005	0.0177	0.0360	0.0140	0.0896
Min	0.0050	0.0050	0.0050	0.0050	0.0050	0.0050	0.0050	0.0050
Max	0.0162	0.0062	0.0797	0.0076	0.0752	0.1118	0.0657	0.4803

LD - limit of detection; X - mean concentration; SD - standard deviation

isomers, the β isomer showed the highest concentrations (Figure 1). Similar proportions were found also in the breast milk samples from different regions of the country (15). When comparing the organochlorine compound levels in the milk collected on the 4th day after delivery with mature milk, worthy of notice is the fact of higher concentrations of total PCBs and β-HCH in mature milk. These variances were statistically significant (p < 0.05) (15). Mean concentrations of these compounds in human breast milk did not differ from those found in human milk in other European countries for example, in France (16), Netherlands (17), and Sweden (18).

The findings indicate the negative relationship between the concentrations of individual DDT metabolites in the placentas and the corresponding concentrations in milk. The concentrations of p,p'-DDE in the women examined were always the highest in milk, and lowest in the placentas. At the same time, p,p'-DDD with marginal concentrations in milk proved to accumulate in large quantities in the placentas (Figure 2). It must be noted that p,p'-DDE is generally the dominant DDT metabolite in human tissues as well as in the environment. The detection of higher concentrations of p,p'-DDD vs. p,p'-DDE in the placentas becomes especially significant when related to the results of research on the DDT metabolite levels in infants where the concentrations of p,p'-DDD were found higher than those of p,p'-DDE. Therefore, the detection of higher p,p'-DDD accumulation in the placentas examined can be supposed to disprove the claim that the placenta is the barrier to this compound. On the contrary: this can indicate that this compound, using unknown mechanisms, selectively accumulates in the placenta to then penetrate to the fetus. However, the confirmation of such correlations requires further research.

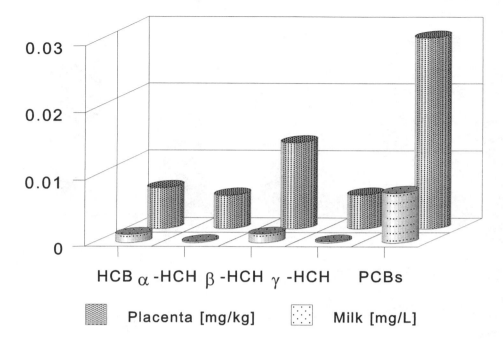

Figure 1. HCB, HCHs, and PCBs in milk and placentas from the same donors

The transportation of organochlorine compounds *via* the placenta varies for mammals, depending on the chemical structure and, especially, the number of chlorine atoms per molecule and their position *(12)*. For example, highly chlorinated PCBs penetrate the placenta more easily than less chlorinated ones. This, however does not explain higher p,p'-DDD concentrations in placentas since the molecules of this compound and p,p'-DDE have the same number of chlorine atoms.

The placenta is an active tissue, both enzymatically *(19)* and hormonally. Lucier has found that the AHH (placental arylhydrocarbon hydroxylase) activity is much higher in the placentas of women exposed to PCBs and PCDFs. Lucier claims that such parameters as the AHH activity, cytochrome P-450 isoenzymes, and epidermal growth factor (EGF) are good indicators of what impact the exposure to these compounds may be.

This makes one wonder if the compounds detected in the placenta can undergo metabolism there and if that metabolism affects the compound levels detected in this tissue. This might explain why the concentrations of p,p'-DDD in the placenta samples we studied proved higher than those of p,p'-DDE.

Relatively high quantities of organochlorine compounds absorbed with breast milk are unlikely to be as detrimental to the infant as the compounds the fetus is exposed to during its prenatal life. The human exposure to organochlorine compounds during the prenatal life is related to disorders in the functioning of many organs *(20)*.

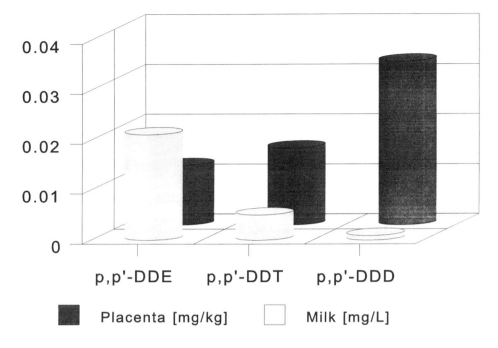

Figure 2. DDT and its metabolites in milk and placentas from the same donors

Disorders in the division and differentiation of the cells during prenatal life may cause fetus deformation. The central nervous system is not fully developed even after birth. The blood-brain barrier is not fully shaped either. Animal studies prove that even small doses can cause behavioral change and delayed psychomotoric development. Any disorders in the sex hormone levels that stem from the exposure to environmental estrogenes or anti-androgenes at the fetal stage adversely affect the operation of sex organs throughout life, e.g., in the sperm production, menstrual cycle, pregnancy, and in the development of the reproductive organs in a male or female direction. Some of these compounds are also anti-estrogenes, thus bringing about disorders in hormonal systems. The negative impact of organochlorine compounds on the child's development is also a result of the compounds' disturbing the metabolism of vitamin A which plays an important role in the development of the fetus and the operation of the immune system. These compounds may also cause a decrease in the size of the thymus whose proper activity is necessary at the fetal stage. In addition, the compounds affect the functioning of the thyroid gland.

The afore-mentioned as well as other possible negative impact of organochlorine compounds on the fetus indicate that man in the prenatal stage is very susceptible to the compounds' exposure. Therefore, the concentrations of these compounds in the environment must be monitored, with the placenta tissue and human milk used as good indicators of any potential threats to the child's body.

Literature Cited

1. Jacobson, J.L.; Jacobson, S.W.; Humphrey, H.E.B. *J. Pediatr.* **1990**, *116*, 38-45.
2. *Environmental Health, PCBs, PCDDs and PCDFs in breast milk: Assessment of Health Risk;* WHO, Copenhagen, 1988; *19*.
3. Kodama, H.; Ota, H. *Arch. Environ. Health* **1980**, *35*, 95-100.
4. Rogan, W.J. *Environ. Health Perspect.* **1995**, *103*, Suppl.6, 19-23.
5. Guo, Y.L.; Lambert, G.H.; Hsu, Ch. *Environ Health Perspect.* **1995**, *103*, Suppl. 6, 117-122.
6. Ando, M.; Saito, H.; Wakisaka, I. *Environ. Res.* **1986**, *41*, 14-22.
7. Bosse, U.; Bannert, N.; Niessen, K.H.; Teufel, M.; Rose, I. *Zentralbl. Hyg. Umwelt.* **1996**, *198*, 331-339.
8. Schecter, A.; Papke, O.; Ball, M. *Chemosphere* **1990**, *21*, 1017-1022.
9. Kanja, L.W.; Skaare, J.V.; Ojwang, S.B.; Maitai, C.K. *Arch. Environ. Health* **1992**, *22*, 367-374.
10. Salama, A.K.; Bakry, N.M.; Abou-Donia, M.B. *Int. J. Occup. Med.. Toxicol.* **1993**, *2*, 383-397.
11. Lanting, C.I.; Huisman. M.; Muskiet, F.A.J.; van der Paauw, C.G.; Essed, C.E.; Boersma, E.R. 18th Symposium on Halogenated Environmental Organic Pollutants, Stockholm, Sweden, 1998; Organohalogen Compounds, *38*, 5-8.
12. Jensen, A.A.; Slorach, S.A.; In *Chemical Contaminants in Human Milk*; CRC Press, Inc. Boca Raton Ann Arbor Boston, 1991.
13. Ahlborg, U.G.; Hanberg, A.; Kenne. K.; In *Risk Assessment of Polychlorinated Biphenyls (PCBs);* Nordic Council of Ministers, Copenhagen. Report NORD 1992; *26*.
14. Ryan, J.J.; Chen-Chin. Hsu; Bojle. M.J.; Guo, Y.L. *Chemosphere* **1994**, *29*, 1263-1278.
15. Czaja, K.; Ludwicki, J.K.; Góralczyk. K.; Strucinski, P. *Bull. Environ. Contam. Toxicol.* **1997**, *58*, 769-775.
16. Bordet, F.; Mallet. J.; Maurice, L.; Borrel, S.; Venant, A. *Bull. Environ. Contam Toxicol.*, **1993**, *50*, 425-432.
17. Koopman-Esseboom, C.; Huisman, M.; Weisglas-Kuperus, N.; Boersma, E.R.; de Ridder, M.A.J.; van der Paauw, C.G.; Tuinstra, L.G.M.T.; Sauer, P.J.J, *Chemosphere* **1994**, *29*, 2327-2338.
18. Vaz, R,; *Food. Addit. Contam.* **1995**, *12*, 543-558.
19. Lucier, G.W.; Nelson, K.G.; Everson, R.B.; Wong, T.K.; Philpot.,R.M.; Tiernan, T.; Taylor, N.; Sunahara, G.I. *Environ. Health Perspect.* **1987**, *76*, 79-87.
20. Bernes, C. In *Persistent Organic Pollutants. A Swedish View of an International Problem*; Swedish Environmental Protection Agency, 1998; Monitor 16.

Author Index

Subject Index

A

Absorption
definition, 196
description, 196
Active excretion, definition, 20
Adipose, dioxin bioaccumulation,
252–262
Adsorption, description, 114
Aerobic conditions, microbial
degradation, 177–180
Age of fish, role on methylmercury
bioaccumulation, 151–162
Agricultural chemicals, chlorinated.
See Chlorinated agricultural
chemicals
Air
occurrence of polychlorinated
biphenyls, 89–90
plant uptake of linear
alkylbenzenesulfonates and di(2-
ethylhexyl)phthalate, 99–110
Air exposure route
description, 19
indirect route, 19
Alcaligenes, degradation of
polychlorinated dibenzo-*p*-dioxins
and dibenzofurans, 176–187
Aldrin, regulation and use, 7
Alkylbenzenesulfonates, linear. *See*
Linear alkylbenzenesulfonates
Anaerobic conditions, microbial
degradation, 180–186
Animal sources of food, dioxins,
245–250
Aryl hydrocarbon hydroxylase
health effects, 267
role of polychlorinated biphenyls,
267–281
Atmospheric concentrations and
behavior

chlordanes, 37–43
nonylphenols, 50–54
polychlorinated biphenyls, 31–37
polycyclic aromatic hydrocarbons,
40, 44–50
Atmospheric semivolatile organic
compound deposition to vegetation
compartment designation for
transport through cuticle, 224
effect of leaf age, 229, 232
effect of leaf surface area, 229–230,
232
effect of lipid composition, 229, 232
effect of lipid content, 228-229
effect of plant architecture, 230
effect of plant species, 228, 231–232
effect of temperature, 227–228, 231
experimental description of
deposition pathways, 219
factors affecting particle-bound
deposition, 220
mechanism of dry gaseous
deposition, 221
mechanism of particle-bound
deposition,, 220–221
one- or two-compartment model of
dry gas deposition, 223
partitioning process for dry gaseous
deposition, 221–223
reasons for interest, 218–219
role of dominant deposition
pathways on octanol–air partition
coefficient, 225–227
role of dry gaseous vs. particle-
bound depostion on octanol–air
partition coefficient, 225, 231
steps in deposition pathways, 219–
222
tortuosity of cuticular waxes during
transport through cuticle, 224–
225

293

B

Bacteria
degradation, 178–180
reductive dechlorination in
sediments, 184–186
Barley, uptake of linear
alkylbenzenesulfonates and di(2-
ethylhexyl)phthalate from sludge-
amended soil, 99–110
Bay shrimp, DDT residues and
dieldrin, 204–216
Beef animals, presence of dioxins,
245–250
Behavior, studies, 28–187
Beijerinckia, degradation of
polychlorinated dibenzo-*p*-dioxins
and dibenzofurans, 176–187
Benzene hexachloride, preparation, 4
Bioaccumulation
butyltin compounds in freshwater
ecosystems, 134–146
definition, 14
depuration routes, 20–21
exposure routes, 14–19
food webs, 21
kinetics, 14, 15*t*
methylmercury on Northern Pike
age and size, 151–162
modulating factors, 22-23
polar bears, 13–14, 24
toxaphene, 164–174
Bioaccumulation factor
definition, 22
use, 22
values, 22
Bioavailability
description, 23
influencing factors, 23
Bioconcentration, definition, 14
Bioconcentration factors, DDT
residues and dieldrin, 215*t*, 216
Biodegradation

sediments, 180–181, 182–183*f*
white rot fungi, 180
Biomagnification, polychlorinated
biphenyl congeners in Lake
Michigan food web, 266–281
Biomagnification factor
calculation, 277
polychlorinated biphenyls, 277–278
use, 22
values, 22
Biota, occurrence of polychlorinated
biphenyls, 91–92
Biota–soil accumulation factor
use, 22
values, 22
Biotransformation
definition, 14
description, 22
examples, 22–23
Biphenyls, polychlorinated. *See*
Polychlorinated biphenyls
Breast milk
dioxin bioaccumulation, 252–262
persistent organochlorine compound
concentrations, 284–290
polychlorinated biphenyl
concentrations, 94
Brevibacterium, degradation of
polychlorinated dibenzo-*p*-dioxins
and dibenzofurans, 176–187
Bromine, discovery, 3
Burkholderia, degradation of
polychlorinated dibenzo-*p*-dioxins
and dibenzofurans, 176–187
Butyltin compounds in freshwater
ecosystems
chemical analysis procedure during
monitoring, 142–143
concentrations of mussel tissues
during monitoring, 144–146
concentrations of sediments during
monitoring, 144
experimental description, 16

I

K

chlorinated, resistant desorption
  kinetics, 112–120
halogenated, *See* Halogenated
  organic chemicals
Organic contaminants, hydrophobic.
  See Hydrophobic organic
  contaminants
Organic pollutants, persistent. *See*
  Persistent organic pollutants
Organic substances, depuration, 20
Organochlorine–diene insecticides.
  See Diene–organochlorine
  insecticides
Organotin compounds
  classification, 134–135
  environmental occurrence, 135
  uses, 135

P

Particle(s), depository for
  contaminants, 112–113
Particle-bound deposition,
  atmospheric semivolatile organic
  compounds to vegetation, 220–221
Passive excretion, definition, 20
Pelagic biota, persistence of DDT
  residues and dieldrin, 212, 213–
  214*t*
Pentachlorophenol-treated wood, role
  in dioxins in food, 245–250
Persistence
  analytical procedure for petroleum
    hydrocarbon mixture in soil
    profile during leaching, 74
  bioconcentration factors for DDT
    residues and dieldrin off pesticide
    processing plant, 215*t*, 216
  component redistribution dynamics
    for petroleum hydrocarbon
    mixture in soil profile during
    leaching, 77–78, 81–83

epibenthic and pelagic biota for
  DDT residues and dieldrin off
  pesticide processing plant, 212,
  213–214*t*
experimental description for
  petroleum hydrocarbon mixture in
  soil profile during leaching, 71
experimental materials for
  petroleum hydrocarbon mixture in
  soil profile during leaching, 71,
  72*t*, 73*f*, 74*t*
experimental procedure for
  petroleum hydrocarbon mixture in
  soil profile during leaching, 72
GC/MS analysis procedure for DDT
  residues and dieldrin off pesticide
  processing plant, 207
intertidal mussel content of DDT
  residues and dieldrin off pesticide
  processing plant, 209, 211–212
kerosene redistribution dynamics for
  petroleum hydrocarbon mixture in
  soil profile during leaching, 75,
  76f
near-surface water content of DDT
  residues and dieldrin off pesticide
  processing plant, 209, 210*t*
previous studies for petroleum
  hydrocarbon mixture in soil
  profile during leaching, 71
problem for petroleum hydrocarbon
  mixture in soil profile during
  leaching, 71
rain vs. irrigation water type effects
  for kerosene redistribution for
  petroleum hydrocarbon mixture in
  soil profile during leaching, 75,
  77f
sample collection and preparation
  procedure for DDT residues and
  dieldrin off pesticide processing
  plant for biota, 205–207

temperature/season effect on
atmospheric concentrations and
behavior, 44, 47–50
transport vectors for atmospheric
concentrations and behavior, 50
Preferential biomagnification,
polychlorinated biphenyl
congeners in Lake Michigan food
web, 266–281
Pseudomonas, degradation of
polychlorinated dibenzo-*p*-dioxins
and dibenzofurans, 176–187

R

Reductive dechlorination, bacteria in
sediments, 184–186
Regulation, polychlorinated
biphenyls, 86–87
Resistance desorption kinetics of
chlorinated organic compounds
from contaminated soil and
sediments
constrictivity factor, 116
desorption experimental procedure,
115
experimental materials, 114
first-order rate constants, 116
fractional desorption, 116, 117*f*
irreversible adsorption equation, 120
limiting factors, 114
mass of sorbed compounds on
sediment, 116
previous studies, 113–114
requirements for prediction of
chemical fate, 120

second-order rate constant, 118–120
sediment experimental procedure,
114–115
Soxhlet extraction procedure, 115–
116

Respiratory surfaces, role on uptake,
16–17
Reversible compartment, description,
59
Risk control, polychlorinated
biphenyls, 125, 131

S

San Francisco Bay, California,
persistence of DDT residues and
dieldrin off pesticide processing
plant, 204–216
Sediment
biodegradation, 180–181, 182–183*f*
contaminated, resistant desorption
kinetics of chlorinated organic
compounds, 112–120
contamination problem, 58–59
irreversible sorption modeling of
hydrophobic organic
contaminants, 58–68
occurrence of polychlorinated
biphenyls, 90
Sediment exposure route
description, 18
influencing factors, 18
Selective persistence, toxaphene,
164–174
Semivolatile organic compounds
atmospheric, deposition to
vegetation, 218–232
sources, 218
Sewage sludge
organic contaminants, 99–100
plant uptake of linear
alkylbenzenesulfonates and di(2-
ethylhexyl)phthalate from
amended soil, 99–110
use, 99